Studies in Big Data

Volume 96

The series "Studies in Big Data" (SBD) publishes new developments and advances in the various areas of Big Data- quickly and with a high quality. The intent is to cover the theory, research, development, and applications of Big Data, as embedded in the fields of engineering, computer science, physics, economics and life sciences. The books of the series refer to the analysis and understanding of large, complex, and/or distributed data sets generated from recent digital sources coming from sensors or other physical instruments as well as simulations, crowd sourcing, social networks or other internet transactions, such as emails or video click streams and other. The series contains monographs, lecture notes and edited volumes in Big Data spanning the areas of computational intelligence including neural networks, evolutionary computation, soft computing, fuzzy systems, as well as artificial intelligence, data mining, modern statistics and Operations research, as well as self-organizing systems. Of particular value to both the contributors and the readership are the short publication timeframe and the world-wide distribution, which enable both wide and rapid dissemination of research output.

The books of this series are reviewed in a single blind peer review process.

Indexed by SCOPUS, EI Compendex, SCIMAGO and zbMATH.

All books published in the series are submitted for consideration in Web of Science.

More information about this series at http://www.springer.com/series/11970

G. P. Obi Reddy · Mehul S. Raval · J. Adinarayana ·
Sanjay Chaudhary
Editors

Data Science in Agriculture and Natural Resource Management

Springer

Editors
G. P. Obi Reddy
Division of Remote Sensing Application
ICAR-National Bureau of Soil Survey
and Land Use Planning
Nagpur, India

J. Adinarayana
Centre of Studies in Resources Engineering
Indian Institute of Technology Bombay
(IITB)
Mumbai, India

Mehul S. Raval
School of Engineering and Applied Science
Ahmedabad University
Ahmedabad, India

Sanjay Chaudhary
School of Engineering and Applied Science
Ahmedabad University
Ahmedabad, India

ISSN 2197-6503 ISSN 2197-6511 (electronic)
Studies in Big Data
ISBN 978-981-16-5846-4 ISBN 978-981-16-5847-1 (eBook)
https://doi.org/10.1007/978-981-16-5847-1

This Springer imprint is published by the registered company Springer Nature Singapore Pte Ltd.
The registered company address is: 152 Beach Road, #21-01/04 Gateway East, Singapore 189721, Singapore

To my beloved parents and family members for their constant encouragement and support.

by G. P. Obi Reddy

In dedication to my wife—Hemal, son—Hetav, and in memory of mother—Charu and father—Shirish for their constant encouragement and support through thick and thin.

by Mehul S. Raval

To my parents and family members for their constant encouragement and support.

by J. Adinarayana

To my parents and family members for their strong support throughout.

by Sanjay Chaudhary

Foreword

The world of Data Science is rapidly changing due to fast-emerging technology changes in the fields of computers, data acquisition systems, and digital technologies. The last two decades witnessed tremendous developments in Information and Communication Technologies (ICT), including the Internet, cloud computing, sensors technology, computer processing, and storage and dissemination systems, which opens new avenues in digital data acquisition, processing, visualization, and disseminating valuable information of the planet earth to the general users, planners, and policy makers. This digital revolution, accompanied by the fast emerging of remote sensing platforms, provides an unprecedented amount of geospatial data on the status of planet earth resources, especially on natural resources, agriculture, and their dynamics, to develop various Earth Observation (EO) applications.

The edited volume on 'Data Science in Agriculture and Natural Resource Management' addressed the principles and applications of data science and emerging challenges in agriculture and natural resource management. The volume contains three important sections, namely, 'Data Science-Principles, Concepts and Applications, 'Data Science—Applications in Agriculture', and 'Data Science—Applications in Natural Resources Management.' The chapters in these sections dealt with specialized areas contributed by eminent experts from research institutes and universities of India and abroad with suitable illustrations and tables. The edited book is unique in the field of 'Data Science Applications in Agriculture and Natural Resource Management.' The book provides advanced knowledge on data science and added the latest knowledge in the field to benefit the global readers. The authors from renowned international organizations/Universities/Industries like International Crops Research Institute for the Semi-Arid Tropics (ICRISAT), International Center for Agricultural Research in the Dry Areas (ICARDA), national organizations like ICAR-National Bureau of Soil Survey and Land Use Planning (ICAR-NBSS&LUP), Indian Institute of Technology Bombay (IITB), Ahmedabad University, Pandit Deendayal Energy University, University of Agricultural Science, Bangalore; and industrial entities such as NTT Data, International Business Machines (IBM) Corporation and Amnex Infotechnologies contributed the chapters to the edited volume with suitable case studies.

This book explored various facets of data science, the big data revolution, and EO applications in agriculture and natural resource management. It stimulates new ideas in data-driven research, new applications by integrating emerging data science techniques and robust prediction models in the emerging fields of cloud computing, artificial intelligence (AI), and deep learning, and ICT applications for geo-smart agriculture and sustainable natural resource management. I am sure the book realizes the aspirations and needs of geospatial, data science, agricultural, natural resources, and environmental scientists/faculty/students from traditional universities, agricultural universities, technological universities, research institutes, and academic colleges worldwide. It also helps planners, policymakers, and extension scientists plan and sustain natural resources for climate-resilient agriculture.

In summary, this book will be an added asset in the field of contemporary data science applications, which depicts the chapters ranging from fundamentals to the advanced mix of high-end information technology and agriculture and natural resource management. As these fields are dynamic, I wish the Editors will bring more editions of the influential book.

July 2021

Prof. Seishi Ninomiya
International Field Phenomics
Research Laboratory
The University of Tokyo
Tokyo, Japan

Preface

Agriculture and natural resources are the core and essential building blocks for the development of any society. It is important to efficiently use them for fulfilling growing food demand. In recent years, we are experiencing tremendous technology growth and the evolution of innovative applications using Data Science (DS). The technology is helping us to capture various types of data related to farming practices to harvesting to post-harvesting and supply chain.

Data Science is a genuinely multi-disciplinary area having an intersection of Computer Science, Mathematics, and Artificial Intelligence (AI). The problems in DS require analyzing discrete structured and unstructured high dimensional data sets, consisting of a collection of observables and their associated responses for predicting responses for previously unseen observations. Each farm and natural resource has a great source of data. DS has all the potential to make the best use of available voluminous, dynamic, and real-time data to generate actionable information. It enables us to make efficient decisions for the best utilization of natural resources and improve the yield of agricultural products.

We feel a significant need to present various aspects of 'Data Science in Agriculture and Natural Resource Management (DSANRM) in the form of a book. It will be helpful for students aiming to undertake projects in the DSANRM. At the same time, it will be a comprehensive quality resource for practitioners, researchers, decision-makers, and policymakers associated with this domain.

With the above aims, this book is organized into three sections:

1. Data Science—Principles, Concepts, and Applications
2. Data Science—Applications in Agriculture
3. Data Science —Applications in Natural Resource Management

The Part I comprises four chapters that focus on DS principles and concepts used in DSANRM. The Chapter "Data Science—Algorithms and Applications in Earth Observation" by Reddy and Kumar N. introduces DS and highlights its interdisciplinary nature. It uses Earth Observations technologies to discuss tools used in DS at various stages—collection, storing, processing, analysis, and visualization.

It shows how DS has been enriched with techniques derived from Machine Learning (ML) and Data Mining (DM). It introduces supervised algorithms for regression, classification, and various clustering algorithms. It discusses the use and challenges of ML algorithms in EO studies like drought monitoring. The Chapter "Emerging Technologies—Principles and Applications in Precision Agriculture" by Mishra, dives deeper and brings out all essential aspects of technology-driven Precision Agriculture (PA). It covers the complete gamut of on-field sensing using the Internet of Things (IoT), remote sensing using Unmanned Aerial Vehicles (UAVs), and satellites. It shows how edge computing and cloud computing can use Big Data Analytics (BDA) and ML to carry out the predictive analysis.

The Chapter "Data Science: Principles and Concepts in Modeling Decision Trees", by Divakaran presents a brief overview of statistical learning techniques like Linear and Non-linear Regression, Support Vector Machines (SVM), Decision Trees (DT), and Neural Networks (NN) for prediction problems in both regression and classification settings. Then, the study of Decision Trees is motivated by highlighting their computational efficiency and higher model interpretability. The chapter provides vital insights into the form of (i) the workings of standard decision tree-based statistical learning methods for making predictions; (ii) the understanding of these methods; and (iii) the use of Decision Trees in the context of Precision Agriculture (PA). Gandhi, in Chapter "Deep Reinforcement Learning for Agriculture: Principles and Use Cases" brings in another dimension of ML—Reinforcement Learning (RL). The chapter introduces RL concepts like Q learning and then details the connection between Deep Neural Networks (DNN) and RL. The chapter presents two fascinating case studies on—yield maximization and fruit detection. The yield case study shows how the RL agent can find the optimal strategy for setting the greenhouse parameters to maximize yield for cherry and tomatoes. In the second case study, RL is applied to identify products from the crop images. The chapter shows the architecture of the RL, implementation details, and results. The chapter is an essential milestone in linking theory and application, following Section II of the book.

Part II comprises a five-chapter dedicated to showing the use of DS in agriculture. Chapter "Computer Vision and Machine Learning in Agriculture" by Raval, Chaudhary, and Adinarayana link the sensing by computer vision (CV) and analysis by ML. Using PA as a domain, the chapter shows how CV and ML can be seamlessly integrated. It provides crucial discussions on public datasets and state-of-the-art algorithms for weed control, fruit detection, and plant phenotyping. The chapter ends with discussions on challenges and future paths. Chapter "An Architecture for Quality Centric Crop Production System by Kumar, Hiremath, and Chaudhary develops a recommendation system for farmers. The system helps farmers to map specific crop attributes with consumer requirements. The recommender uses endogenous and exogenous data obtained through IoT sensors and AI. Interestingly, the architecture is generic and can be extensible to a wide range of crops. Kumar J. et al. pens Chapter "Crop Classification for Precision Farming Using Machine Learning Algorithms and Sentinel-2 Data". The chapter focuses on remote sensing-based crop monitoring for small farm holding. It uses Sentinel—2 satellite images to delineate farm boundaries and perform crop classification using Random Forest (RF) classifier and Open—source Geographic Information system. The method is generic and applied to different sizes of farms and crops.

Chapter with the title "Machine Learning Approaches and Sentinel-2 Data in Crop Type Mapping," by Panjala, Gumma, and Teluguntla, uses Google Earth Engine (GEE) for crop monitoring. The chapter showcases evaluation of several ML supervised learning algorithms and spectral matching techniques using Sentinel—2 data to classify several crops. The chapter compares classifier results with agricultural statistics and shows that the RF classifier is the best crop classification. The last chapter of Part II, "Big Data Analytics for Climate-Resilient Food Supply Chains: Opportunities and Way Forward," written by Twarakavi et al., is about the reliable forecast of food production. The authors argue that current methods for forecasting food production levels are rudimentary and not scalable. The chapter develops and validates methodology by combining weather forecasts, remote sensing, scalable machine-learning methods, and cloud computing to estimate production risk at regional levels.

While Part II is focused on agriculture, Part III brings attention to Natural Resource Management with five chapters. Chapter "Machine Learning Algorithms for Optical Remote Sensing Data Classification and Analysis" by Reddy and Kumar A., introduces advanced ML algorithms for remote sensing analysis and classification. The RF, SVM, and Classification and Regression Tree (CART) were applied on GEE platform to derive land-use-land-cover (LULC) classes. The chapter demonstrates the immense potential for advanced ML algorithms in remote sensing data classification. Chapter "Geo-Big Data in Digital Augmentation and Accelerating Sustainable Agroecosystems"—presents ongoing digital augmentation efforts for rice fallows in India. The authors, Krishna and Biradar, argue about the need for augmentation and digitization to accelerate an agroecological transition. The chapter shows the capabilities and limitations of mapping rice-based farming systems using Sentinel series data during the *Kharif* season in a monsoon climate. The chapter shows a novel approach for multi-scale mapping (at state or region level) to micro-scale mapping (at district/village level) to assist informed decision-making.

Chapter "Transforming Soil Paradigms with Machine Learning" by Kumari S. et al.—brings attention to the most important natural resource—soil. The chapter shows that ML and BDA can be effectively used to predict, identify, and classify soil resources. Primarily, it shows the use of Digital Soil Mapping (DSM) and ML algorithms to predict soil properties. The soil-related studies continue in Chapter "Remote Sensing and Machine Learning for Identification of Salt-affected Soils". The authors Kumar N. et al. showcase ML techniques in identifying salt-affected soils (SAS) in Indo-Gangetic plains in India. The chapter studies salinity indices and principal components on images of the Landsat—8 satellite. The results were validated with field observations and GEE data. The final chapter, 14, "Geoportal Platforms for Sustainable Management of Natural Resources" by Reddy, uncovers Geo-portals' role in disseminating geospatial information. The chapter weaves the exciting development of the Bhoomi geoportal and its services. It provides Web Map Services (WMS) to access, query, visualize, and disseminate information. The chapter also highlights fundamental challenges in the development of such portals, such as cross-domain thematic applications.

Nagpur, India
Surat, India
Mumbai, India
Ahmedabad, India
July 2021

G. P. Obi Reddy
Mehul S. Raval
J. Adinarayana
Sanjay Chaudhary

Acknowledgements

Writing or editing a book with excellence is an asymptotic process towards perfection—there is always a scope to improvise. The editors are eternally grateful to all the authors of the book—Nirmal Kumar, Shakti Mishra, Srikrishnan Divakaran, Ratnik Gandhi, Vikaskumar, Deepak Hiremath, Jay Prakash Kumar, Dimple Singhania, Sanket Nayankumar Patel, Mihir Dakwala, Pranay Panjala, Murali Krishna Gumma, Parthasaradhi Teluguntla, Navin Kumar C. Twarakavi, Kamal Das, Mohamed Akram Zaytar, Fred Otieno, Jitendra Singh, Bruno Silva, Komminist Weldemariam, Shantanu Godbole, K. C. Arun Kumar, Gopal Krishna and Chandrashekhar Biradar, Kumari Sweta, S. Dharumarajan, Amar Suputhra, M. Lalitha, R. Vasundhara, B. Kalaiselvi, B. N. Shashi Kumar, A. Sathish, Rajendra Hegde, M. S. S. Nagaraju, and R. K. Naitam. The book would not have seen a timely outcome without their expert contribution and real time responses to reviewers' comments.

Big thanks to Professor Seishi Ninomiya, International Field Phenomics Research Laboratory, Graduate School of Agriculture and Life Sciences, the University of Tokyo for penning Foreword and words of advice.

The editors express their gratitude to all associated with 8th International Conference on Big Data Analytics conference (BDA2020) conference and co-located Virtual Workshop on Data Science for Agriculture and Natural Resource Management (DSANRM2020), where the seeds of the book were sown.

The editors would like to thank leadership at their parent organizations—ICAR-National Bureau of Soil Survey and Land Use Planning (ICAR—NBSS&LUP), Nagpur, India; Ahmedabad University, Ahmedabad, India; and Indian Institute of Technology Bombay (IITB), Mumbai for providing time and freedom to work on this book project.

The editors also acknowledge the support from Prof. Dr. Janusz Kacprzyk Series Editor "Big Data" of Springer, and Jennifer Sweety Johnson and Suresh Dharmalingam from Springer Nature for project coordination of the book.

Nagpur, India
Surat, India
Mumbai, India
Ahmedabad, India
July 2021

G. P. Obi Reddy
Mehul S. Raval
J. Adinarayana
Sanjay Chaudhary

Contents

About the Editors

G. P. Obi Reddy holds Ph.D. and is working as Principal Scientist in the field of remote sensing and GIS applications at Division of Remote Sensing Application, ICAR-National Bureau of Soil Survey and Land Use Planning, Nagpur, India. He has significantly contributed in the field of remote sensing along with GIS applications in digital terrain modelling, landforms mapping, soil-landscape modelling, land degradation assessment, agroecology and development of soil information systems. He is instrumental in the design and development of the Land Resource Information System (LRIS) of India and Bhoomi geoportal. On ICAR deputation, he visited Sri Lanka, The Netherlands, Nepal and South Africa. He has published 102 research articles in reputed national/international journals, 7 books, 53 book chapters and 15 technical bulletins. Currently he is acting as a Chairman, Data Content Standards of DST-NSDI and Member, ISO/TC 211/WG 04 on "Geospatial Services". He is Recipient of Indian National Geospatial Award-2007 and National Geospatial Award for Excellence-2013 from Indian Society of Remote Sensing, Dehradun, Outstanding Scientist Award-2016–17 from ICAR-NBSS&LUP, Nagpur and Fellow of Indian Society of Soil Survey and Land Use Planning, Nagpur.

Mehul S. Raval holds Ph.D. and is Associate Dean—Experiential Learning and Professor at Ahmedabad University, India. His research interest includes computer vision, and he has contributed to problems in surveillance, medical imaging, biometrics and agriculture. His academic pursuits involve visits to under Sakura Science Fellowship (2015) to Okayama University, Japan, Argosy visiting Associate Professor at Olin College of Engineering, MA, the USA (2016), and Sacred Heart University, the USA (2019). He serves as Member, technical program committee, for leading national and international conferences, workshops and symposiums. Currently, he chairs IEEE Computational Intelligence Society—Gujarat Chapter and served in IEEE Gujarat section under various capacities. He has received research grants from Board of Research in Nuclear Science and DST—Govt. of India. He has published 79 scholarly works in journals, magazines, conferences, workshops at the national and international stage. He reviews IEEE, ACM, Springer, Elsevier, IET and other leading publishers and has supervised three Ph.D. students.

J. Adinarayana holds Ph.D. and is working as Teaching/Research Faculty Member from 1986 and currently Institute Chair Professor at Centre of Studies in Resources Engineering (CSRE), IIT Bombay, India. His areas of expertise include agro-informatics in contemporary agriculture. As a Team Leader, he led various interdisciplinary national and international R&D projects on digital agriculture. He is the Immediate Past President of Asia- Pacific Federation for Information Technology in Agriculture (APFITA) Board and the Current President of INSAIT. He served as an Editorial Board Member of 'Regional Geoderma' (Elsevier) & CIGR journals. Currently, he is a Member, CIGR Board on Technical Section VII (IT Systems); Vice-Chair of Agriculture and Open & Sharing Data Working Groups of the Asia-Pacific Advanced Network (APAN), Expert Committee Member in various national organizations, including ICAR, ISRO, MoA&FW and DST. He is Recipient of JSPS Invitation Fellowship, DST Young Scientist Project, INSA Visiting Scientist/Faculty awards, Visiting and Adjunct Faculty/Scientist at Univ. of Tokyo (2013), Univ. of Bonn (2007), Univ. of Aston/UK (1991); Edith Cowan Univ./Perth (2010-17). He guided about 12 Ph.D.s and more than 30 M.Techs and presented research results in more than 100 research papers/book chapters in International Journals/publishers.

Sanjay Chaudhary holds Ph.D. and is Professor at the School of Engineering and Applied Science and Dean of Students of Ahmedabad University. During 2001 to 2013, he was Professor as well as Dean (Academic Programs) at Dhirubhai Ambani Institute of Information and Communication Technology (DA-IICT), Gandhinagar, India. His research areas are cloud computing, blockchain technology, big data analytics and ICT applications in agriculture and rural development. He has authored eight books and nine book chapters. He has published more than 125 research papers in international conferences, workshops and journals. He has received research grants from leading organizations including IBM, Microsoft and Department of Science and Technology, Govt. of India. He is Vice President of INSAIT. He has guided seven Ph.D. students and more than 32 M.Tech. students. He holds a doctorate degree in computer science from Gujarat Vidyapith.

Part I
Data Science—Principles, Concepts and Applications

Data Science—Algorithms and Applications in Earth Observation

G. P. Obi Reddy and Nirmal Kumar

Abstract Data Science (DS) is a fast-evolving interdisciplinary field and it encompasses various scientific approaches, procedures, and systems to abstract information, and insights from large datasets. The vital components of DS are identifying the reliable data sources, data curation, model building, planning and evaluation, results communication, and visualization. DS tools are being widely used for data collection, storage, extraction, cleaning, analysis, and visualization. In recent times, DS is enriched by various techniques like Artificial Intelligence (AI), Machine Learning (ML), Data Mining (DM), predictive analysis, pattern recognition, and visualization. Decision tree algorithms, Support Vector Machines (SVM), K-Nearest Neighbor (KNN), Logistic Regression (LR), Neural networks are some of the important ML algorithms. In recent years, Earth Observation (EO) satellites generated voluminous spatial data, and integration of these datasets with ground-based observational data significantly enhanced the capabilities in mapping, monitoring, and forecasting various earth system processes. ML algorithms have vast potential in monitoring earth resources to provide timely information for site-specific interventions, and implementation. This chapter aims to illustrate various DS tools, algorithms, and the potential applications of ML algorithms in important EO applications like drought monitoring, vegetation monitoring and assessment, digital soil mapping, soil moisture monitoring, Soil Organic Carbon (SOC) assessment, crop type mapping, and precision agriculture. The challenges associated with DS applications in EO are also discussed in the chapter.

Keywords Data Science · Data engineering · Earth observation · Machine learning

---

G. P. O. Reddy (✉) · N. Kumar
ICAR-National Bureau of Soil Survey and Land Use Planning, Amravati Road, Nagpur 440033, India
e-mail: GPO.Reddy@icar.gov.in

G. P. O. Reddy et al. (eds.), *Data Science in Agriculture and Natural Resource Management*, Studies in Big Data 96, https://doi.org/10.1007/978-981-16-5847-1_1

1 Introduction

Earth Observation (EO) technologies and applications play a pivotal role in achieving global challenges like no poverty, zero hunger, and climate change (Andries et al. 2019). EO technologies aimed to gather and record information about a point of interest with the help of sensors usually mounted on a satellite without contacting the object physically (Campbell and Wynne 2011). In recent years, digital data produced through EO technologies are playing a key role in monitoring, and analyzing various physical, chemical and biological processes of planet Earth (Yang et al. 2013). In general, EO data acquired in the form of images collected by various sensors in different time scales are cost-effective and time-efficient to provide solutions at global, regional, and local scales (Roy et al. 2017; Reddy 2018a). The United Nations recognized the potential of EO data to supplement the official statistics generation processes, and complement traditional sources of various environmental data (UN 2017). Time-series remote sensing technologies have demonstrated their potential in EO especially in mapping, and monitoring of earth resources to realize sustainable development (Reddy et al. 2017; Reddy and Singh 2018). To address various emerging issues, there is a need to apply new approaches, and techniques to effectively process, and analyze voluminous EO data received from the various ground, air, and space-borne sensors. In recent years, a host of data science-based Machine Learning (ML) algorithms like supervised, and unsupervised learning, K-means, neural net processing, and pattern recognition have enabled us to transform the field of EO research and applications.

Data Science (DS) covers various scientific approaches, procedures, and systems, which are used to abstract information, and insights from large datasets (Dhar 2013). DS applies mathematical, and statistical concepts and computer tools for processing, extract insights and information from big data. The distributed processing, monitoring of workflows, visualization, and performance monitoring are core parts of the DS domain. DS techniques help in finding the useful patterns, and relationships within the data. In recent years, many organizations adopted DS tools for collection, storing, and processing of data as a part of their regular operations, and services. Artificial Intelligence (AI) and ML are related to each other and are often used interchangeably in DS. AI is the superset; it has ML and Deep Learning (DL) as subsets. AI describes machines that can execute tasks resembling those of humans through artificially modeling human intelligence. ML is a subsection of AI that device means by which systems can automatically learn, and improve from experience. ML algorithms play a vital role in effectively handling EO data as these datasets are huge in quantity, and formats. The focus of ML is the depiction of the input data, and simplification of patterns for use from big data.

In the past three decades, the advanced capabilities in computing and sensor technology have tremendously transformed the field of remote sensing, data collection systems, processing, and producing valuable products, and services. The important portals like USGS Earth Explorer and Google Earth Engine (GEE) (Gorelick et al. 2017), Copernicus Sentinel Hub (https://scihub.copernicus.eu/), and ISRO-Bhuvan (https://bhuvan.nrsc.gov.in/home/index.php) provide public access to obtain enormous EO resources data to the scientific communities to build various EO applications. Integration of data obtained from space-based EO satellites and field-based observations is a formidable combination to develop complex, and comprehensive data-driven EO applications. In addition, multi-resolution, and multi-temporal EO data divulge various planet earth processes, and features, which are not able to observe through conventional methods. Thus, the frequency and coverage of large quantities of EO data provide voluminous data in the form of images, and maps, such large datasets could not be easily generated through ground-based technologies alone. Advances made in spectral, spatial, and temporal resolutions of remote sensing sensors technology, and ever-growing number, and diversity of EO satellites, implementation of open data access policies, and ever-increasing demand for innovative applications prompted to adopt innovative approaches, and platforms to store, manage, process, and analyze voluminous EO data (Guo 2017; Dhu et al. 2017; Baumann et al. 2019). Data generated through EO technologies has immense potential in developing the applications and services in managing the planet's resources and environmental monitoring. However, processing large quantities of EO data, and converting it into useful information necessitates robust computational resources and it remains a major challenge to build tailor-made applications to meet the user requirements (Giuliani et al. 2017). This chapter aimed to discuss various components, tools, techniques, and potential applications of DS in emerging EO like drought monitoring, vegetation monitoring and assessment, digital soil mapping, soil moisture monitoring, SOC assessment, crop mapping, and precision agriculture. The performance of the RF algorithm in SOC assessment, and challenges associated with DS applications in EO were also discussed in the chapter.

2 Data Science and Its Components

DS principles and concepts help to mine both structured and unstructured large datasets in order to recognize the patterns, trends, and extract insights. This is an inter-disciplinary field and it includes statistics, inference, computer science, predictive analytics, ML algorithms, and new technologies, which immensely helps to infer insights from big data. Big data in DS undergo numerous processes like data discovery, development, warehousing, cleaning, classification, analysis, and validation for visualization of patterns, and insights. (Ngiam and Khor 2019). Big data implies voluminous datasets that have been generated to develop possible applications and provide solutions through the extraction of value, and knowledge (Gandomi and Haider 2015). The main concerns in DS are efficient capture, storage, extraction, process, analysis, and visualization of the information from the enormous datasets. The important components of DS can be grouped into data discovery, data preparation, model planning, model building, model operationalization, results communication, and discovery/visualization (Fig. 1 and Table 1).

Fig. 1 Data science and its components (*Source* https://www.edureka.co/blog/what-is-data-science/)

3 Data Science: Tools

In DS, tools can be defined as an equipment or device that can be used to carry out a specific task. However, technique follows detailed scientific procedures to accomplish a particular task. Data scientists often follow operational methods/techniques on the data by using various devices as tools. The combination of tools and techniques often are used in data acquisition, refinement, manipulation, analysis, and visualization. DS tools are used at various stages of data processes such as data collection, storing, processing, analysis, and visualization. DS tools can be categorized according to important steps involved in the processing data like data acquisition, storage, extraction, cleaning, analysis, and visualization (Table 2).

Table 1 Various components of Data Science (DS) and their brief description

DS components	Description
Discovery	In DS, discovery includes problem identification, which is a fundamental and at the same time one of the important components. It includes the finding of relevant data, and their formats from numerous sources to accomplish basic necessities, priorities, and objectives of the task. While framing the research problem, the project objectives, manpower requirement, technology, timelines, budget, data, and the end goal are to be clearly identified
Data preparation	It includes data cleaning, reduction, integration, and transformation. It refers to identifying the data gaps, incorrect, unrelated parts of the data to replace, modify, update, or delete the unwanted data. After performing these tasks, the cleaned data can be used for further process, and analysis. In this phase, data engineering plays a key role in data acquiring, storing, retrieving, and transforming the data, and developing metadata of the data
Model building	Model building is a critical stage and its selection depends on the project objectives, nature, and volume of data. In this stage, depending on the model, method, or technique selected, model variables are to be differentiated. In model building, datasets are to be developed for training as well as validation purposes. Association, classification as well as clustering methods can be adopted in building the robust models by using popular model building tools
Model planning	In this phase, the relationship between the input variables needs to be clearly determined depending on the methods, and techniques adopted. The tools like exploratory data analytics, and visualization can be used to comprehend the relations between variables. The potential tools like SQL Analysis Services, R, SAS, and Python can be used in the model planning process
Model evaluation	Methods for evaluating the performance of the adopted model can be divided into two categories, namely holdout, and cross-validation, which uses a test dataset to evaluate the model performance, its accuracy, and other characteristics. At this stage, if needed the required changes can be made in the model to get the best results. In case the desired accuracy is not obtained, the model building process is to be revisited, select the appropriate model, and then perform the model evaluation to obtain the best result as per the objectives
Visualization	In DS, data visualization is an important phase and assumes greater importance, where representing data in a visual context to easily understand the importance and limitations of data. Various data visualization tools enhance the capabilities in the representation of data in a graphical form. Visualization is also an important component in data discovery, and decision-making process. Data visualization helps to translate information into easily understandable formats such as a map or graph to abstract the insights, identify the patterns, trends, and outliers in large datasets

(continued)

Table 1 (continued)

DS components	Description
Results communication	It is an important component of DS to communicate the results of the study to the intended audience to reach the goal set in the initial phase. The core objective of this phase is to bring out final technical reports by focusing on the highlights of the study. In results communication, the technical details of the task completed need to be furnished for its full deployment, implementation, and findings

4 Data Science: Techniques

The important DS techniques are AI, ML, natural learning process, data modeling, probability and statistics, predictive analysis, contextual analysis, DM, pattern recognition, and visualization (Fig. 2). Suitable DS techniques to be used to extract important information from heterogeneous structured or unstructured data.

4.1 Artificial Intelligence (AI)

AI applies the principles of science and engineering in making smart machines specifically with intelligent computer programs (McCarthy et al. 2020). It enables the machines with the ability to mimic human behavior, particularly with cognitive functions. In recent times, the scope of AI has increased many folds as the intelligence of smart machines embedded with ML capabilities created huge impacts in various fields (Davenport et al. 2020). AI gives machine thinking abilities to achieve higher efficiency in smart ways in a time-efficient and cost-effective manner. AI can make systems self-dependent and has the ability of decision making by using different algorithms. AI has the ability to augment the automation process, business outcomes, cost reduction, and revenue generation.

4.2 Internet of Things (IoT)

IoT provides adaption properties to the machine and it helps the machine to communicate with the other machine to achieve stabilization among different machines connected in the network. IoT plays an important role in application development by using data analytics tools. IoT architecture consists of perception, network, middleware, application, and business layers (Atzori et al. 2010). IoT enables the remote handling of the equipment and it increases the operating range of the network. IoT has the ability to transform the present world to realize more efficient industries, automated machinery control, smart farming, and smarter cities. IoT applications

Table 2 Data science tools and their utilities

Data science tools	Utilities
Data acquisition tools	Data can be acquired through various equipment/devices such as satellites, in situ measurements, online portals and surveys, personnel interviews, etc. Data generated through different sources need to be cleaned and converted into a usable format to perform various data analysis tasks. "CREODIAS" as a cloud platform (https://creodias.eu/) facilitates to perform operations like search, view, and process satellite products, and develop applications in EO. "WEkEO" (https://www.wekeo.eu/) provides Copernicus data covering Sentinel satellites data in various fields EO. It also offers cloud computing packages with integrated service delivery tools to develop business solutions
Data storage tools	Data storage tools are used widely for storing large amounts of data through shared computers. These tools offer a platform to integrate servers to access data seamlessly. "Microsoft Excel" is extensively used to handle small to medium size of datasets. "Apache Cassandra" tool uses SQL (Structured Query Language) and CSL (Cassandra Structure Language) to connect with the database and retrieve required data from the stored data in different servers. "Hadoop" has the capabilities to store a large volume of data and perform various data processing tasks in a distributed architecture. "Apache Hadoop" facilitates the storage of high volume of data among clusters of systems and performs various computations with ease. "Hive" as a data warehouse architecture facilitates to quickly query the data stored in several databases, and file systems
Data extraction tools	Data extraction tools or web scraping tools allow the users to extract information, and data automatically from websites. "OctoParse" (https://www.octoparse.com/) provides output in structured spreadsheets for easy to use and perform further operations. "Content Grabber" can be used to extract data from web sources and provide structured data as output. Web scraping tools can also be made into functional Application Programming Interfaces (APIs) for use by researchers as APIs create interfaces for programmers to easily access, and process published data
Data cleaning tools	These tools are often integrated with databases and are time-saving in searching, sorting, and filtering data (Blei and Smyth 2017). "Data Cleaner" tool works to improve the quality of data through the removal of data redundancy and transforming it into a single record. "OpenRefine" tool cleans data before performing the data transformation task to convert it into the desired form
Data analysis tools	These tools perform various data analysis tasks including data modeling to derive meaningful information from the input dataset and it helps in the decision-making process to determine the problem. "R programming language" is widely used in performing statistical computations, and graphics. "Python" programming language is also used in application development and the outputs in CSV formats. Similar data analysis tools like SAS, Jupyter, R Studio, MATLAB, and Excel were also widely used in data analysis

(continued)

Table 2 (continued)

Data science tools	Utilities
Data visualization tools	The important data visualization tools are R, Jupyter, and Tableau. "Python" provides data visualization facilities with a wide range of graphical representations of data. "Tableau" software provides various user-friendly data visualization capabilities to visualize the results in various forms. "Orange" is another data visualization tool and it supports various visualization tasks like data extraction and analysis. DataWrapper, Qlik, and Gephi are some of the popular data visualization tools that support data visualization with the excellent support of graphical representation

Fig. 2 Data science techniques (*Source* Sood and Rinehart 2016)

are being widely used in big data analytics, and cloud computing (Baseca et al. 2019). The capabilities of IoTs were widely applied in various fields like management, monitoring, control, and unmanned machinery systems (Ojha et al. 2015). Security services such as privacy, and authentication are some of the crucial factors for modern technologies to realize the full potential of IoT services (Atzori et al. 2017).

4.3 Data Mining (DM)

DM is a subdomain of AI and it denotes the extraction of hidden analytical information from big data (Weiss and Indurkhya 1998). DM techniques originated from the classical methods of statistics, databases, parallel computing, pattern recognition, AI, and visualization. It extracts the hitherto unknown, predictive information, hidden patterns from the available big data. It also helps in sorting the particular information from a huge dataset, and provides proper information to the particular machines. The important DM techniques like clustering, patterns recognition and classification, association, regression, outlier detection, and prediction are being widely used in various applications. Knowledge Discovery in Databases (KDD) approach of DM is useful to identify the underlying relationships, and characterize the knowledge in a given dataset (Hira and Deshpande 2015).

4.4 Data Visualization

Data visualization elucidates the significance of data by representing it in a visual context for its easy understanding. In general, the complex patterns, trends, and correlations in big data are difficult to detect and comprehend, however, by applying data visualization techniques, which can be transformed in the form of graphs, charts, scattered plots, gauges, etc. Innovative data visualization tools offer advanced functionalities and allow data analytics at several levels with the support of human intelligence, and reasoning capabilities. In agriculture, with the use of appropriate visualization techniques, it is possible to find out the patterns, connections, or similarities in the observed datasets. GPS-enabled tractors and field sensors generate voluminous spatial data that are not so easy to analyze and understand. However, the deployment of visualization techniques helps to transform them in a graphical format to easily understand and optimize the resources.

4.5 Machine Learning (ML)

ML algorithms were grouped into three categories, which include supervised, unsupervised, and reinforcement learning (RL) (Fig. 3). The supervised algorithms need output values in the training data (Berry et al. 2019), on the other hand, unsupervised algorithms need the input values in the training dataset to find out the hidden patterns in the dataset through clustering or dimension reduction algorithms (Baştanlar and Özuysal 2014). Whereas, while building the systems, RL algorithms learn from the interface with the environment by using rewards, and punishment rulesets, where every event is random (Osband et al. 2020; Busoniu et al. 2008). ML algorithms were widely applied in various fields of EO like smart agriculture, crop monitoring, yield

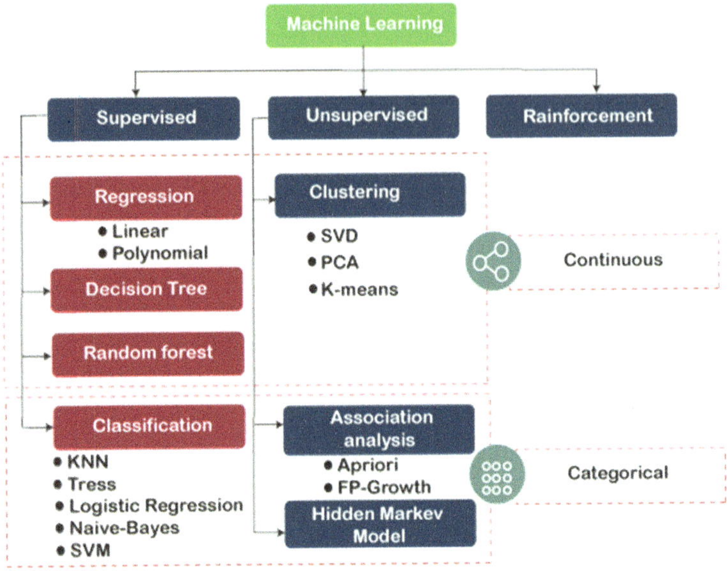

Fig. 3 Important categories of machine learning algorithms (*Source* https://www.edureka.co/blog/machine-learning-algorithms/)

prediction, disease detection, prediction of soil properties, soil and water management, etc. ML algorithms have the ability to overcome human limitations in terms of speed, accuracy, reliability, consistency, and transparency in performing the assigned tasks. The absence of quality and consistency, and any inherent biases within the dataset always pose hindrances in realizing the full potential of ML algorithms.

4.5.1 Supervised Machine Learning

Supervised learning involves training a system as per the labeled data to predict the output from every input as it trains the system according to its expectations. In this process, trained systems give predictions in classification and detect fault (Cui et al. 2019). This method helps to approximate a function between the input, and output data. Here, the system learns the training dataset's classifiers thereafter automatically applies this classification to an unknown dataset. Regression analysis (Linear and polynomial regression), Decision Tree (DT) algorithms, Random Forest (RF), and classification algorithms like K-Nearest Neighbor (K-NN), Logistic regression (LR), Naive Bayes Classifier (NBC), and Support Vector Machine (SVM) are the examples of supervised ML algorithms.

Regression Analysis

It is used to estimate the relationship among the variables, which have a reason, and result relationships (dependent and independent variable). However, multilinear regression models work with one dependent variable, and more than one independent variable. Linear regression works with straight regression lines and predicts the dependent variable outcomes based on the independent variables. Polynomial regression works with curved regression lines to predict the best fitting data points. LR works with success/failure probabilities, and binary variables to predict possible outcomes. The applications of these techniques are directly associated with discrete, and continuous distribution of variables. Regression analysis is a widely used technique to determine the correlations between several variables based on experimental or observed data (Lei et al. 2016). Selecting the right regression technique to perform the task depends on the type of dataset.

Decision Tree

DT learning is a popular classification technique used in ML applications. The DT algorithm offers support in decision-making, and it defines the logical structure, associated uncertainties, and constructive results of the decisions (Khader et al. 2013). DT algorithms contain a hierarchy of both internal as well as external nodes, which are linked by branches. In the model, a serial tree represents a logical model as a binary tree constructed using a training dataset to foresee the value of a target variable by making use of predictor variables. In the DT model, each node generates its child nodes until either the subgroups are too small to undertake the similar meaningful division or further splitting is not possible to produce statistically significant subgroups. Some sections in the sample may have outcomes in a big tree, and some of the links may generate outliers. Such branches are required to be removed. DT models are efficient in multi-feature extraction, and removing outliers without overfitting. DT algorithms like Chi-square–Automatic–Interaction–Detection (CHID), and Classification and Regression Tree (CART) are widely used to build decision trees.

Random Forest (RF)

As a "tree-based" classifier and extension to the CART algorithm, RF improves the overall model performance (Breiman 2001) and is applied to the prediction of discrete as well as categorical variables. In the RF algorithm, many trees are created, however, while growing the trees no pruning takes place. In RF for every individual tree, a subset of the prediction variables is only used. In the RF algorithm, the input dataset determines the number of predictors to be used in building each tree as well as the number of trees to be built in the forest (Liaw and Wiener 2018). In the RF model, each tree is built depending upon the sample dataset selected from the total dataset by adopting the bootstrap process (Efron and Gong 1983). RF algorithm has many advantages in the analysis of large datasets such as temporal satellite images.

Classification Algorithms

Classification algorithms used in ML utilize the input training data to predict the likelihood or probability of the data that follows one of the predetermined categories. The important classification algorithms like K-Nearest Neighbors (K-NN), LR, Naive Bayes, and SVM are briefly defined in Table 3.

4.5.2 Unsupervised Algorithms

Unsupervised algorithms recognize the patterns in input data without the participation of a target attribute. These algorithms use all the variables in the analysis as inputs, hence, these algorithms perform well in clustering, and association data mining tasks. Such algorithms perform without supervision and have the ability to continue independently with training based on unlabeled data input (Cui et al. 2019). These algorithms work well in creating the labels in the input data and subsequently use the same in the implementation of supervised learning tasks (Hofmann 2001). Unsupervised clustering algorithms have the ability to identify inherent groupings within the unlabeled data and assign labels to individual data values (Marshland 2015). Whereas, unsupervised association mining algorithms perform well in identifying the rules that precisely characterize the relationships between attributes. Clustering algorithms, a priori algorithms, combinatorial algorithms, and Self-Organizing Maps (SOM) come under this category.

Clustering Analysis

In the field of DM, clustering algorithms used to classify the input data into different groups depends on the resemblances, traits, and behavior of data (Jain et al. 1999). Clustering involves the categorization of data based on their similar properties. Clustering predictive models are a widely used method of data analysis, which segregates the dataset into subsections of similar items (Cui et al. 2019). Clustering sorts the data into different clusters, where the data elements have high similarity within the clusters but are different from the data elements in the other clusters. Singular Value Decomposition (SVD), Principal Component Analysis (PCA), and K-Means are the important ML clustering algorithms, which are briefly described below (Table 4).

Table 3 Important classification algorithms and their functions

Classification algorithms	Functions
K-NN	K-NN is the supervised algorithm used in DM and ML, where the learning is based on "how similar" is a data from other. It accepts the similarity that exists between the new case and existing cases and it places the new case into the group that is most similar to the existing groups. The algorithm first stores the entire available data and then classifies a new data point depending on the similarity. As a nonparametric algorithm, K-NN doesn't make any hypothesis on the primary data. The potential of K-NN algorithm can be leveraged in mining unstructured datasets with respect to their distribution
LR	It is basically a statistical technique, which is similar to linear regression (Hosmer and Lemeshow 2000). LR equation predicts an outcome for a binary variable *Y*, from one or more response variables *X*. Since the model doesn't firmly need continuous data, the response variable can be either categorical or continuous. The LR model is best suitable to define and test the hypotheses about relationships between a categorical outcome variable, and one/more categorical/continuous predictor variables. The core advantage of LR is that it can indeed provide probabilities, and extend to multi-class classification difficulties (Hastie et al. 2009)
Naive bayes	These classifiers are relying on Bayes's probability theorem, and describe the relationship of conditional probabilities of statistical quantities. When Naive Bayes classifier is used to create clusters, it basically uses conditional probability principle to assign data values to classes (Dey 2016; Pavithra and Jayanthi 2018). Bayesian classifiers follow a rational approach to update the likelihood of hypotheses in the context of new evidence (Berry 1996)
SVM	Cortes and Vapnik (1995) introduced SVM as a supervised learning algorithm and it performs binary classification. Through a hyperplane, it classifies datasets into different classes. In the process, it reduces the classes and maximizes the margin between classes to provide unique distinctions. SVM performs well with nonlinear patterns as it converts the original data into new features. It also works well with low-dimensional datasets as it has robust capabilities to generalize the information (Li and Zhao 2009). The challenges associated with SVM algorithm performance are the selection of a suitable kernel, and optimum kernel parameters

Association Analysis

Association Analysis is basically a rule-based technique widely used to draw relationships between large datasets. It derives significant rules in big data to measure some interestingness. Association rules are highly popular with DM processes in categorical databases. This technique has grown its importance for marketing campaigns, finding hidden correlations in big data, bioinformatics, website performances, and countless other applications in almost every economic sector. "A priori algorithm" is commonly used in extracting the frequent itemsets from big data and discovering the

Table 4 Important clustering ML algorithms and their functions

Clustering algorithms	Functions
SVD	SVD is used as a robust algorithm for the decomposition of orthogonal matrices, where the main function of SVD is its relationship with the rank of matrices, and its capability to estimate the matrices from an allotted rank (Swathi et al. 2017). SVD has the ability to reduce large datasets comprising a number of values to fewer datasets with significant values. In this process, it preserves the correlation of variables of the original dataset. SVD is commonly used in matrix rank estimation, canonical correlation analysis, image compression, image recovery, and spectral clustering
PCA	As a dimensionality reduction algorithm, PCA is being widely used to make visualizations of the complex datasets through identification of the patterns in the datasets and make correlations of variables. PCA projects these identified correlations on a similar dimensional subspace. PCA is being widely used in all forms of analysis in extracting relevant data from complex datasets. PCA was widely used to detect changes in time-series data as it has the ability to enhance even subtle modifications. PCA method initially calculates the covariance matrix, eigenvalues, and eigenvectors for input data, subsequently gets the percentage of the total dataset variance, then a series of new data is obtained by multiplying the eigenvector for the original input data (Lasaponara 2006)
K-Means	MacQueen (1967) proposed the K-Means algorithm as an unsupervised ML algorithm for clustering. Partition-based clustering aimed to cluster big datasets, wherein a number of objects are considered first, then partitioned the objects into a number of groups, where each group comprises similar data points (Neha and Vidyavathi 2015). It partitions the dataset based on the mean value of the given dataset and iteratively processes until no more partitions are available. K-Means algorithm partitions the dataset with an assumption that the number of clusters is fixed. The optimum number of clusters is to be chosen to achieve the better performance of this algorithm. (Shafeeq and Hareesha 2012)

association rule to notice the information. It makes multiple iterations in the database to find out frequent itemsets, in this process the algorithm uses k-itemsets to make k + 1-itemsets (Rao and Gupta 2012). The main limitation of the Apriori algorithm is that it consumes more time to clutch a vast number of candidate sets with frequent itemsets.

4.5.3 Reinforcement Learning

As a new type of learning algorithm, RL algorithms have gained importance in recent times. RL is the learning of a mapping from situations to actions to optimize a reward, or reinforcement signal. In the RL algorithm, an agent learns through trial-and-error communication with its situation where every event is random (Busoniu et al. 2008). This algorithm attempts a cluster of different things and rewards when it does something well. In this algorithm, the user doesn't give the machine examples of correct input–output pairs, however, it provides a technique for the machine for quantification of its performance.

5 Data Science Applications in EO

EO data offer a wealth of data and information for various EO applications, however, the user needs various tools and techniques to transform this raw data into useful information (Müller et al. 2015). Though EO data is widely used in various EO applications, a better understanding of the processes in transforming the raw data into useful information is critical, where ML techniques significantly contribute to ease the process and computations. Emerging technology like cloud computing offers access to a wide range of robust computing resources often at lower or no prices. Cloud-based geospatial platform of Google Earth Engine (GEE) (https://earthengine.google.com/) offers the most common time-series remote sensing data like Landsat, Sentinel, and MODIS for free of cost and can be processed without downloading the datasets for various EO applications (Pekel et al. 2016). In recent years, ML algorithms were widely used in various EO applications like Neural Networks in the estimation of soil moisture (Li et al. 2020), Support Vector Machines in prediction accuracy of soil properties (Deissa et al. 2020), Decision Trees in land cover classification (Phiri et al. 2020), and Random Forests in the mapping of coastal plain wetlands (Zhang et al. 2019). The potential of ML techniques in some of the important EO applications like drought monitoring studies, vegetation monitoring and assessment, digital soil mapping, soil moisture monitoring, prediction of soil organic carbon with a case study, crop type mapping and monitoring, and precision agriculture were discussed below.

5.1 Drought Monitoring Studies

Drought is one of the important natural hazards that occurs in all climatic regions, especially in arid and semi-arid regions of the planet (Sivakumar et al. 2014; Sandeep et al. 2021; Kumar et al. 2021). ML algorithms like RF and DT were used in drought monitoring as they provide valuable inputs in the decision-making process and minimize the subjectivity in analysis and obtain the critical inputs (Tyralis et al. 2019). RF algorithm works in conjunction with several CART models trained with bootstrap aggregation (Breiman 1996). Kuswanto and Naufal (2019) used ML algorithms of CART and RF to classify the droughts and reported that both algorithms provide high efficiency in the prediction of droughts. PCA eliminates redundancy by converting the original feature space axes, and shows drought information in a new model space without correlation (Lasaponara 2006). Alizadeh and Nikoo (2018) study the drought events by using data fusion techniques with high-resolution remotely sensed data at feature level and reported that ML approaches of RF, Support Vector Regression (SVR), and Artificial Neural Networks (ANN) showed excellent performance in analysis and obtaining the drought information.

5.2 Vegetation Monitoring and Assessment

ML algorithms are being widely used in conjunction with remotely sensed data in EO applications especially in Land Use/Land Cover (LULC) classification and monitoring of vegetation dynamics (Abdi 2019). The information generated through time-series satellites data is being widely used in various EO applications and created detailed maps in the field of vegetation monitoring (Reddy et al. 2020) and land-use changes. RF has shown its potential in LULC classification and obtained the best classification outputs from geo-big data computations (Belgiu and Drăut 2016). DL algorithms were used in remote sensing data analysis including scene classification, LULC analysis, and object detection (Chen et al. 2014). Srivastava et al. (2012) evaluated the robustness of the ANN and SVM algorithms in the study of LULC dynamics and reported that the ANN performs better as compare to SVM. Whereas, Keshtkar et al. (2017) advocated that the SVM classifier performed better as compared to RF in the classification of various types of land cover. However, the selection of an appropriate algorithm for a specific task is important, keeping in view the merits of ML algorithms and large input parameters needed to obtain accurate results in vegetation monitoring and assessment (Breiman et al. 1984).

5.3 Digital Soil Mapping

Soil mapping process involves soil resource inventory, classification, and mapping (Kempen et al. 2012). In soil mapping, the information on environmental variables like climatic conditions, lithology/parent material, landforms, and vegetation was considered in establishing soil–landscape relationship (Reddy et al. 2013). Digital Soil Mapping (DSM) approach showed its potential in generating spatial information on soil parameters (McBratney et al. 2003). DSM techniques offer to generate soil properties distribution maps by applying numerical models, which takes into account various soil environmental covariates in deriving the heterogeneity in soil properties (Lagacherie and McBratney 2007). In DSM, tree-based ML algorithms and the simplest version of regression tree, and their variants were widely used (Taghizadeh-Mehrjardi et al. 2016). Many authors used RF algorithms in DSM to derive various soil properties (Tziachris et al. 2019; Vaysse and Lagacherie 2015). Quantile regression forest was used in mapping the uncertainty involved in the prediction of various soil parameters (Vaysse and Lagacherie 2017). A number of studies used ML algorithms like neural networks (Lamichhane et al. 2019), ANN (Dai et al. 2014), and SVM (Forkuor et al. 2017) in the mapping of soil properties.

5.4 Soil Moisture Monitoring

Monitoring of soil moisture dynamics assumes greater importance in crop planning, precision agriculture, sustainability of cropping systems, and mitigation of climate change. Conventional methods follow in situ measurements to estimate Soil Moisture Content (SMC), however, these methods are tedious, and time-consuming to cover larger areas in different time scales (Schmugge et al. 1980). Obtaining SMC estimates from EO satellites (Petropoulos et al. 2015) is necessary to acquire at various spatio-temporal scales to monitor the soil moisture dynamics in near real-time especially in precision agriculture. Adab et al. (2020) demonstrated RF, SVM, and ANN algorithms in the estimation of soil moisture, and reported that blending of surface reflectance properties with supplementary geospatial data improves the accuracy in estimation. Ahmad et al. (2010) compared the SVM algorithm with the estimates obtained from ANN, and Multivariate Linear Regression (MLR) algorithms and reported that SVM shows better performance in soil moisture estimation as compared to ANN and MLR algorithms.

5.5 Prediction of Soil Organic Carbon (SOC)

SOC is a critical soil property in soil health management, water-holding capacity, nutrient availability, and plant growth (Chen et al. 2019a). The soil forming factors like climate, organisms, relief, parent materials, and time determine the spatial variability of SOC at scales (Jenny 1994). In recent times, the use of ML algorithms in the prediction of SOC has significantly increased in the context of the availability of huge remotely sensed imageries, large datasets on climatic parameters, and Digital Elevation Models (DEM) at various resolutions in the public domain (Padarian et al. 2019). Time-series remote sensing data provides cost-effective, time-efficient, and reproducible spatial datasets for quantification of spatial variability in SOC (Gehl and Rice 2007). In SOC prediction, many authors used various multispectral optical satellite sensors data products, which include MODIS (Chen et al. 2019b), Landsat (Chen et al. 2019a), SPOT (Liu et al. 2015), and combined Landsat and MODIS (Emadi et al. 2020). Numerous ML algorithms like ANNs (Were et al. 2015; Zhao et al. 2010), SVR (Ballabio 2009), and multivariate adaptive regression splines (Rossel and Behrens 2010) were applied in SOC prediction. ML algorithms were proven their merits as compared to the traditional linear and geostatistical methods in the prediction of SOC, because of their higher capabilities in estimation at unsampled points.

5.5.1 Prediction of Soil Organic Carbon (SOC)- A Case Study

In order to assess the performance of RF algorithm in soil organic carbon prediction and to compare the results obtained from geostatistical models of Ordinary Kriging (OK), and Regression Kriging (RK), a case study from a heterogeneous landscape of Chhattisgarh plateau, Central India, has been selected. In the study, various terrain, bio-climatic, and productivity input variables were derived from DEM at 30 m of Shuttle Radar Topography Mission (SRTM), WorldClim, and Moderate Resolution Imaging Spectroradiometer (MODIS)-based Normalized Difference Vegetation Index (NDVI) at 250 m, respectively. Under terrain variables, elevation, aspect, slope, Topographic Wetness Index (TWI), Topographic Position Index (TPI), multiresolution ridge top flatness, multiresolution valley bottom flatness, and mean slope position were computed. Under bio-climatic variables mean annual temperature, maximum temperature of warmest month, minimum temperature of coolest month, temperature seasonality, mean annual precipitation, precipitation of wettest month, precipitation of driest month, and precipitation seasonality were generated. The important productivity variables like mean NDVI, mean NDVI of monsoon months, and mean NDVI of winter months were considered in the models. The collected random soil samples were grouped into calibration dataset and validation dataset. Initially, SOC was predicted by using the geostatistical techniques of OK and RK. The performance of all three models was evaluated by using the validation dataset. The results obtained in the prediction of SOC through the RF algorithm were compared with the results obtained from OK and RK geostatistical methods (Chen et al. 2019b). The results of the study indicate better performance of RF algorithm over the geostatistical methods of OK, and RK in predicting, and explaining the spatial variability of SOC in heterogeneous landscape (Fig. 4). The correlation analysis among the predicted SOC by using OK, RK, and RF models is depicted through Corrplot (Fig. 5). The analysis indicates that RF algorithms perform better than RK and OK with correlation coefficient (R^2) of 0.989, 0.664, and 0.561, respectively. The study demonstrates that RF is the robust ML algorithm for faster, more efficient, and better data interpretation in the estimation of spatial variability of SOC. However, the predictive power of the model output also depends on the set of input variables (Aertsen et al. 2010).

5.6 Crop Type Mapping and Monitoring

Satellite-based crop type mapping and monitoring assume a greater importance in agriculture, and ensures global food security. Pelletier et al. (2016) reported that the pixel-based RF algorithm is highly useful in classifying high-resolution datasets in land cover analysis, and mapping. Teluguntla et al. (2018) followed the RF algorithm by using time-series Landsat data on the GEE platform in producing cropland at 30-m and reported that the RF algorithm has the capability to precisely map croplands at cost-effective and time-efficient at various resolutions. GEE platform is being widely employed in the analysis of multi-temporal satellite data in crop type mapping by

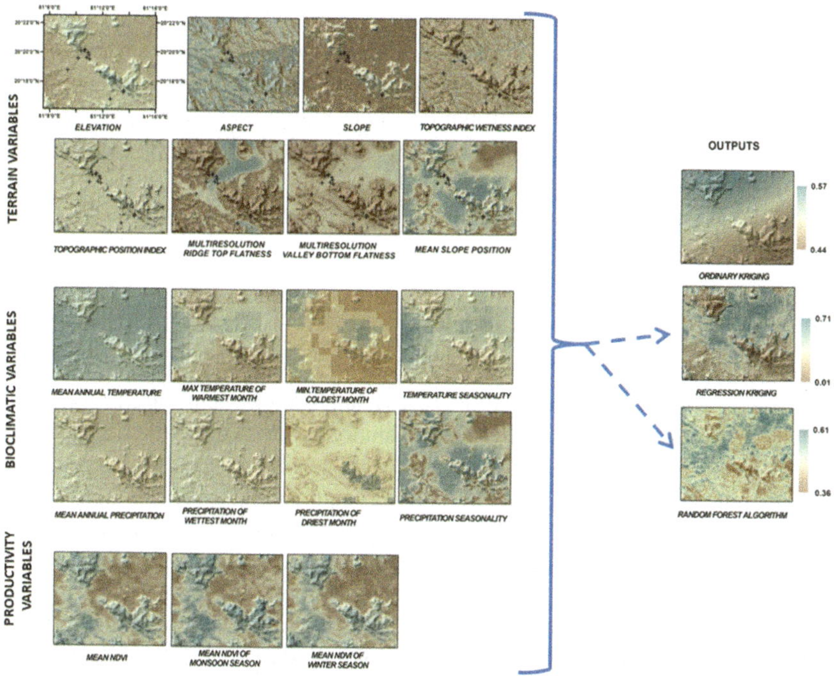

Fig. 4 SOC estimation using ordinary kriging, regression kriging, and RF algorithm

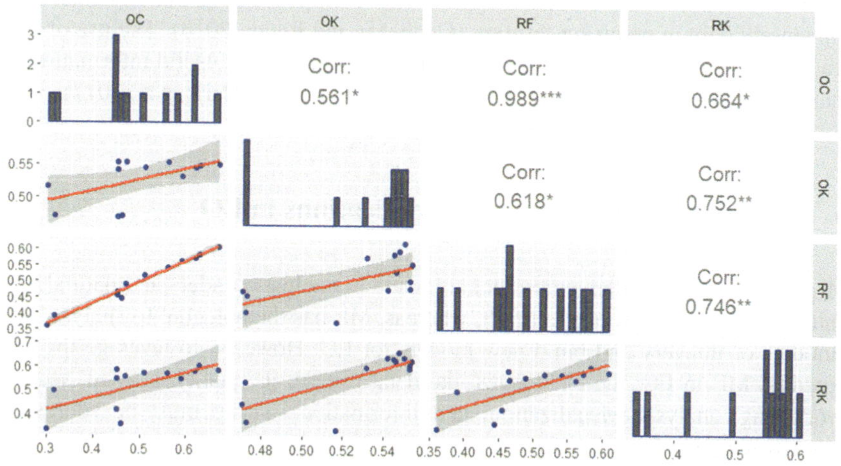

Fig. 5 Correlation matrix between observed and predicted SOC through OK, RF, and RK models

using ML algorithms (Wang et al. 2019). Gumma et al. (2020) used Landsat 30-m data on the GEE platform by adopting RF algorithm in cropland mapping in south Asia and reported that RF algorithm gives better overall accuracy in producing agricultural croplands areas. Moumni and Lahrouni (2021) used high-resolution fused satellite data with ML algorithms of ANN, SVM, and Maximum Likelihood (ML) in mapping crop types in irrigated areas, and reported that fused images of optical, and microwave satellites improves the accuracy in crop type mapping and classification.

5.7 *Precision Agriculture (PA)*

PA is one of the efficient applications of IoT applications being used in agricultural surveys (Ojha et al. 2015). IoT-based solutions were demonstrated in advance farming, monitoring of crop fields in conjunction with sensors, and automating the irrigation system. IoT enables agricultural automation, increased agricultural production (Bu and Wang 2019), and secure food supply chain. The application of IoT (Gubbi et al. 2013) enables the farmers to adopt a data-driven approach in collecting precise information from field sensors in improving farm yields and mitigate the risks. The application of Wireless Sensors Network (WSN) in PA assists the farmers to make precise, and informed decisions (Kodali et al. 2014). Besides the traditional data collections systems, advanced sensing technologies with Geographic Information System (GIS), Global Positioning System (GPS) (Reddy 2018b), ML algorithms, Variable Rate Technology (VRT), intelligence irrigation systems, etc., immensely help in precision farming (Gadhekar and Reddy 2018). The use of IoT leads to generate large-scale data and such big data support as useful information in managing the farms more effectively in smart farming (Kamilaris et al. 2017).

6 Challenges in Data Science Applications in EO

ML and EO are complementary technologies. EO helps to understand natural and anthropogenic changes on the Earth, whereas, ML has the potential to analyze vast amounts of imagery and build new models for EO. However, to leverage the full potential ML in EO, the challenges like data capture, its quality, storage, access, processing, analysis, visualization, and information sharing must be addressed effectively (Table 5).

7 A Way Forward

Though the currently available DS algorithms and techniques address the applications in EO, some corners are remaining to be addressed. However, the rapid developments

Table 5 Some of the challenges associated with DS applications in EO

Challenges	Description
Data capture	Capturing data is a gigantic task as the demand and complexity of datasets are increasingly growing. Capturing and creating valuable data involves huge investments to ramp up the infrastructure and manpower (Chen et al. 2014). The challenges associated with the modern big data acquisition process are its collection, integration, storing, and management
Data quality	Analysis of big data from various sources by using suitable ML algorithms helps to understand the customer needs and improve service quality, and predict the risks. However, challenges associated with the use and analysis of big data from different sources to generate value from big data depend on accurate and high-quality data
Data storage	Often users face difficulty in storing ever-increasing big data. Hadoop Distributed File System (HDFS) became a foundational data management platform for large data, and it also supports data analytics, predictive analysis, DM, and ML. Storage and management of voluminous data include aggregation from different sources, and cleaning to use in analysis and encode the data for security purposes (Chen et al. 2014)
Data access	Open data access policies of major scientific EO data enable the users to analyze time-series satellite and allied resources at various scales in developing the applications in EO (Cerbaro et al. 2020). The innovative cloud platforms like EarthServer and GEE can be used to access services to analyze voluminous data
Processing and analyzing	Recently, open data cubes provide a new solution to organize, store, manage, and analyze voluminous EO data (Giuliani et al. 2017). Different data cubes like EarthServer, Swiss Data Cube (SDC) (Giuliani et al. 2017), and Digital Earth Australia (Dhu et al. 2017) provide the services to meet the needs of EO data users to utilize the products of regional to global scale. EO data cubes provide a wide variety of datasets, infrastructures, and user-friendly interfaces to the users
Data visualization	The core aim of data visualization is to depict information, and knowledge efficiently by applying suitable visualization techniques. DM algorithms play a significant role to discover the patterns, and extract hidden patterns and values from large datasets (Oussous et al. 2018). The high dimensions of big data always pose challenges in effective analysis and visualization (Chen et al. 2014). The advanced DM algorithms can be deployed to effectively visualize the trends, and patterns in big data

(continued)

Table 5 (continued)

Challenges	Description
Information sharing	Information sharing becomes important, it includes accessibility and Web-based geoportals. Sharing of interdisciplinary research, and new developments among the experts of different fields enhance the utility of products, services, and information generated. It becomes more pertinent when integrating EO data with non-EO data and it enhances the scope to build new ML models to address the contemporary issues in various applications and it enhances the interpretations (Guo et al. 2017)

and progress being made in the field of EO and DS cope with the new and emerging challenges. These advanced developments augment the DS domain to discover and produce more robust ML algorithms in the years to come.

- The potential of advanced ML algorithms should be leveraged especially in the field of natural resources management, precision agriculture, and climate change studies.
- There is a need to optimize the technological advancements in digital infrastructure, especially in high-end computing and storage capabilities like cloud computing.
- Instead of exploring, and accessing the raw EO data, the use of ML models helps to fully utilize the ready-to-use products, and services provided by online platforms to develop the applications in addressing many critical issues of EO.
- The potential of data cubes could be utilized effectively to perform the analysis on time-series datasets in various applications of EO.
- The institutional resources and computational infrastructure need to be strengthened to translate the voluminous EO data and services offered by cloud computing EO platforms into information by using suitable ML algorithms.
- To develop sustained applications in EO monitoring by using advanced ML algorithms, the continuity of various EO satellite missions to be ensured to provide continuous inputs to the various models and applications.
- There is a need to optimize the multisource EO data and ground observation data with robust data assimilation techniques for effective data processing, analysis, validation, and visualization by using suitable ML algorithms.
- To improve the data quality, and output of ML algorithms, automated data entry, data quality at the source, filling data gaps, creating robust business rules, workflows, and Standard Operating Procedures (SOPs) are essential.

8 Conclusions

In the chapter, DS tools, techniques, and applications in emerging EO were discussed. The important components of DS like data discovery, data preparation, model building, model planning, model evaluation, results communication, and visualization play an important role in the effective execution of any project. The tools of DS such as data acquisition, storage, extraction, cleaning, analysis, and visualization were discussed. The important techniques of DS like AI, ML, DM, predictive analysis, contextual analysis, pattern recognition, and visualization were deliberated. Though satellite observations provide a wealth of data on a regional to a global scale, the concern remains in efficiently capturing, storing, extracting, processing, analyzing, and visualization of information from these vast datasets. DS techniques have immense potential to identify and extract critical information from a variety of structured/unstructured data. The important applications of ML algorithms in EO, which includes drought monitoring, vegetation monitoring and assessment, digital soil mapping, soil moisture monitoring, prediction of SOC with a case study, crop type mapping and monitoring, and PA were discussed in the chapter. Despite the potential of ML algorithms, the computational resources and expert knowledge assume greater importance in effectively interpreting and visualizing the results. To leverage the full potential of DS principles, algorithms, and applications in EO, the challenges associated with data capturing, quality, storage, access, processing, and analyzing, information sharing, and visualization must be addressed effectively.

References

Abdi A (2019) Land cover and land use classification performance of machine learning algorithms in a boreal landscape using Sentinel-2 data. GISci Remote Sens 1–20

Adab H, Morbidelli R, Saltalippi C, Moradian M, Ghalhari GAF (2020) Machine learning to estimate surface soil moisture from remote sensing data. Water 12:3223

Aertsen W, Kint V, van Orshoven J, Özkan K, Muys B (2010) Comparison and ranking of different modelling techniques for prediction of site index in Mediterranean mountain forests. Ecol Model 221:1119–1130

Ahmad S, Kalra A, Stephen H (2010) Estimating soil moisture using remote sensing data: a machine learning approach. Adv Water Resour 33(1):69–80

Alizadeh MR, Nikoo MR (2018) A fusion-based methodology for meteorological drought estimation using remote sensing data. Remote Sens Env 211:229–247

Andries A, Morse S, Murphy R, Lynch J, Woolliams E, Fonweban J (2019) Translation of earth observation data into sustainable development indicators: an analytical framework. Sustain Dev 27:366–376

Atzori L, Iera A, Morabito G (2010) The internet of things: a survey. Computer Netw 54(15):2787–2805

Atzori L, Iera A, Morabito G (2017) Understanding the internet of things: definition, potentials, and societal role of a fast-evolving paradigm. Adhoc Netw 56:122–140

Ballabio C (2009) Spatial prediction of soil properties in temperate mountain regions using support vector regression. Geoderma 151:338–350

Baseca CC, Sendra S, Lloret J, Tomas J (2019) A smart decision system for digital farming. Agron 9(5):216

Baştanlar Y, Özuysal M (2014) Introduction to machine learning. Methods Mol Biol 1107:105–128

Baumann P, Misev D, Merticariu V, Datacubes BPH (2019) Towards space/time analysis-ready data. Service-oriented mapping. Springer, Cham, pp 269–299

Belgiu M, Drăgut L (2016) Random Forest in remote sensing: a review of applications and future directions. ISPRS J Photogramm Remote Sens 114:24–31
Berry D (1996) Statistics: a bayesian perspective. Duxbury Press, Belmont, CA
Berry M, Mohamed A, Yap BW (2019) Supervised and unsupervised learning for data science. Springer, Cham, p 271
Blei DM, Smyth P (2017) Science and data science. Proc Natl Acad Sci USA 114(33):8689–8692
Breiman L (1996) Bagging predictors. Mach Learn 24:123–140
Breiman L (2001) Random forests. Mach Learn 45(1):5–32
Breiman L, Friedman J, Stone CJ, Olshen RA (1984) Classification and regression trees. CRC Press, Boca Raton, FL, USA
Bu F, Wang X (2019) A smart agriculture IoT system based on deep reinforcement learning. Future Gener Comput Syst 99:500–507
Busoniu L, Babuska R, Schutter BD (2008) A comprehensive survey of multiagent reinforcement learning. IEEE Trans Syst Man Cybernetics Part C Appl Rev 38(2):156–172
Campbell JB, Wynne RH (2011) Introduction to remote sensing. Guilford Press, New York, USA
Cerbaro M, Morse S, Murphy R, Lynch J, Griffiths G (2020) Information from earth observation for the management of sustainable land use and land cover in Brazil: an analysis of user needs. Sustain 12:489
Chen D, Chang N, Xiao J, Zhou Q, Wu W (2019a) Mapping dynamics of soil organic matter in croplands with MODIS data and machine learning algorithms. Sci Total Environ 669:844–855
Chen L, Ren C, Li L, Wang Y, Zhang B, Wang Z, Li L (2019b) A Comparative assessment of geostatistical, machine learning, and hybrid approaches for mapping topsoil organic carbon content. ISPRS Int J Geo-Information 8:174
Chen Y, Lin Z, Zhao X, Wang G, Gu Y (2014) Deep learning-based classification of hyperspectral data. IEEE J Sel Top Appl Earth Observ Remote Sens 7:2094–2107
Cortes C, Vapnik V (1995) Support-vector networks. Mach Learn 20(3):273–297
Cui Q, Gong Z, Ni W, Hou Y, Chen X, Tao X, Zhang P (2019) Stochastic online learning for mobile edge computing: learning from changes. IEEE Commun Mag 57:63–69
Dai F, Zhou Q, Lv Z, Wang X, Liu G (2014) Spatial prediction of soil organic matter content integrating artificial neural network and ordinary kriging in Tibetan Plateau. Ecol Indi 45:184–194
Davenport T, Guha A, Grewal D, Bressgott B (2020) How artificial intelligence will change the future of marketing. J Acd Mark Sci 48:24–42
Deissa L, Margenotb AJ, Culmanc SW, Demyan MS (2020) Tuning support vector machines regression models improve prediction accuracy of soil properties in MIR spectroscopy. Geoderma 365:114227
Dey A (2016) Machine learning algorithms: a review. Int J Comp Sci Infor Tech 7(3):1174–1179
Dhar V (2013) Data science and prediction. Commun ACM 56(12):64–73
Dhu T, Dunn B, Lewis B, Lymburner L, Mueller N, Telfer E, Lewis A, McIntyre A, Minchin S, Phillips C (2017) Digital earth Australia—unlocking new value from earth observation data. Big Earth Data 1:64–74
Efron B, Gong G (1983) A leisurely look at the bootstrap, the Jackknife and cross-validation. Amer Stat 37:3648
Emadi M, Taghizadeh-Mehrjardi R, Cherati A, Danesh M, Mosavi A, Scholten T (2020) Predicting and mapping of soil organic carbon using machine learning algorithms in Northern Iran. Remote Sens 12:2234
Forkuor G, Hounkpatin OKL, Welp G, Thiel M (2017) High resolution mapping of soil properties using remote sensing variables in south-western Burkina Faso: a comparison of machine learning and multiple linear regression models. PLoS One 12:e0170478
Gadhekar SR, Reddy GPO (2018) Precision agriculture-geospatial technologies and applications. Agromet Publishers, Nagpur, p 274
Gandomi A, Haider M (2015) Beyond the hype: big data concepts, methods, and analytics. Int J Infor Manag 35:137–144

Gehl RJ, Rice CW (2007) Emerging technologies for in situ measurement of soil carbon. Clim Chang 80:43–54

Giuliani G, Dao H, De Bono A, Chatenoux B, Allenbach K, De Laborie P, Rodila D, Alexandris N, Peduzzi P (2017) Live monitoring of earth surface (LiMES): a framework for monitoring environmental changes from earth observations. Remote Sens Environ 202:222–233

Gorelick N, Hancher M, Dixon M, Ilyushchenko S, Thau D, Moore R (2017) Google earth engine: planetary-scale geospatial analysis for everyone. Remote Sens Environ 18–27

Gubbi J, Buyya R, Marusic S, Palaniswami M (2013) Internet of things (IOT): a vision, architectural elements, and future directions. Future Gen Comp Sys 29(7):1645–1660

Gumma MK, Thenkabai PS, Teluguntla PG, Oliphant A, Xiong J, Giri C, Pyla V, Dixit S, Whitbread AM (2020) Agricultural cropland extent and areas of South Asia derived using Landsat satellite 30-m time-series big-data using random forest machine learning algorithms on the google earth engine cloud. Gisci Remote Sen 57:302–322

Guo H (2017) Big earth data: a new frontier in earth and information sciences. Big Earth Data 1:4–20

Guo H, Zhen L, Hao J, Changlin W, Jie L, Dong L (2017) Big earth data: a new challenge and opportunity for digital earth's development. Int J Digital Earth 10(1):1–12

Hastie T, Tibshirani R, Friedman J (2009) The elements of statistical learning, 2nd edn. Springer Verlag

Hira S, Deshpande P (2015) Data analysis using multidimensional modeling, statistical analysis and data mining on agriculture parameters. Procedia Comp Sci 54:431–439

Hofmann T (2001) Unsupervised learning by probabilistic latent semantic analysis. Mach Learn 42:177–196

Hosmer DW, Lemeshow S (2000) Applied logistic regression, 2nd edn. Wiley

Jain A, Murty M, Flynn P (1999) Data clustering: a review. ACM Comput Surv 31(3):264–323

Jenny H (1994). Factors of soil formation: a system of quantitative pedology; courier corporation. North Chelmsford, MA, USA

Kamilaris A, Kartakoullis A, Prenafeta-Boldú FX (2017) A review on the practice of big data analysis in agriculture. Comput Electron Agric 143:23–37

Kempen B, Brus DJ, Stoorvogel JJ, Heuvelink GBM, Vries F (2012) Efficiency comparison of conventional and digital soil mapping for updating soil maps. Soil Sci Soc Am J 76:2097–2115

Keshtkar H, Voigt W, Alizadeh E (2017) Land-cover classification and analysis of change using machine-learning classifiers and multi-temporal remote sensing imagery. Arab J Geosci 10:154

Khader AI, Rosenberg DE, McKee M (2013) A decision tree model to estimate the value of information provided by a groundwater quality monitoring network. Hydrol Earth Syst Sci 17:1797–1807

Kodali RK, Rawat N, Boppana L (2014) WSN sensors for precision agriculture. In: Region 10 symposium. IEEE, pp 651–656

Kumar KCA, Reddy GPO, Masilamani P, Turkar SY, Sandeep P (2021) Integrated drought monitoring index: a tool to monitor agricultural drought by using time-series datasets of space-based earth observation satellites. Adv Space Res 67(1):298–315

Kuswanto H, Naufal A (2019) Evaluation of performance of drought prediction in Indonesia based on TRMM and merra-2 using machine learning methods. Methods X 6:1238–1251

Lagacherie P, McBratney AB (2007) Spatial soil information systems and spatial soil inference systems: perspectives for digital soil mapping. In: Lagacherie P, McBratney AB, Voltz M (eds) Digital soil mapping: an introductory perspective, vol 31. Amsterdam, Elsevier, pp 3–22

Lamichhane S, Kumar L, Wilson B (2019) Digital soil mapping algorithms and covariates for soil organic carbon mapping and their implications: a review. Geoderma 352:395–413

Lasaponara R (2006) On the use of principal component analysis (PCA) for evaluating Interannual vegetation anomalies from SPOT/VEGETATION NDVI temporal series. Ecol Model 194:429–434

Lei Y, Jia F, Lin J, Xing S, Ding SX (2016) An intelligent fault diagnosis method using unsupervised feature learning towards mechanical big data. IEEE Trans Indus Electron J 36:3137–3147

Li X, Zhao H (2009) Weighted random subspace method for high dimensional data classification. Stat Interface 2(2):153–159
Li Y, Yan S, Chen N, Gong J (2020) Performance evaluation of a neural network model and two empirical models for estimating soil moisture based on Sentinel-1 SAR data. Prog Electromagn Res C 105:85–99
Liaw A, Wiener M (2018) Classification and regression by random forest. R-News 2(3):18–22
Liu S, An N, Yang J, Dong S, Wang C, Yin Y (2015) Prediction of soil organic matter variability associated with different landuse types in mountainous landscape in southwestern Yunnan province, China. CATENA 133:137–144
MacQueen J (1967) Some methods for classification and analysis of multivariate observations. In: Proceedings of the fifth Berkeley symposium on mathematical statistics and probability. IEEE, pp 281–297
Marshland S (2015) Machine learning: an algorithm perspective. CRC Press, Boca Raton, FL
McBratney AB, Santos MLM, Minasny B (2003) On digital soil mapping. Geoderma 117:3–52
McCarthy J, Minsky ML, Rochester N, Shannon CE (2006) A proposal for the Dartmouth summer research project on artificial intelligence. AI Mag 27:12
Moumni A, Lahrouni A (2021) Machine learning-based classification for crop-type mapping using the fusion of high-resolution satellite imagery in a semiarid area. Scientifica Article ID 8810279
Müller H, Rufin P, Griffiths P, Barros Siqueira AJ, Hoster P (2015) Mining dense Landsat time series for separating cropland and pasture in a heterogeneous Brazilian savanna landscape. Remote Sens Environ 156:490–499
Neha D, Vidyavathi BM (2015) A survey on applications of data mining using clustering techniques. Int J Comp Appl 126(2):7–12
Ngiam KY, Khor IW (2019) Big data and machine learning algorithms for health-care delivery. Lancet Oncol 20:e262–e273
Ojha T, Misra S, Raghuwanshi NS (2015) Wireless sensor networks for agriculture: the state-of-the-art in practice and future challenges. Comp Electron Agri 118:66–84
Osband I, Doron Y, Hessel M, et al (2020) Behaviour suite for reinforcement learning. Int Conf Learn Rep
Oussous A, Benjelloun FZ, Lahcen AA, Belfkih S (2018) Big data technologies: a survey. J King Saud Univ Comp Infor Sci 431–448
Padarian J, Minasny B, McBratney AB (2019) Using deep learning for digital soil mapping: a review aided by machine learning tools. Soil 5:79–89
Pavithra D, Jayanthi AN (2018) A study on machine learning algorithm in medical diagnosis. Int J Adv Res Comp Sci 9(4):42–46
Pekel JF, Cottam A, Gorelick N, Belward AS (2016) High-resolution mapping of global surface water and its long-term changes. Nature 540:418–422
Pelletier C, Valero S, Inglada J, Champion N, Dedieu G (2016) Assessing the robustness of random forests to map land cover with high resolution satellite image time series over large areas. Remote Sens Environ 187:156–168
Petropoulos GP, Ireland G, Barrett B (2015) Surface soil moisture retrievals from remote sensing: current status, products & future trends. Phys Chem Earth Parts a/b/c 83:36–56
Phiri D, Simwanda M, Nyirenda V, Murayama Y, Ranagalage M (2020) Decision tree algorithms for developing rulesets for object-based land cover classification. ISPRS Int J Geoinfor 9:329
Rao S, Gupta R (2012) Implementing improved algorithm over APRIORI data mining association rule algorithm. Int J Comp Sci Tech 489–493
Reddy GPO (2018b) Global positioning system: principles and applications, In: Reddy GPO, Singh SK (eds) Geospatial technologies in land resources mapping, monitoring and management. Geotechnologies and the environment, vol 21. Springer, Cham, pp 63–74
Reddy GPO (2018a) Satellite remote sensing sensors: principles and applications, In: Reddy GPO, Singh SK (eds) Geospatial technologies in land resources mapping, monitoring and management. Geotechnologies and the environment, vol 21. Springer, Cham, pp 21–43

Reddy GPO, Kumar N, Sahu N, Srivastava R, Singh SK, Naidu LGK, Chary GR, Biradar CM, Gumma MK, Reddy BS, Kumar JN (2020) Assessment of spatio-temporal vegetation dynamics in tropical arid ecosystem of India using MODIS time-series vegetation indices. Arab J Geosci 13(15):1–13

Reddy GPO, Nagaraju MSS, Ramteke IK, Sarkar D (2013) Terrain characterization for soil resource mapping in part of semi-tract of Central India using high resolution satellite data and GIS. J Indian Soc Remote Sen 41(2):331–343

Reddy GPO, Patil NG, Chaturvedi A (2017) Sustainable management of land resources-an Indian perspective. Apple Academic Press Inc., Canada, p 796

Reddy GPO, Singh SK (2018) Geospatial technologies in land resources mapping, monitoring, and management. Geotechnologies and the environment, vol 21. Springer, Cham, p 638

Rossel RV, Behrens T (2010) Using data mining to model and interpret soil diffuse reflectance spectra. Geoderma 158:46–54

Roy PS, Behera MD, Srivastav SK (2017) Satellite remote sensing: sensors, applications and techniques. Proc Natl Acad Sci India Sect A Phys Sci 87(4):465–472

Sandeep P, Reddy GPO, Jegankumar R, Kumar KCA (2021) Monitoring of agricultural drought in semi-arid ecosystem of Peninsular India through indices derived from time series CHIRPS and MODIS data sets. Ecol Indic 21:107033

Schmugge T, Jackson T, McKim H (1980) Survey of methods for soil moisture determination. Water Resour Res 16:961–979

Shafeeq BM, Hareesha KS (2012) Dynamic clustering of data with modified K-means algorithm. In: International conference on information and computer networks (ICICN 2012), IPCSIT, vol 27. IACSIT Press, Singapore

Sivakumar MVK, Stefanski R, Bazza M, Zelaya S, Wilhite D, Magalhaes AR (2014) High Level meeting on national drought policy: summary and major outcomes. Weather Clim Extrem 3:126–132

Sood AK, Rinehart M (2016) Data Science as a tool for cloud security: cloud generation visibility, detection and protection. ISACA J 4:1–5

Srivastava PK, Han D, Rico-Ramirez MA, Bray M, Islam T (2012) Selection of classification techniques for land use/land cover change investigation. Adv Space Res 50:1250–1265

Swathi HR, Sohini S, Surbhi GG (2017) Image compression using singular value decomposition. IOP Conf Ser Mater Sci Eng 263(4):5–8

Taghizadeh-mehrjardi R, Toomanian N, Khavaninzadeh AR, Jafari A, Triantafilis J (2016) Predicting and mapping of soil particle-size fractions with adaptive neuro-fuzzy inference and ant colony optimization in central Iran. Eur J Soil Sci 67:707–725

Teluguntla P, Thenkabail PS, Oliphant A, Xiong J, Gumma MK, Congalton RG, Yadav K, Huete A (2018) A 30-m landsat-derived cropland extent product of Australia and China using random forest machine learning algorithm on google earth engine cloud computing platform. ISPRS J Photogramm Remote Sen 144:325–340

Tyralis H, Papacharalampous G, Langousis A (2019) A brief review of random forests for water scientists and practitioners and their recent history in water resources. Water 11:910

Tziachris P, Aschonitis V, Chatzistathis T, Papadopoulou M (2019) Assessment of spatial hybrid methods for predicting soil organic matter using DEM derivatives and soil parameters. CATENA 174:206–216

UN (United Nations) (2017) Earth observations for official statistics: satellite imagery and geospatial data task team report. https://unstats.un.org/bigdata/task-teams/earth-observation/UNGWG_Satellite_Task_Team_Report_WhiteCover.pdf

Vaysse K, Lagacherie P (2015) Evaluating digital soil mapping approaches for mapping GlobalSoilMap soil properties from legacy data in Languedoc-Roussillon (France). Geoderma Reg 4:20–30

Vaysse K, Lagacherie P (2017) Using quantile regression forest to estimate uncertainty of digital soil mapping products. Geoderma 291:55–64

Wang S, Azzari G, Lobell DB (2019) Crop type mapping without field-level labels: random forest transfer and unsupervised clustering techniques. Remote Sens Environ 222:303–317

Weiss SM, Indurkhya N (1998) Predictive data mining: a practical guide. Morgan Kaufmann Publishers, Inc.

Were K, Bui DT, Dick ØB, Singh BR (2015) A comparative assessment of support vector regression, artificial neural networks, and random forests for predicting and mapping soil organic carbon stocks across an Afromontane landscape. Ecol Indic 52:394–403

Yang J, Gong P, Fu R, Zhang M, Chen J, Liang S, Xu B, Shi J, Dickinson R (2013) The role of satellite remote sensing in climate change studies. Nat Clim Chang 3:875–883

Zhang M, Chen F, Tian B, Liang D (2019) Multi-temporal SAR image classification of coastal plain wetlands using a new feature selection method and random forests. Remote Sens Lett 10:312–321

Zhao Z, Yang Q, Benoy G, Chow TL, Xing Z, Rees HW, Meng FR (2010) Using artificial neural network models to produce soil organic carbon content distribution maps across landscapes. Can J Soil Sci 90:75–87

Emerging Technologies—Principles and Applications in Precision Agriculture

Shakti Mishra

Abstract Technology is playing a vital role in bringing about transformational changes and progress in the agricultural sector. With the advent of the Internet of Things (IoT), Remote Sensing, Global Positioning System (GPS), Aerial photography, it has become much easier to monitor crop health, identify in-season nutrient deficiencies, and other issues causing crop stress and reduced yield. Precision agriculture is an information and technology-based system that uses various sources of data such as soil, crop, pest, humidity, yield, nutrients, temperature, moisture for sustainability, optimum profitability, and protection of the environment. Precision Farming helps to better understand the condition of the crops via considering the environmental and economic considerations with the help of remotely sensed data and prevent crop damage by being proactive in the required time frame by applying the right amount of pesticide, irrigation water, and nutrients. The typical components of Precision Farming include background data, record-keeping systems, data analysis, and decision-making process, specialized implementation equipment, and evaluation and revision. Significant benefits of precision agriculture include field scouting, precise application of nutrients in environmentally sensitive areas, detailed soil analysis, variable rate irrigation, etc. This chapter broadly covers various components of precision-based farming and how the underlying technologies are integrated to give a predictive analysis of yield production, sustainable irrigation solutions, crop disease management, and weed identification.

Keywords Data analysis · Internet of things (IoT) · Machine learning · Precision agriculture · Sensors

S. Mishra (✉)
Pandit Deendayal Energy University, Gandhinagar 382007, India
e-mail: Shakti.mishra@sot.pdpu.ac.in

G. P. O. Reddy et al. (eds.), *Data Science in Agriculture and Natural Resource Management*, Studies in Big Data 96, https://doi.org/10.1007/978-981-16-5847-1_2

1 Introduction

The importance of agriculture in the Indian context can be inferred from the fact that over 60% of the population is involved directly or indirectly in some agricultural activity or the other. Not just it provides the basic bread and butter to the majority of Indian households, it also drives the economy. The agricultural sector is plagued by various factors leading to inefficiency at each level of crop production to delivery at the consumer level. An analysis at the broader level points to inadequacies like conventional farming methods, low-level implementation of mechanization, insufficient monitoring mechanisms, lack of precise information on weather, and techniques to combat the natural phenomenon.

The unprecedented invention in technology has seen a major shift from the traditional way of farming to precision-based farming. From mechanization to the green revolution, the field of agriculture has seen tremendous growth in adopting technologies to increase production, better ways of dealing with soil management, yield prediction, and livestock farming. In the late 1990s, with the advent of Global Positioning System (GPS) based location-tracking technology, which is a worldwide radio-navigation system, computer-based data analysis brought an agricultural revolution. In traditional methods of farming, growers apply the same amount of fertilizers, water, and pesticide at the predefined period in the number of years or as per crop requirement. However, this may lead to uneven production and yield across the farmlands (Blackmore 1994). The term precision farming includes improved management technologies such as yield monitoring and mapping, soil sensing and mapping, Geographic Information Systems (GIS), variable rate application, satellite-based positioning, remote sensing, field and crop scouting, livestock management, and automatic steering. Now, it has become much easier to monitor and control the environmental parameters that may affect the productivity of crops and livestock. Equipped with data from soil sensors, GPS-equipped harvesting machines like tractors, pumps, sheds, and external sources such as satellite-based weather forecasting, local weather channels, precision agriculture enable farmers to better manage key resources including seed, fertilizer, and pesticides, and increasing productivity. Internet of Things (IoT) has enabled the interconnection of various sensors, actuators, and other devices. Data received from these devices has been stored on cloud storage and further analyzed to infer useful interpretations using various distributed data management platforms and frameworks such as big data and machine learning. Data-driven agriculture leverages technologies in a number of stages to enhance the existing land fertility and yield. The aim of precision agriculture is to proliferate production, improve the economy, and to achieve environmental sustainability. Artificial Intelligence (AI), predictive analytics, connected and cost-effective sensors, and other emerging technologies could further increase crop production, improve the methods of irrigation, soil nutrients, pesticides, and other inputs, and build sustainability and pliability across the agricultural sector.

In India, there have been efforts from government organizations to leverage AI and Machine Learning (ML) for crop cutting experiments, an assessment method to estimate the yield of a crop during a harvesting cycle, monitoring of crop and soil health assessment using drones and intelligent decision-making by an app-based system, where farmers do not need to install any sensors in the fields but require phones that can receive text message about sowing advisories.

2 Lifecycle of Precision Based Farming

The lifecycle of Precision Agriculture (PA) includes background data, record-keeping systems, data analysis and decision-making process, specialized implementation equipment, and evaluation and revision. Observation and background data collection requires the usage of technologies such as IoT where sensor-based systems are used to capture the data. This typically includes measuring soil moisturizing level, soil temperature, and soil pH levels (Bhatnagar and Chandra 2020) using ground sensors. UAV sensors are used to identify crop size, shapes, leaf wetness to predict crop health; location sensors are used for the longitude and altitude of the plants. The data captured by these sensor devices are used to analyze to find out useful patterns about field and crop (Fig. 1).

Fig. 1 Lifecycle of precision farming

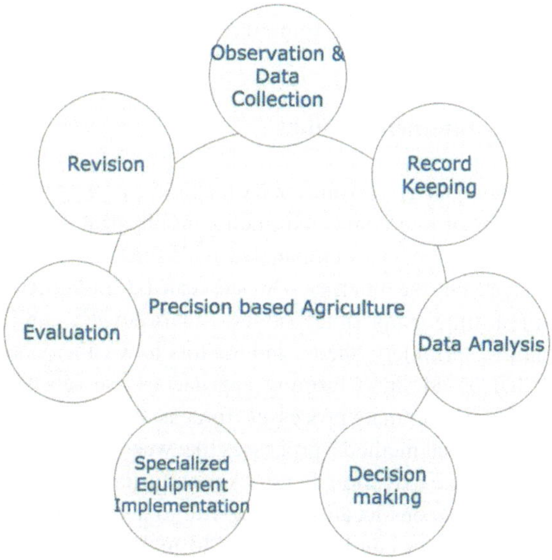

2.1 Observations and Data Collection

Traditional methods of data collection use manual procedures to characterize soil and yield attributes. The local practices of such prediction depend heavily on having the correct sampling process and obtaining variograms. Obtained data is also considered to be spatially dependent in all interpolation methods, which means that sampling should be at an interval that falls within the correlation range of spatial variation (Kerry et al. 2010). There are various factors to be considered for measuring the sampling density, such as objectives, field variability, and costs. They can range widely to a more elaborative coverage of the field. Traditionally, samples are collected for entire fields or segments of fields to provide average values. There are different methods for the soil sampling: (a) simple random (locations are randomly selected), (b) stratified random (fields are divided based on some characteristics and then randomly selected), (c) grid sampling (involves laying a grid and samples are collected at regular intervals across the landscape), (d) stratified systematic (to reduce the bias introduced by grid sampling), judgmental sampling (based on the problem like crop health, low yield, etc.) (Shibusawa and Haché 2009; Ferguseon et al. 2000).

Remote sensing and unmanned aerial vehicle technology have also been used prominently to capture the spatial data and imagery data. Topographical mapping of farm fields is proven beneficial for collecting the imagery data that could be analyzed for pesticide spraying in designated areas, observing soil properties, monitoring the crop nutrient deficiencies, weeds, crop diseases, and other crop conditions. Technologies like IoT, Remote Sensing, Light Detection And Ranging (LIDAR), UAV, and GPS are used to acquire data from agricultural fields.

2.1.1 Internet of Things

IoT emerged as a revolutionary technology for agricultural applications. It refers to a network of interconnected intelligent physical devices, data, people, and processes. IoT enables devices empaneled with sensors to connect to and interact with other devices via the Internet. Devices can be monitored from distance and controlled in real time using smartphones. Smart devices can include anything from motors, pumps, sprinklers, sheds, and tractors to weather stations and computers.

IoT-based smart farming approaches use sensor-based networks to collect the field data, automate processes, forecast events, and real-time improvisation. Unlike conventional methods, processes like weed detection, climatic condition monitoring, variable rate irrigation, livestock monitoring can be better handled by placing sensors at advantageous locations (Saiz-Rubio and Rovira-mas 2020). Data collected from sensor devices can help farmers to make intelligent decisions, helping them farm smoother, safer, and adapt more swiftly to changing environmental conditions. Precision farming uses robots, drones, computer imaging, and remote sensors to capture the data along with analytical algorithms to predict future lines of action. The key benefits of connecting physical devices on farms through IoT technologies are as follows:

- Distant monitoring of farm environments, saving time and labor on daily farm checks.
- Improved decision-making through machine learning algorithms and data analytics.
- Deep insights of market demand while capturing real-time data across the supply chain.
- Sustainable solutions to ensure less wastage, convenience to outreach markets.
- Improved productivity while adding new capabilities and adapting to emerging technologies (Table 1).

Benefits of Internet of Things

The thing is any device capable of creating a connection to the Internet within the context of IoT. These things are also called "Smart devices" as they can perform autonomous computing and connect for data exchange. The integration of technological resources into the agricultural production process offers several benefits. Some of these resources include, among others, temperature, moisture, humidity, pressure, water stress, soil concentration, unmanned aerial equipment, video cameras, GPS, and communication networks (Navarro et al. 2020). IoT has the considerable potential to address and resolve irrigation, crop health monitoring, soil management, and

Table 1 Precision agriculture lifecycle with associated technology platforms

Stages	Task	Technology platform available
Observation and background data collection	• Soil monitoring, sampling, and mapping • Crop monitoring and mapping • Global Satellite Positioning (GNSS) Remote sensing Field/Crop scouting	Internet of things Aerial photography
Data storage and decision-making	• Crop map or plan • Maps of inputs • Fertility, soil tests • Irrigation map • Yield history	Cloud computing Edge computing Fog computing Geographical information System
Data analysis	• Better productivity • GIS based Agricultural mapping software • Financial analysis	Big data Artificial intelligence Machine learning
Application technologies	• Variable rate application • Section control • Agricultural robots	Robotics Drones Autonomous tractors Automatic watering Seeding robots

productivity. The real-time data about soil, air quality, water stress, etc. can help farmers in making an optimal decision about planting and harvesting crops, thereby increasing the overall yield and revenue.

Drip irrigation is an optimal method to water the crops when soil moisture reaches a set level without manual intervention. The same measures can also be applied to humidity, light, and temperature control on farm fields. The smart irrigation system greatly improves efficiency and reduces labor costs, allowing farmers to ease out the process. Sensor devices are capable of monitoring any significant change in temperature, weather, humidity, nutrient in soil, moisture as well as the health of each crop in real time. In extreme weather conditions, remote sensing, predictive analytics, and innovative methods and tools have been proven to be useful. With the advancements in sensor technology, cloud, fog, and edge computing, it has been possible for many farmers to transmit the data from fields to expert labs and get the analysis done in real time. Subject matter experts and eminent agricultural scientists across the globe may connect to farmers and provide useful suggestions about essential variables like soil, crops, livestock, and weather (Aune et al. 2017).

Site-specific treatment such as wireless devices, IoT framework, and capturing data from satellites or drones, helps farmers to reach the data in a faster and precise manner. It also allows for timely farming operations such as seed sowing, pesticides application, weeding and crop harvesting, the use of fertilizers, pesticides, and irrigation according to the soil, rainfall and targeted crop protection procedures. Process automation achieves better control over the production process and maintains the higher quality of crop and yield while enhancing growth capacity through farm automation (Table 2).

Sensor Types

Different types of Sensors Output (Navarro et al. 2020)

The raw data produced by sensor devices becomes executable while analysis. Different processes as per sensors output involved in precision farming are as follows:

- **Yield Monitoring** and Mapping systems are fixed on yield harvesting mean such as tractors, maize harvesters use longitudinal and latitudinal coordinate data from GPS sensors. This data is then merged with the GPS data to create crop maps.
- **Variable Rate Fertilizer** application instruments use crop maps and monitor plant health. Automatically controlled variable rate controllers use real GPS locations, though they may be human-controlled too.
- **Weed Detection** generates maps by quickly identifying the location with a datalogger device and GPS receptor. Precise information can be obtained by overlapping the weed pattern with crop maps, and fertilizer maps.
- **Variable Spraying** controllers turn pesticide and nutrient spray on and off, and modify the quantity of spray applied. This is helpful in site-specific nutrient treatment.
- **Landscape and Geographical mapping** can be logged using high-precision GPS and turns out to be beneficial while interpreting crop maps and weed maps.

Table 2 Sensor type used in IoT environment for smart farming (Schriber 2021)

Sensor type	Functions	Brands
Optical sensors	• It uses light to measure soil properties. • It analyses the amount of light on the growing plants that help to regulate the moisture levels in damp conditions.	Vishay Ambient Light Sensor like AEQ 101 Osram Opto Semiconductor
Location sensors	• It uses signals from GPS satellites to determine three-dimensional analysis, latitude, longitude, and altitude.	GPS-integrated circuits like the NJR NJG1157PCD-TE1
Electrochemical sensors	• It provides pH and soil nutrient levels. • Sensor electrodes work by detecting specific ions in the soil.	NPK Sensor using ISE and ISFET
Mechanical sensors	• It measures soil compaction or "mechanical resistance." • It records resistive forces that roots exert to absorb water.	Honeywell FSG15N1A
Airflow sensors	• It measures soil air permeability. • It records various types of soil properties, including gaseous substances, optimum pressure, structure, humidity, and moisture-holding capacity.	Bosch MAF, Honeywell Airflow Sensor Awm5104vc

Geographical information about fields, existing infrastructure, and wetlands, marshes can be precisely located to help in farm planning.

- **Direction Systems** can accurately position a moving equipment or vehicle using GPS. These systems replace legacy equipment for spraying or seeding. In the future, autonomous vehicles will likely be put into use for smart agriculture.

Connectivity

The need to transmit data between many agricultural facilities and devices poses a considerable challenge for adopting smart farming. The connection between these devices should be reliable enough to resist bad weather conditions and ensure non-disruptive operations. Smart agriculture with several sensor devices is connected via many connectivity options. Choosing a suitable connectivity method is a crucial decision to make. The significant characteristics for the same are following:

- Range: In typical IoT environments, sensor devices collect the data and send it across to actuators devices for further processing. The proximity between sensing devices and devices sending data to the cloud or network requires line of sight or not?

- Bandwidth: Transmitting large amounts of data requires high bandwidth. The optimal amount of data needs to be sent over the connection.
- Power: Data-intensive environments often require low-energy solutions. Sensors, key elements in IoT networks are battery based and transmit small amounts of data over longer distances.
- Cost: While scaling up to a large IoT environment, cost-effective solutions are required to implement for network connectivity.

Orchestrating the balance between these key characteristics has emerged as a critical component of Smart Farming. Different available connectivity options are:

Wi-Fi: Wi-Fi can be used to connect data from instruments in the farm directly to smartphones. It is considered to be a high-speed Internet that allows Wi-Fi-compatible devices to interface with the Internet. There are three essential elements operating into Radio signals, Antenna, and Router. Wi-fi signals may be relayed across many acres. Sensor devices measure soil humidity, plant health, weather conditions, and relay information back to farmers over Wi-fi.

GPRS/2G/3G: GPRS is a packet-oriented data service for GSM-based cellular phones where two devices can communicate if both are in the same GSM service area. GPRS outperforms the range limitation challenge of wireless devices and thus making it suitable for periodic monitoring applications. Enhanced Data rates for Global Evolution (EDGE), an advanced version of GPRS, offers a high data rate with low latency processing without any hardware/software changes in the GSM core networks (Ojha et al. 2015).

Satellites: Reliable Internet connectivity in the field is a must requirement to push data from one place to another in Agriculture 4.0, often termed, as connected farms for better data analysis (Jarman and Dimmock 2018). Satellite Earth Observation is a remote sensing tool emphasized on collecting satellite imagery. But this does not only provide spatial imagery but also is very useful in providing fast and reliable Internet and voice communications across the world in areas that cellular and fiber connectivity cannot reach. Satellite communications can fill the gap to enable farmers to connect data-producing devices in the field to business applications (Tompkins 2019).

LoRa and LoRaWAN: LoRaWAN is a protocol built upon LoRa as recommended by the LoRa Alliance. LoRA uses radio wave transmission that can be received by any of the gateways available. This is quite useful for sending updates a few times per hour. This technology improves and streamlines day-to-day operations in smart agriculture. It is suitable for a number of applications from livestock monitoring to measure soil moisture (SemTech 2017).

Bluetooth and Bluetooth Low Energy (BLE): Bluetooth is a radio-frequency standard to save battery power by keeping transmission power extremely low. Bluetooth is wireless, inexpensive, and automatic. Bluetooth protocol is similar to Wi-Fi but constrained to usage scenarios within a short range of about 30 ft./10 meters. The sensor node gathers data from multiple sensors and transmits the data via Bluetooth to a hub or other Internet devices such as a workstation. In the IoT environment, gateways nodes are preferred with devices that have an Internet connection and can interact with Bluetooth Low Energy (BLE) devices. In agricultural fields where the weather conditions may lead to interrupted power supplies, a Bluetooth low energy device enables to use beacons that can be powered by a small coin cell battery.

Mesh Technology: Sensor nodes are placed throughout the farm and collect sensor values. Each sensor node is not connected to the Internet directly; rather, it communicates with other sensor nodes in a mesh network. An edge node communicates all information to the central server and database. Sensor nodes collect information once every two hours, and are otherwise put in deep sleep to conserve power. The sensor nodes are designed to last long periods of time without being connected to a power source. There's a main router that connects directly to the modem, and a series of satellite modules, or sensor nodes, mobile devices, smart pumps, which are placed around the farm fields as part of a single wireless network, and share the same SSID and password.

LRWPAN: LR-WPAN is a collection of Low-Rate Wireless Personal Area Networks (LR-WPAN) standards. These standards are used in high-level communications protocols such as ZigBee. It provides cost-effective and low-speed communication to power-restricted devices. LR-WPAN offers fast transmission of small packets of data from sensors and to actuators, minimizing power utilization and reducing design complexity.

HSDPA-4G/5G: High-Speed Downlink Packet Access, HSDPA, is based on the 3G networks and used for fast data transfer rates and increased data capacity. It supports HD video live streaming and faultless connectivity.

2.1.2 Remote Sensing Technologies

Remote sensing technologies have become cost-effective solutions for capturing field data. Aggregation of field-level data is useful in applying Variable-Rate Technology (VRT) that provides the capability to modify the rate of soil, nutrients, and pesticides for site-specific application. Remote sensed imagery could be used for identifying soil properties, classification of crop species, detection of crop water level, detection of weeds and identification of crop diseases, and mapping of crop yield.

Figure 2, depicts some of the parameters affecting the quality of the remote-sensed imagery. Different issues that must be carefully evaluated in the remote-sensed images for agricultural decision-making, including the accuracy of the image relative to the ground location, earth rotation (also called geometric precision), to the extent that the image depicts spatial and spectral resolutions features in the ground and the quality of spectral information such as sun position, cloud information, etc., represented in acquired images.

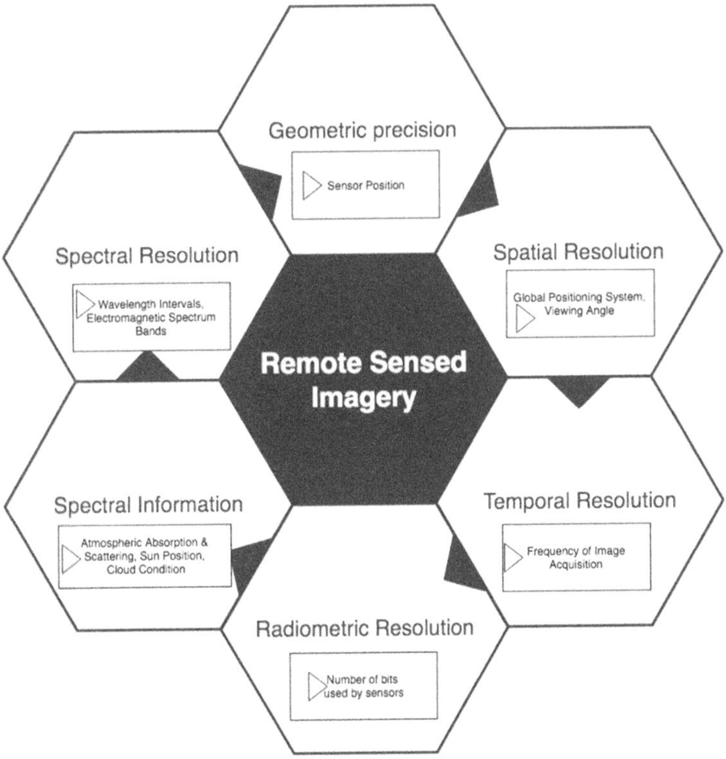

Fig. 2 Quality parameters for Remote-Sensed Imagery (Khanal et al. 2017)

Image Resolution in Remote Sensing

While analyzing images, different types of resolutions are named as: spatial, temporal, spectral, and radiometric. Spatial and spectral are particularly substantial as they impact the ability to extract ample information from an image (Table 3).

2.1.3 Unmanned-Aerial Vehicle (UAV) Technology

In collaboration with GIS, GPS technology forms widely adopted precision agriculture practices allowing fine-grain monitoring and mapping of yield and crop. This also helps in gaining an adequate understanding of their crops using robust data analytics. Crop Scouting is one operation that can be made more accessible by using drone data to plan and improve waterways and evolving pesticide applications accurately. Drone devices are unmanned aircraft that can fly autonomously. It can also be considered as a flying robot that can be remotely controlled through software. The flight plans are software controlled and work in conjunction with sensors and GPS systems. The typical component includes a high-resolution camera, antenna, receiver, flight

Table 3 Different types of image resolution in remote sensing (Khanal et al. 2017)

Resolution	Definition	Types	Advantages/Challenges
Spatial tesolution	Determines the size of the smallest identifiable features in an image	• High spatial resolution: small objects can be detected, which in turn displays features in detail. • Low spatial resolution, the size of a pixel is high. Multiple features are represented by a single pixel	• Different spatial resolution images acquired by manned-aircraft or UAVs typically and satellite image, • Less stable camera position • Images need to be converted into similar spatial resolution
Spectral resolution	Refers to an ability of a sensor to differentiate between wavelength intervals in the electromagnetic spectrum	• Multispectral images deal with fewer bands • Hyperspectral images deal with thousands of fine wavelength intervals to provide detailed information.	• Hyperspectral images capture variability in crop and soil conditions in agricultural sectors. • It includes the high cost of the sensor, high image storage requirement, and the complexities associated with image processing. • Sensors with the high spectral resolution are useful in distinguishing features that are often not easily detectable by sensors with broad wavelength ranges.
Temporal resolution	Signifies the frequency at which images are collected over the same area (e.g., field)	• Images from manned aircraft and UAVs can have higher temporal resolution due to flexibility in scheduling flight plans. • Satellite images may have lesser temporal resolutions due to fixed revisit cycles of satellites.	• Capturing images at frequent intervals in the crop-growing season to detect possible in-season nutrient and water levels.
Radiometric resolution	Reflects a sensor's ability to identify or discriminate very slight differences in reflected or emitted energy	• The higher the radiometric resolution, the more sensitive it is to detecting small differences in information represented in images.	• Images of higher radiometric resolution for detecting differences in features such as crop canopy or soil variations.

controller, speed controller, altimeter, and accelerometer. Various kinds of sensors such as the Hyperspectral sensor mounted on UAV are helpful in identifying vegetation, crop health, water levels, and ground structure. Thermal sensors are beneficial for livestock management, Heat stress detection, security and protection of fields from wild animals, etc.

The data produced by drone devices are usually high-resolution in nature and henceforth, UAVs are particularly useful for the careful monitoring of large areas of farmland, considering topographical factors such as slope and elevation, to identify crop disease, weeds, soil conditions, troubleshoot irrigation systems and yield monitoring. The drone technology has also proven useful in gaining insight into plant development and population, as more precise data can help with cultivation decisions, retreating and trimming activity, and the improvement of crop models (Merriam 2020). The advantages of smart drone systems are the optimal use of inputs, seed, fertilizers, and water, identification of weeds, pests, and fungi using high-resolution imagery, thus, saving time in crop scouting, improving variable rate prescription in real time, and maximizing operations efficiency. This leads to precise recommendations, prescriptions, and increased yield from the field. However, the usage of drone devices comes with a lot of challenges. Some of them may be enlisted as:

- **Limited Battery**: The advantages of aerial vehicles may be shortened due to battery life issues. Drones or quadcopters do need a significant amount of energy to fly, as they are required to rotate at high revolutions per minute to generate lift.
- **Limited payload and flight endurance**: Any additional source of power and payload mechanism increases the cost of devices.
- **Security Concerns**: Drone devices fitted with video cameras may violate the regulatory laws and wrongly shoot in unauthorized locations.
- **High Cost**: Drone devices, equipped with high-end cameras to capture images for large fields live-in streaming facility may cost very high.

2.1.4 Tethered Eye

Due to the number of operational issues in UAV devices such as reduced battery power, high capital expenditure, deployment, and operational expertise with lack of live-in media streaming capabilities, an alternative aerial imagery platform, which has low capital investment, low operational costs, and high resolution has been proposed by Microsoft (Swamy et al. 2019). This technology uses the existing imaging devices for capturing the images like smartphones, cameras. This system is named as TYE (for Tethered eYE). It uses lighter-than-air gas (such as helium) filled balloons to provide lift to an existing imaging device such as a camera, smartphone. The balloon is held using a tether at heights from 50 ft. to 150 ft. depending on the required resolution.

2.2 *Data Storage and Decision-Making*

Precision agriculture has tremendous potential to expand agricultural operations through better data-driven decision-making. The rising use of precision farming has given birth to several sensor devices deployed across farms. These sensor devices are responsible for collecting data periodically and transmitting this data to gateway devices for analysis. Technology solutions that help manage the data and its processing need to be highly scalable and reliable. This part of precision analysis needs to be carried out without setting up massive data centers or much investment in high-end servers. The cloud-computing paradigm offers a solution to this requirement.

2.2.1 Cloud Computing

Introduction

In the cloud computing paradigm, computing is transformed into a model consisting of various services. These services have been commoditized and offered in a manner similar to traditional utilities such as water, electricity, gas, and telephony. Such a computing model, enables users to access applications and services from anywhere in the world on demand. Organizations are attracted by the opportunity that not only reduces the expenditure cost but also eliminates the maintenance effort. Apart from being known for its cost-effective, hassle-free implementation, and minimal to zero maintenance costs, cloud computing can accelerate the penetration of the Internet among agriculture technology.

Cloud Service Model

Cloud computing model encompasses different service layers. Unlike traditional data centers, in cloud computing, these services are virtualized and delivered on-demand. The cloud services are classified into three categories:

Infrastructure as a Service (IaaS): IaaS offers the dynamic provisioning of infrastructural IT resources such as Desktop, Server, Networking, Voice and Data Communications, Security, and Storage as services to various customers. Example: Amazon EC2, GoGrid, and Eucalyptus.

Platform as a Service (PaaS): Platform as a Service can be referred to as delivering the computing and application platform over the web. The PaaS provider offers the hardware, operating system, software upgrades, security, and everything related to the hosting of an application. Examples: Windows Azure, Google AppEngine.

Software-as-a Service (SaaS): SaaS refers to the sharing of software applications to many users. The software runs on SaaS provider sites and users need not worry about the security, installation, and up-gradation aspects. Example: Salesforce.com, Google Apps.

Cloud Deployment Model

Virtualization and network are the underpinnings for all kinds of cloud services. A collective point of view on cloud computing envisions that there are many different types of clouds such as public, private, virtual, hybrid, community clouds, and many different services such as infrastructure, platform, software, database, storage, an application that would be delivered through cloud marketplace. Depending upon the user accessibility/usage of cloud services, different cloud solutions can be provided. Various cloud deployment models and their brief descriptions are shown in Table 4.

Apart from the cloud architectural models as shown in Table 4, there are few other types of clouds also such as Virtual Private Clouds and Inter-cloud.

Table 4 Cloud computing deployment models with examples

Cloud type	Public cloud	Private cloud	Hybrid cloud	Community cloud
Definition	The Cloud Service Provider (CSP) develops applications and services and makes them available for the public on a pay-as-usage model basis Mostly deployed off-customer premises and mix applications transparently from different customers on shared architecture	Private clouds are proprietary to the particular organization Designed and managed by an IT department within an organization, highly controlled over cloud services and infrastructure Generally built from reasonable commodity hardware, and may be in a collocated facility or in an existing data center	A hybrid cloud is a composition of two or more clouds (public, private, or community) that remain unique in themselves but are bound together by standardized or proprietary technology that enables interoperability	Community cloud falls in between public and private cloud It is designed for a community of organizations that have common privacy, security, and regulatory considerations/goals
Example	Salesforce.com, Amazon Elastic Compute Cloud (EC2), IBM's Blue Cloud, Sun Cloud, Google AppEngine, and Windows Azure Services Platform	Eucalyptus, Elastra VMware, and Microsoft.	Merging of internally built or private cloud infrastructure with already deployed security threat intelligence network	Health Community Cloud Automation (HCCA) was initiated by Parallels and Microsoft Corp

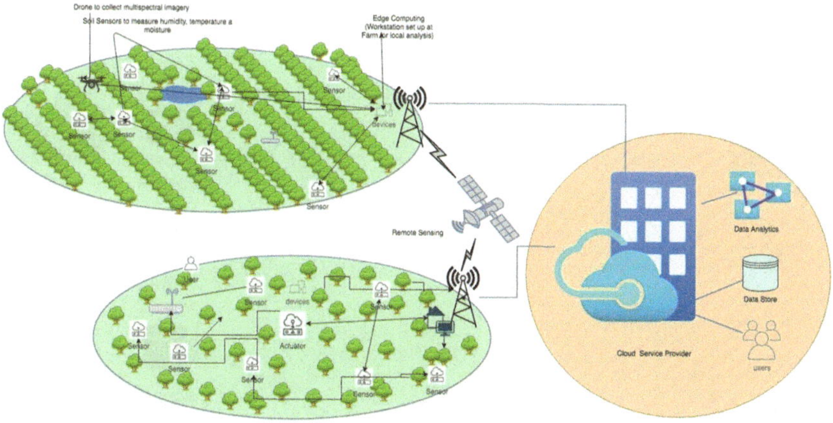

Fig. 3 Typical cloud computing deployment scenario for precision farming

Cloud Computing in Precision Farming

Cloud-based decision support enables us to process data efficiently at a reasonable cost, and to translate the data to better decisions and actions in a field. Figure 3 demonstrates a typical scenario where sensor devices, UAV devices have been deployed across farm fields and crop scouting has been done using remote sensing. The data received from satellite images and sensor devices is stored in a cloud platform. Cloud computing infrastructure deployed at Cloud Service Provider (CSP) offers the services like infrastructure, platform, or software. Microsoft Farm beats (Kapetanovic et al. 2019; Vasishta et al. 2017) is a technology that helps to collect data from several IoT devices such as sensors, cameras, and drones to produce detailed analysis for farmers. Farm Beats is also used for a variety of farming applications such as yield storage monitoring, generating precision maps of crop fields, and livestock monitoring.

Cloud Computing Challenges

Cloud computing paradigm presents several opportunities for smart farming and also poses new challenges. Some of these challenges are listed out as follows:

- In smart farms, a number of sensors and data sources are deployed for different activities, each device with its own data format and semantics. Hence, cloud-based solutions need to handle different data types and formats;
- How to analyze the data for different applications such as irrigation, nutrient deficiency, or crop scouting?
- To control field devices safely from the cloud.

A cloud-based decision support system (Fig. 4) is used to integrate data generated from different sensors and data sources to measure field variability. To handle a variety of field devices, Software-Defined Network (SDN) may also be used. Figure 5

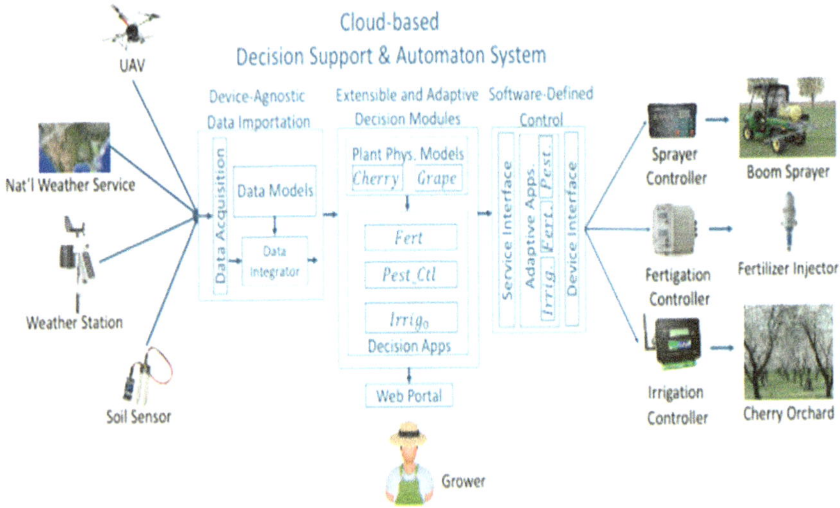

Fig. 4 Cloud-based decision support system using software-defined network (Tan 2016)*

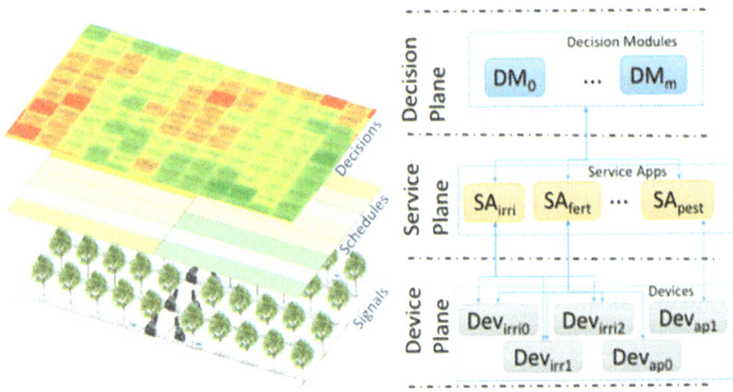

Fig. 5 Software-defined control for smart irrigation system (Tan 2016) *Images re-printed with written permission from IFAC-paperOnline and Elsevier

illustrates software-defined control, using a 3-zone irrigation system. A decision made by the decision plane specifies a distribution of water volume in an orchard (Tan 2016). SDN control structures the adaptation in a three-layer architecture: (1) a decision plane integrating device-independent decisions that require actions at a high level, (2) translating decisions to schedules for each device, and (3) the device plane executing the decision schedules.

2.2.2 Edge Computing

With the increasing use of IoT devices and the arrival of fast wireless networks, it has become much easier to bring the data and compute together. Collecting the data from hundreds of devices scattered across farmland and processing this data in real time to gain key insights require real-time computing power. Faster key networking technologies leverage the creation, processing, and handling of real-time data such as image processing, weather forecasting, yield prediction. IoT devices generate enormous amounts of data continuously. Transmitting all this data to the cloud required huge bandwidth and loss of quality too. With this additional operational cost, it is highly imperative to store the data collected by sensors at the local source of processing and storage. Edge computing hardware and services address this problem by deploying a local center for processing and storage.

As per Gartner, Edge computing is a "part of distributed computing topology in which information processing is located close to the edge-when things and people produce or consume that information." Any edge gateway can process data from edge devices and send only relevant data back through the cloud, reducing the need for higher bandwidth. Edge devices (gateways) are deployed in close proximity to farms, the data collected from soil sensors, such as humidity level, temperature, time, moisture content, water level, along with crop images processed at the edge devices and suitable decision, like for this example, increasing/decreasing the water drip rate can be taken at farm level (O'Grady et al. 2019). However, once local decisions and subsequent controlling instructions are given, for further processing this analysis can be combined with other applications data, such as nutrient deficiency, or pesticide spraying in the cloud. Figure 6 represents edge devices positioned at farms and the transmission of data between cloud, and local processing.

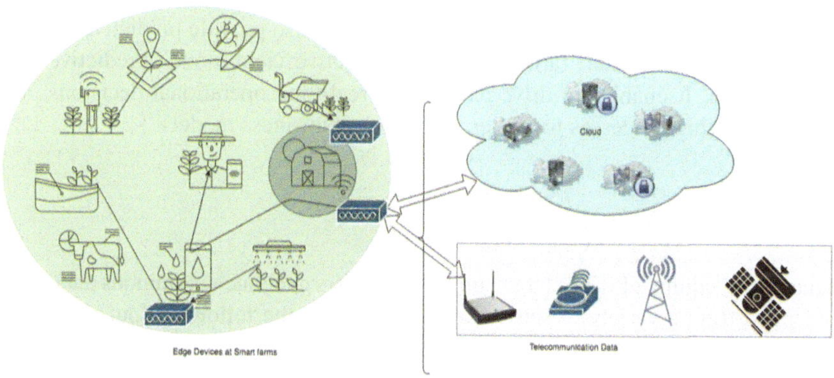

Fig. 6 Data transmission with edge computing and cloud computing for precision farming

2.2.3 Fog Computing

Fog Computing aims at moving the processing abilities closer to the data source, proving useful to the smart grid, realizing more rapid M2M (Machine-to-Machine) communications, and even in the precision farming context, shifting the decisions near the place where data are collected (Bonomi et al. 2012). Both fog computing and edge computing offer the same set of functions in terms of pushing both data and intelligence to analytic platforms that are situated either on the computation side or close to where the data originated from such as a local computing center at farmer's place. Fog nodes perform data mining and data analysis on a large volume of heterogeneous data, collected from various sensor devices and other IoT devices. This, in turn, achieves real time and fast processing for decision-making (Peng and Zhang 2018). Henceforth, in most of the literature, it is found out that fog computing and edge computing are actually the same concept. Both of these two technologies leverage the computing capabilities inside a local network to carry out computation tasks that would generally have been carried out in the cloud in order to save network bandwidth, reduce latency and security.

2.3 Data Analysis

Precision Farming provides insight into the data collected from sensor-equipped fields and imagery received from UAV and remote sensing devices. Correct measures need to be taken to analyze this data. Big Data is a key technology, which majorly focuses on processing large amounts of data—both structured and unstructured. This analysis is particularly helpful in identifying hidden patterns, hidden correlations between the number of collected parameters, finding relevant information, undiscovered trends, and associations. Big data analytics also provide predictive insights to future models of smart farming such as predictive yield model, predictive feed intake model. It enables to drive data-driven real-time operational decisions, and reinvent farming processes for faster, and efficient business models.

2.3.1 Big Data Analytics

Smart farms equipped with a large number of sensors generate voluminous data. Data generated from a variety of sensors are processed in the following four sequential steps:

- A large amount of unstructured data is generated by IoT devices, across the field is collected in the big data system. This IoT-generated big data is characterized by its huge volume, velocity, and variety.
- Big data system is actually a shared distributed database that stores a huge amount of big data files.

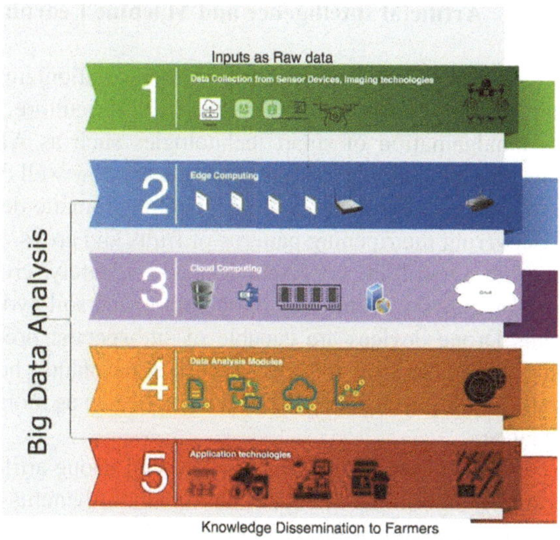

Fig. 7 Big data analytics framework for smart farms

- Stored data is analyzed using tools like framework Hadoop MapReduce or Spark on computing platforms.
- The reports of analyzed data are generated and visualized.

Figure 7 shows the process of collecting the raw inputs from sensor devices and subsequently processing this data at edge devices located at local area networks in farms. Further data is analyzed in cloud computing servers using data analysis modules. Derived analysis and inferences are integrated with application technologies and application-level decision-making happens (Wolfert et al. 2017).

Application of Big Data in Agriculture

- **Improving farm productivity**: Big data analysis helps farmers with information on changes in weather, temperature, airflow, soil nutrients, and other factors that affect crop yield. With all this data, the farmers are able to make precise and reliable decisions, thus improving farm productivity.
- **Site-specific Pesticide application**: By analyzing the crop health and soil conditions, big data helps farmers to provide reliable usage of pesticides in specific sites This reduces the unnecessary usage of pesticides that could be detrimental to the yield quality.
- **Yield prediction**: Yield prediction entails the use of technology and combining data collected from different IoT devices to analyze information on weather, chemicals, soil health, weeds, vegetation, and irrigation, to improve decision-making by stakeholders. Big data analysis helps farmers to easily decide on where and when to plant seeds, weed detection, water stress, chemicals application, nutrient deficiency, and when to harvest.

2.3.2 Artificial Intelligence and Machine Learning

AI and Big Data Analytics are seen revolutionizing rural farmlands into smart connected farms, also called as Precision Agriculture. This became possible through an amalgamation of smart technologies such as AI, Big Data, Cloud, IoT, and Machine Learning. Transformed farms are now well connected through IoT devices and empower the features ranging from automatic detection of drought patterns to discovering the ripening patterns of fruits and crops. Drone devices are extensively used in agriculture for research analysis, safety, crop scouting, terrain scanning, spatial analysis, monitoring temperature, and soil hydration, identifying yield problems Drone devices are capable of site-specific précised spray pesticides on the diseased plants from the vast expanse of farmlands, help add micro and macronutrients, check physical properties from soil such as moisture, chemical properties, pH balance.

Precision Agriculture empowered with strong artificial intelligence and machine learning algorithms facilitate the maximum benefits mentioned above. It includes the classification of disease after analyzing leaf conditions, by sending corrective measures and so on. The other one includes automatically controlling the water level by analyzing the soil moisture. The powerful statistical analysis seems endless and gives tremendous opportunities in the field of precision farming (Figs. 8 and 9).

Machine learning presents key algorithms and models that can help to draw results from the agricultural field. Combining data received from sensor devices and performing suitable machine learning techniques opens the path to optimize pesticide usage in the field, adopting the ways to increase the yield and managing the livestock. A typical process includes cleaning the data, handling missing values, removing noisy elements, and standardizing, normalizing, and de-duplicating the data values. Further steps include extracting suitable features, integrating data from multiple devices and databases, and creating new attributes from the data.

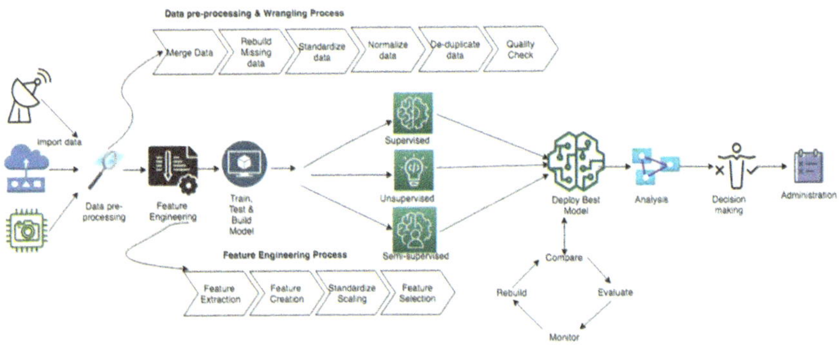

Fig. 8 Major steps in applying machine learning concepts for precision farming

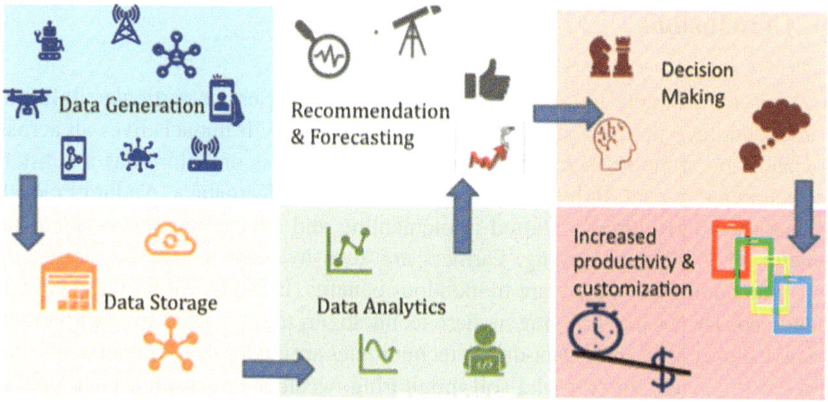

Fig. 9 Data-driven agricultural process leads to better yield and productivity

The next step is to leverage a suitable machine learning algorithm (supervised, semi-supervised, unsupervised) to train the model. The final step includes validating the trained model and evaluating the results by testing the accuracy of the models.

3 Case Study

In India, the University of California, Berkeley has been roped into launch an experiment called "Smart Villages" in Mori Village of East Godavari District of Andhra Pradesh. In this project, the emphasis was upon easing out the process of agriculture with the help of the integration of technologies. The open innovation allows simple integration of technologies offered by several industries in three phases: Pre-Harvest, Growth and Management, Post-Harvest. Pre-harvesting phases include farm analysis and advisory in real time. This has been proven very useful for Shrimp farmers for getting the critical parameters like pH-level and DO-level with respect to the pond water. With the knowledge enrichment of pre-harvest phases, farmers find farm controlling much easier using smart sensors and actuators in the growth and management stage. To handle power failure and optimized energy usage, high-quality solar panels and reliable bio-energy solutions offer sustainable solutions. Digital technology solutions improved the supply chain process and increased the profits for the farmers. The survey conducted showed that 86% of farmers are interested in the implementation of new technologies for smart farming (Darwin and Chesbrough 2017).

4 Conclusions

Agriculture is largely untouched by modern technologies, in particular, Information Technology until recently carries this vast potential. It impacts lives all across; is sizable; is arduous; lacks shared information; success or failure has a substantial economic impact. In India, Gujarat, Uttar Pradesh, Karnataka, Andhra Pradesh, Maharashtra have already started implementing and educating farmers about the benefits of intelligent farming. Farmers are keen to adapt themselves as per the newer technologies as there are tremendous benefits. In the recent past, the government made strides into bringing modern technologies to rural areas and exploit them to solve these core issues. Data-driven technologies are emerging to help in improving efficiency concerning crop and soil monitoring, weather forecasting, yield prediction, predictive agricultural analytics, markets, and supply chain efficiency. The cloud computing infrastructure with the use of data-led technologies, IoT, and AI enables the development of precision agriculture and strengthens the farmers in practicing smart and precision-based farming, smart irrigation, intelligent fertilizer application, and disease/pest diagnosis/detection, intelligent spraying, and harvesting. Machine learning and soft computing methods with pattern recognition through image and video (drone cameras, satellite imagery) data processing are widely used worldwide and provide accurate imagery, which researchers and scientists use for deeper analysis. This, in turn, offers huge financial stability and sustainable solutions to stakeholders.

References

Ai Y, Peng M, Zhang M (2018) Edge computing technologies for internet of things: a primer. Digit Commun Netw 4(2):77–86

Aune JB, Coulibaly A, Giller KE (2017) Precision farming for increased land and labour productivity in semi-arid West Africa-A review. Agron Sustain Dev 37:16

Bhatnagar V, Chandra R (2020) IoT-based soil health monitoring and recommendation system. In: Pattnaik P, Kumar R, Pal S (eds) Internet of things and analytics for agriculture, Vol 2. Studies in Big Data, Vol 67. Springer, Singapore

Blackmore S (1994) Precision farming: an introduction. Outlook Agri 23(4):275–280

Bonomi F, Milito R, Zhu J, Addepalli S (2012) Fog computing and its role in the internet of things. In: Proceedings of the first edition of the MCC workshop on mobile cloud computing, Helsinki, pp 13–16

Darwin S, Chesbrough H (2017) Smart farming in Andhra Pradesh, India: An Innovation Approach. Berkeley Haas

Ferguson RB, Hergert GW (2000) EC00-154 Precision agriculture: soil sampling for precision agriculture. Historical Materials from University of Nebraska-Lincoln Extension. 708 https://digitalcommons.unl.edu/extensionhist/708

Jarman M, Dimmock J (2018) Satellites for agriculture Accessed on May 8th, 2021 https://projectblue.blob.core.windows.net/media/Default/Imported%20Publication%20Docs/SatellitesForAgriculture1825_181217_WEB.pdf

Kapetanovic Z, Chandra R, Chakroborty T, Nelson A (2019) FarmBeats: improving farm productivity using data-driven agriculture, 2019 SIAM conference on computational science and engineering

Kerry R, Oliver MA, Frogbrook ZL (2010) Sampling in precision agriculture. In: Oliver MA (eds) Geostatistical applications for precision agriculture

Khanal S et al. (2017) Remote sensing in precision agriculture accessed on May 9th, 2021 https://ohioline.osu.edu/factsheet/fabe-5541

Li T (2016) Cloud-based decision support and automation for precision agriculture in orchards. IFAC-PapersOnLine 49(16):330–335

Merriam P (2020) The next era of crop scouting: putting drones to work Accessed on May 09, 2021 https://www.precisionag.com/in-field-technologies/drones-uavs/the-next-era-of-crop-scouting-putting-drones-to-work/

Navarro E, Costa N, Pereira A (2020) A systematic review of IoT solutions for smart farming. Sensors 20:4231

O'Grady MJ, Langton D, O'Hare GMP (2019) Edge computing: a tractable model for smart agriculture?. Artif Intell Agric 3:42–51

Ojha T, Misra S, Raghuwanshi N (2015) Wireless sensor networks for agriculture: the state-of-the-art in practice and future challenges. Comput Electron Agric 118:66–84

Revolutionizing Smart Agriculture Using Semtech's LoRA Technology, SemTech White paper, Oct, 2017 Accessed on May 8th, 2021 https://www.semtech.com/uploads/technology/LoRa/WP-SEMTECH-LORA-SMART-AGRICULTURE.pdf

Saiz-Rubio V, Rovira-Más F (2020) From smart farming towards agriculture 5.0: a review on crop data management. Agronomy 10: 207

Shibusawa S, Haché C (2009) Data collection and analysis methods for data from field experiments. In: Ting KC, Fleisher DH, Rodriguez LF (eds.) System analysis and model. of food and agri. EoLSS

Sjaak W, Lan G, Cor V, Marc-Jeroen B (2017) big data in smart farming a review. Agric Syst 153:69–80

Steven Schriber (2021) Smart agriculture sensors: helping small farmers and positively impacting global issues, Too. Accessed on May 8th, 2021 https://www.mouser.in/applications/smart-agriculture-sensors/

Swamy VN, Kumar A, Patil R, Jain A, Kapetanovic Z, Sharma R, Vasisht D, Chandra R (2019) Low-cost aerial imaging for small holder farmers. In: Proceedings of the 2nd ACM SIGCAS conference on computing and sustainable societies, COMPASS '19, www.3gpp.com. Accessed on May 8th, 2021

Tompkins S (2019) Enabling the connected farm—the importance of satellite communications Accessed on May 9th, 2021 https://www.inmarsat.com/en/insights/enterprise/2019/enabling-the-connected-farm-the-importance-of-satellite-communications.html

Vasisht D, Kapetanovic Z, Won J, Jin X, Chandra R, Kapoor A, Sinha SN, Sudarshan M, Stratman S (2017) Farmbeats: an IoT platform for data-driven agriculture. In: Proceedings of the 14th USENIX conference on networked systems design and implementation (NSDI'17). USENIX Association, USA, 515–528

Data Science: Principles and Concepts in Modeling Decision Trees

Srikrishnan Divakaran

Abstract Data Science is a fast-emerging multidisciplinary area at the intersection of Computer Science, Mathematics, and Artificial Intelligence, with applicability in diverse fields like Agriculture, Business, Education, Finance, and Healthcare for solving problems like Annotating Genome Sequences, Automation in Manufacturing Industry, and Medical Image Analysis. These problems require analyzing discrete structured and unstructured high-dimensional data set consisting of a collection of observables and their associated responses for predicting responses for previously unseen observations. This chapter begins with a brief overview of statistical learning techniques like Linear and Non-linear Regression, Support Vector Machines (SVM), Decision Trees (DT), and Neural Networks (NN) for prediction problems in both regression and classification settings. Then, the study of Decision Trees is motivated by highlighting their computational efficiency and higher model interpretability in comparison with other earlier mentioned statistical learning techniques. This is followed by the introduction of basic terms and concepts necessary for understanding DT models and their associated statistical learning methods in both supervised and unsupervised learning contexts. Finally, key insights are provided into (i) the workings of some standard decision-tree-based statistical learning methods for making predictions; (ii) the understanding of statistical guarantees provided by these methods; and (iii) the use of Decision Trees in the context of Precision Agriculture.

1 Introduction

Data Science is a fast emerging multidisciplinary area at the intersection of Computer Science (data structures, algorithms, parallel and high-performance computing), Mathematics (Discrete Mathematics, Linear Algebra, Probability and Statistics, Information Theory, Modeling, and Simulation, Optimization), and Artificial Intelligence (AI) and Machine Learning (ML), with applicability in diverse fields like agriculture (Diawara 2019), Business, Finance, and Healthcare for solving problems

S. Divakaran (✉)
Ahmedabad University, Ahmedabad, India
e-mail: srikrishnan.divakaran@ahduni.edu.in

G. P. O. Reddy et al. (eds.), *Data Science in Agriculture and Natural Resource Management*, Studies in Big Data 96, https://doi.org/10.1007/978-981-16-5847-1_3

like Annotating Genome Sequences (Abril et al. 2019), Automation in Manufacturing Industry (Dietz 2019), and Medical Image Analysis (Gupta 2018) to mention a few. Many of these applications involve analyzing discrete structured and unstructured high dimensional input data sets consisting of a collection of observables (inputs) and their associated responses (outputs) for predicting responses for previously unseen observations (test data). This involves using the given input data set to: (i) define models that either explicitly or implicitly characterize the nature of the relationship between the observables and their associated responses in terms of model parameters; (ii) use statistical learning methods to learn the model parameters and predict the responses for test data (previously unseen observations), and (iii) provide statistical guarantees on their predictions.

More precisely, given a data set consisting of the n observables $X = (x_1, x_2, \ldots, x_n)$ and their corresponding responses $Y = (y_1, y_2, \ldots, y_n)$, where each observation $X_i = (X_{i1}, X_{i2}, \ldots, X_{ip})$ consists of p factors, statistical learning techniques help in characterizing the relationship between X and Y and use this characterization to develop methods for predicting responses to test data along with some statistical guarantees on their predictions. If the response Y is a continuous variable, then the prediction process is commonly referred to as regression, whereas it is referred to as classification when the output variable is discrete.

Linear and Non–linear Regression, Logistic Regression, Decision Trees (DT), Support Vector Machines (SVM), and Neural Networks (NN) are some of the widely used statistical learning techniques for making predictions and providing statistical guarantees on its prediction accuracy. Linear Regression is used when the response is continuous and has a linear relationship with the factors of the observables. In this technique, the relationship between the observable and its associated response is characterized by a linear function $y = a_0 + a_1x_1 + a_2x_2 + \cdots + a_px_p$, where $a_0, a_1, \ldots, a_p$ are the coefficients (model parameters) of $x_1, x_2, \ldots, x_p$, the p factors of the observable, determined by finding the best fit for the input data set. However, in situations where the relationship between the response and the observable is not linear or the factors of the observable are not independent (i.e. there is an interaction between the factors), the prediction methods based on linear regression are not accurate and hence provide weak statistical guarantees on its predictions. A very natural and immediate generalization of linear regression is non–linear regression. Here, the functional form can be specified as $y = f(x_1, x_2, \ldots, x_p)$, where f can be any polynomial, exponential, logarithmic, trigonometric, or power function of the factors $x_1, x_2, \ldots, x_p$. This technique is more powerful and has wider applicability than linear regression but is not well suited when there is an interaction between the factors. In addition, in many emerging domains specifying the functional form of f is extremely challenging because of the insufficient theoretical framework for both characterization and subsequent statistical analysis of data.

Logistic Regression is used in classification problems (response values are discrete and can be categorized into a few categories) by fitting a non–linear S-shaped logistic function to the data. Logistic Regression works well when the response can be modeled as a binary variable (i.e. the number of categories is 2) but as the number of categories increase, the models are unwieldy and their prediction accuracy is often poor. Also, like Linear Regression, this technique does not work well when there is an interaction between the factors.

SVM is used for modeling linear and non–linear prediction problems in both classification and regression settings. In this technique, a line or hyper–plane is used to separate data into classes and is very effective when there is a linear separation boundary. However, in many practical problems where the separation boundary is highly non–linear, SVM is better suited to serve as a starting point for applying other non–linear statistical learning techniques.

NN is a powerful technique for modeling non–linear relationship through highly interconnected layered networks that simulate neurons that accept inputs from a subset of neurons from the previous layer and produces an output that is typically fed as inputs to another subset of neurons in the next layer. The output of a neuron is modeled using a non–linear function of its weighted inputs. The first layer corresponds to the input and the final layer corresponds to the eventual output. Some neurons may send feedback to neurons in the earlier layer in the network. NNs are trained by an iterative process where the weights applied to each input at each neuron are adjusted to optimize the desired output. This technique behaves like a black box when it comes to the interpretability of its predictions and hence is extremely difficult to extend these NN models to new situations and also provide statistical guarantees on its predictions.

DT is a simple but effective technique for modeling complex relationships between an observable and its response for prediction problems that arise in both regression and classification settings. For a given dataset, the Decision Tree is constructed by essentially partitioning the dataset into subsets recursively into smaller and smaller subsets where the subsets have a hierarchical relationship that is captured by a tree. The predictions on test data are made based on which subset the test data falls into. This method also works when there are missing features, when there is a mix of categorical and numerical features and when different features have different scales. In the classification settings, DT (Classification Trees) works well even when the response has more than two discrete values. In addition, like NN, it can model non-linear relationships between the response and observable and interactions between factors. In addition, the DT unlike NN is computationally very efficient, their predictions are highly interpretable and often are very easy to implement. For a succinct comparative study of DT versus other statistical learning techniques we refer the interested reader to DTREG: a predictive modeling software (DTREG 2021; Cheesinglee 2016).

Example 1: Figure 1a and b (Cheesinglee 2016) highlight the superior classification of Decision Trees when compared to Logistic Regression and SVM when the two classes are separated by a non-linear boundary.

In this chapter, we overview the key concepts and techniques in the design and analysis of statistical models (James et al. 2013; Leo Breiman et al. 1984) and methods for prediction problems involving discrete high-dimensional data that are (i) hard to characterize; and (ii) traditional optimization techniques are not helpful in designing computationally efficient methods that can provide performance guarantees. In particular, the focus is on DT-based statistical learning methods that are

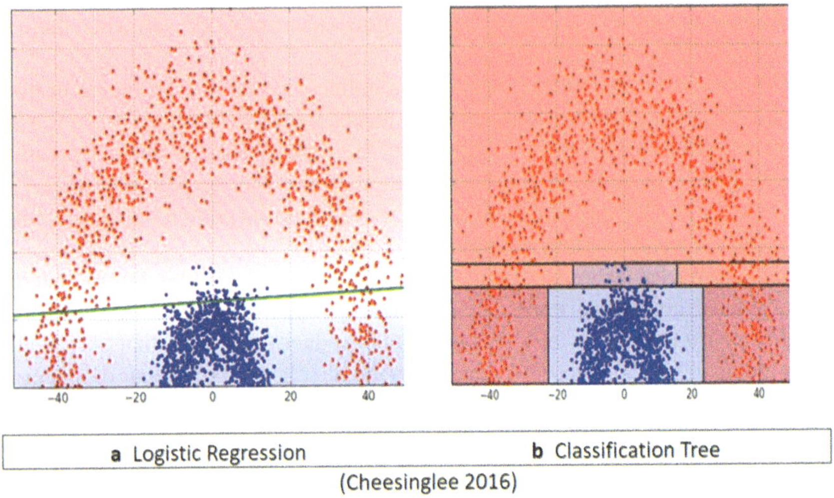

Fig. 1 **a** Logistic regression. **b** Classification tree

simple (highly interpretable), computationally efficient, and also provide statistical guarantees on its predictions. Many of these DT models and related methods solve key problems in computer vision (Rosebrock 2012; Nielson 2015; Prince 2012) with applications in Agriculture Monitoring and Management (Bisen 2020; Diawara 2019). In this chapter, we attempt to provide key insights into (i) the workings of some standard DT-based statistical learning methods for making predictions; (ii) understanding the statistical guarantees provided by these methods; and (iii) the use of DTs in the context of Precision Agriculture. For a comprehensive treatment of models and methods for prediction and inference problems, we refer the reader to Hastie et al. (2009), James et al. (2013) for a machine learning perspective, to Mitchell (1997) for a machine learning perspective and Goodfellow (2016), Han et al. (2019), Nielson (2015) for a deep learning perspective. We also refer the reader to prediction and inference problems that arise in Computer Vision (Rosebrock 2012; Nielson 2015). The rest of the chapter is organized as follows: In Sect. 2, we overview the basic terms and definitions necessary for understanding our models, statistical methods, and their analysis. In Sect. 3, we present DT-based statistical learning models and methods for prediction problems that occur in both classification and regression settings in the supervised learning context. In Sect. 4, we present statistical learning models and methods for clustering problems that arise in the unsupervised learning context. In Sect. 5, we highlight the applicability of DT in precision agriculture. In Sect. 6, we summarize the learnings of this chapter and its future scope.

2 Preliminaries

In this section, we overview the basic terms, definitions, and concepts for understanding (i) statistical learning models and methods for prediction; and (ii) the statistical basis of these methods for providing guarantees on their predictions. For a comprehensive introduction from a statistical learning perspective, we refer the reader to James et al. (2013), Hastie et al. (2009). For an excellent introduction to these techniques for prediction/inference problems in Computer Vision, we refer the reader to Rosebrock (2012), Prince (2012), Deep Learning and Neural Nets, we refer the readers to Goodfellow et al. (2016), Nielson (2015).

From a statistical learning perspective, the relationship between the n inputs/observables $X = (x_1, x_2, \ldots, x_n)$ and their corresponding outputs/responses $Y = (y_1, y_2, \ldots, y_n)$ can be characterized by $Y = f(X) + \varepsilon$, where f is a fixed but unknown function of X and ε is a random error term independent of X with mean 0. So, for predicting the response for X, we employ a statistical learning method that predicts its response for X to be $\widehat{Y} = \widehat{f(X)}$, where $\widehat{f}$ is an estimate of f. The accuracy of $\widehat{Y}$ as a response for X depends on how accurately $\widehat{f}$ estimates f and the variability associated with ε.

The error in estimating f is called the reducible error and can be reduced by choosing an appropriate statistical learning method, whereas the error due to the random error term is referred to as irreducible error and cannot be reduced because it depends only on the variability of ε. So the goal in prediction is to choose a statistical learning method for estimating f that minimizes the reducible error. In some systems, the inputs (observables) are explicitly related to the output (response) and the goal of modeling is to help characterize the functional form of f and then use statistical learning methods to estimate f along with providing statistical guarantees on its predictions. The modeling and learning techniques employed in these contexts are categorized as supervised learning. However, in some systems, the goal of modeling is to help characterize and determine the relationship among the inputs/observables, and there are no explicitly defined outputs. The modeling and learning techniques employed in these contexts are categorized as unsupervised learning. Also, there are situations where there is missing, noisy, or incomplete data where we may have to consider a hybrid of supervised and unsupervised approaches, and these approaches are categorized as semi-supervised learning. So, more formally, in the case of supervised learning, we are given input set $X = (x_1, x_2, \ldots, x_n)$ and its corresponding outputs $Y = (y_1, y_2, \ldots, y_n)$ (that we usually cannot observe), we would like to determine a function f that provides an estimate for Y by considering a class of functions that are efficiently computable and accurately approximates the relationship between X and Y. In the case of unsupervised learning, we are given the input data sets and we would like to determine a clustering function that partitions the input data set into K different classes, where K may be fixed a priori or determined during the partitioning process. Linear Regression, Logistic Regression, SVM, and DT are examples of supervised learning, whereas Principal Component Analysis (PCA), K-means clustering, NN are examples of unsupervised learning. The problem of determining the function $\widehat{f}$ that provides an estimate for f is fundamental to solving the following two broad categories of statistical learning problems: (1) prediction problems: predict output Y as a function of inputs X, and (2) inference problems: infer the relationship between Y and subsets of X. In the prediction context, we need not know the functional form of f to predict Y but in the inference context, we need to know the functional form of f for inferring the relationship between X and Y. If the output Y is a continuous variable then the prediction process is commonly referred to as regression, whereas it is referred to as classification when the output variable is discrete. For estimating the function f, we need to first build models to either implicitly or explicitly characterize the functional form of f. Then we can employ standard statistical learning methods to (i) determine $\widehat{f}$, an estimate of f; and (ii) provide performance guarantees on its predictions/inferences using $\widehat{f}$.

The models that are employed for statistical learning problems can be categorized broadly into probabilistic and non–probabilistic based on the nature of the relationship between the observations and responses. In both systems, we need to model the inputs/observations X and their corresponding output/response Y as random variables because of the randomness associated with (i) the process of recording the observations; (ii) modeling the relationship between the inputs and outputs approximately either due to inadequate knowledge or computational considerations, and (iii) noise in the system. In non–probabilistic models, the relationship between the observations and the responses can be modeled using functions that are deterministic whereas in probabilistic systems there is non-determinism in the nature of the relationship between the observations and responses. This causes any given input to be associated with a set of possible outputs each with different weights specified by a probability distribution. In this chapter, we will focus only on non-probabilistic models that are used by statistical learning methods for prediction in both supervised and non-supervised settings. For an excellent introduction to probabilistic models for prediction for solving problems in the context of Computer Vision, we refer the reader to (Prince 2012). For building a model, we provide our statistical learning methods with training data, a set of n different data points consisting of pairs of inputs and their corresponding output (based on past knowledge/experience). These statistical methods can be categorized into parametric or non-parametric based on whether we assume about the functional form of f or not. For models built using a parametric method, we assume that the function f can be specified in terms of model parameters and we use the training data to fit or train the model (determine the model parameters). Essentially, in this approach, we reduce the problem of estimating f to one of estimating a set of parameters. Maximum Likelihood (ML), Maximum a-Posteriori (MAP), and Bayesian Methods are some of the popular methods for parameter estimation. The simplest parametric models are those models where the functional form of f is defined in terms of as few parameters as possible. In this case, the model is simple and highly interpretable. However, if the assumed functional form of f is very far from its true functional form then its predictions are poor. This problem can be addressed by increasing the model flexibility by choosing a functional form that is a combination of several functional forms. However, this results in a complex model (less interpretable) that makes good predictions but is computationally expensive, and risks over-fitting (finding non-existing patterns attributable to randomness) when it attempts to fits the training data closely. For models built using a non-parametric method, we do not make explicit assumptions about the functional form of f. Instead, we estimate f by fitting the data points as closely as possible without too many discontinuities or too many wiggles. This approach helps avoid the need for assuming any functional form for f without placing restrictions on the range of possible shapes for f. However, for these methods to be effective, we need far more observations than parametric methods for providing good statistical guarantees on its estimates. The performance of a statistical learning method is a function of the difference between its predicted output $\widehat{f(x_i)}$ and the true output y_i for each observation $x_i \in X$. For the regression setting, the Mean Squared Error (MSE) given by $MSE = (1/n)\sum_{i=1}^{n}(y_i - \widehat{f(x_i)}))^2$ is the most commonly used measure. The MSE will be small if the predicted responses are very close to the true responses, and will be large if, for some of the observations, the predicted and true responses differ substantially. A natural strategy for choosing a statistical method for estimating f is to choose the method whose MSE for the training data (i.e. data used to train the model) is minimum. This MSE is commonly referred to as the training MSE. However, an objective measure of a statistical method's performance is the MSE for unseen data, commonly referred to as test MSE. However, minimizing the training MSE does not guarantee that the test MSE will also be minimized. That is, if the statistical method uses a flexible model (closely follows the training data) then the training MSE will be low but the test MSE may not be low. This is because the statistical method finds patterns in the training data that are due to randomness rather than due to the true relationship between the observable and the response. This is usually referred to as over-fitting the data. Often over-fitting can be spotted when a simpler model yields a lower test MSE than a complicated model. Therefore, to objectively validate the performance of any model we need to validate the model.

2.1 Cross-Validation

Cross-validation (Stone 1974), a simple but effective technique that can be used to validate a model. In this technique, a random subset of the input dataset is used to train the model (training data) and the rest of the input dataset (depending on the validation technique) is set aside as test data. The training data is randomly split further into sub-training and sub-validation sets. The model is fitted using the sub-training set and evaluated using the sub-validation set. This procedure is iterated using different sub-sets and averaging of these results using measures like MSE or RSS helps in estimating the model's performance while providing statistical guarantees on its performance. This technique of cross-validation is computationally efficient and it also does not explicitly require data to be set aside for validation and hence does not affect the training of the model.

Notice that both the training and validation data can be viewed as random samples drawn from the input data set. So, MSE for validation data is a random variable and hence we need to compute it in the expected sense. The expected MSE for an estimate $\widehat{f(x_0)}$, where x_0 is an input selected at random from the input set, can be expressed as the sum of the following three quantities: (i) Variance in the estimate $\widehat{f(x_0)}$; (ii) Bias is the error in the estimate $\widehat{f(x_0)}$; due to the use of a simpler model; and (iii) a random error term that is supposed to be independent of $\widehat{f(x_0)}$ with mean 0 and variance $var(\varepsilon)$. So to minimize expected MSE, a statistical method needs to minimize both the variance and bias in its estimate. The random error term is an irreducible term that cannot be eliminated as it is independent of its estimate. Generally, by fitting the training data closely (using flexible methods—more parameters), the variance in the estimate will increase but the bias will decrease. As we choose more flexible methods, the decrease in bias tends to be faster than an increase in its variance. However, the nature of the tradeoff between the variance and bias determines the level of flexibility of a statistical method that is optimal for minimizing the test MSE. Another tradeoff one needs to consider while building a model is the tradeoff between prediction accuracy and model interpretability. The more restrictive models are simple but their prediction accuracy is limited because they restrict the functional forms that are allowed. A flexible model allows for a wider range of possible functional forms for f and thereby improving the prediction accuracy but is more complex from the perspective of interpretability. For prediction problems, the interpretability of the predictive model is simply not of interest, and hence defining a flexible model may be appropriate. However, in Inference problems, simple interpretable models provide advantages in getting good insights into the inference process.

Now, we will describe the general principle of k-fold cross-validation (Molinaro et al. 2005, Lianne and Justin 2020), a widely used cross-validation technique that can help in selecting a model flexibility based on the amount of bias the user is willing to tradeoff for variance and the computational costs for model validation.

K-Fold Cross-Validation

1. First, split the input data set into training and test sets. Then, we perform the k-fold cross-validation as follows.
2. Randomly partition the training dataset into k roughly equal-sized smaller sets (folds) F_1, F_2, ..., F_k.
3. For i in $[1..k]$
   (a) Train the model using the $k - 1$ fold other than F_i.
   (b) Calculate a model performance using (e.g., RSS or MSE) of the fitted model using the validation fold F_i (ith set).
4. Combine the performance measure from all the k models by computing a single metric such as average. This single measure is used to decide the model performance.

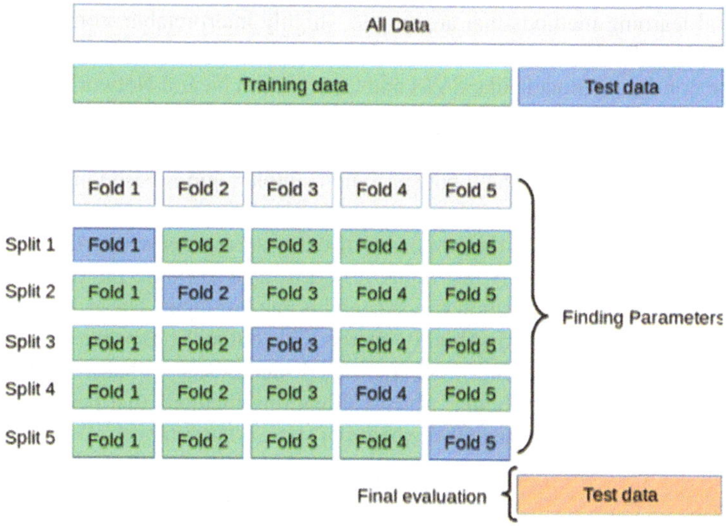

Fig. 2 Fivefold cross-validation process (Lianne and Justin 2020)

Example 2: Figure 2 shows a fivefold cross-validation process where Model one uses fold 1 for evaluation, and folds 2–5 for training. Model two uses fold 2 for evaluation, and the remaining folds for training, and so on.

Notice if $k = n$ then each observation becomes a fold and this type of cross-validation is called Leave-One-Out Cross-Validation (LOOCV). In this case, the computational effort is high because we will have to validate n times and also the performance measures of the n models are highly correlated (due to almost similar training data) resulting in a high variance in its performance. As k gets smaller, the training fold will be smaller in comparison to the validation fold. This will result in higher bias but less variance and less computational effort for validation. So one has to choose k depending on how much the user can tradeoff bias for variance and the computational effort one can expend for validation.

3 Tree-Based Models for Regression and Classification

The input and output spaces of many prediction problems are discrete and high dimensional with several discontinuities, and hence are very difficult to compactly represent and characterize. In addition, from a combinatorial optimization perspective, even special cases of many of these prediction problems are known to be NP-hard. For these problems, the traditional statistical learning methods based on Regression Models, Support Vector Machines, and Convolution Neural Networks are complex—not easily interpretable, computationally expensive, and do not provide strict statistical guarantees on their predictions. So, we present decision tree-based statistical learning methods that are simple—highly interpretable, computationally more efficient than the traditional methods mentioned earlier but are not as accurate as of the non-linear models like SVM and Convolution Neural Networks (CNN). In this section, we first present decision-tree-based statistical methods for prediction problems in both classification and regression settings. These methods make their predictions by segmenting the input space into simple regions based on the training data set. Then for each region, it defines as response a representative value that is some statistical function of the responses of the observations within that region in the training set. Then for any test input, it returns as a response the representative value corresponding to the region which contains that test input. These decision-tree-based methods are intuitive and highly interpretable but at the expense of prediction accuracy. Then, by modifying these tree-based methods, we obtain Bagging and Random Forests, the two popular statistical learning methods that help to achieve better prediction accuracy but at the expense of some interpretability. For a complete treatment of these tree models and methods for prediction, we refer the reader to James et al. (2013); Hastie et al. (2009), and Han et al. (2019).

3.1 Regression Trees

Given the input set $X = (x_1, x_2, \ldots, x_n)$ of n data points and their corresponding outputs (responses) $Y = (y_1, y_2, \ldots, y_n)$, where for $i \in [1..n]$, data point $X_i = (X_{i1}, X_{i2}, \ldots, X_{ip})$ consists of p predictors, we build a regression tree by first partitioning the p-dimensional predictor space into non-overlapping rectangular regions (high dimensional rectangular regions) $R_1, R_2, \ldots, R_K$, where the number K is determined by the partitioning process and not known a priori. Then, for each region choose a representative response (output) to be a statistical function (mean, min, max, and mode) of the responses of the training observations in that region. Then for any test input, it returns the representative response corresponding to the region which contains the test input. Given a training data set (X, Y), the goal of partitioning is to find rectangular regions $R_1, R_2, \ldots, R_K$, that minimizes the Residual Sum of Squares (RSS) $RSS = \sum_{k=1}^{K} \sum_{i \in Rk} (y_i - \widehat{y(R_k)}))^2$, where $\widehat{y(R_k)})$ is the mean response for the training observations within the kth region. This problem of partitioning the p dimensional space based on training data set X, Y with K high-dimensional rectangles to minimize the RSS is known to be NP hard. So, we present a greedy heuristic that initially starts with a single region consisting of the entire input space, recursively chooses from among the regions the region that has more than some minimum number of points and its splitting into two results in the maximum reduction in RSS. If we build the regression tree by fitting the training data very closely then the constructed regression tree will have a low RSS for the training data. However, this tree is likely to be complex (have many splits)—resulting in poor interpretability of its predictions. More importantly, the RSS of this tree for the test data is likely to be high due to the high variance in its predictions caused by over-fitting (fitting the tree close to the training data). For addressing this, we need to prune the regression tree (choose a sub-set of cuts) to build a simpler tree that is more interpretable and has lower variability on its RSS on test data, although at the expense of a slight increase in bias. For building this simpler tree, we need to determine which sub-set of splits we choose from among the exponential number of splits. This problem is NP-hard, so we present a heuristic that is defined in terms of a tunable real number $\alpha \in [0, 1)$ that helps in constructing a pruned tree by restricting the number of splits it considers. This heuristic essentially restricts the number of trees it considers by penalizing a tree by adding to its RSS a penalty that is α times the number of its terminal nodes. This penalty helps the heuristic to prefer trees with fewer terminal nodes. We estimate the performance of these pruned trees for the various values of α, by using K-fold validation (the training set and split it randomly into K nearly equal-sized groups (or folds) and onefold is set aside for testing) to estimate the RSS on test data.

Algorithm Regression Tree Construction (Breiman 1984)

Preprocessing: Divide the n input observations randomly into K groups, or folds, of nearly the same size. Notice that at any given instance, onefold is held back for validation (test data) while the remaining $k - 1$ fold are used for the construction of the regression tree T_0.

1. Use recursive binary splitting to grow a large tree T_0 on the training data, stopping only when each terminal node has fewer than some minimum number of observations.
2. Apply cost complexity pruning to the large tree T_0 in order to obtain a sequence of best sub-trees, as a function of α.
3. Use K-fold cross-validation to choose α. That is, divide the training observations into K folds. For each $k = 1, ..., K$:

   (a) Repeat Steps 1 and 2 on all but the kth fold of the training data.
   (b) Evaluate the mean squared prediction error on the data in the left-out kth fold, as a function of α. Average the results for each value of α, and pick α to minimize the average error.

4. Return the sub-tree from Step 2 that corresponds to the chosen value of α.

For each value of α, there corresponds a sub-tree $T_\alpha \subset T_0$ such that $\sum_{m=1}^{|T|}\sum_{Xi\in Rm}(y_i - \widehat{y(R_m)}))^2 + \alpha|T_\alpha|$ is as small as possible. Here $|T|$ indicates the number of terminal nodes of the tree T_α, R_m is the rectangle (i.e. the subset of predictor space) corresponding to the mth terminal node, and y_{Rm} is the predicted response associated with $R_{m,}$ that is, the mean of the training observations in R_m. The tuning parameter α controls a trade-off between the subtree's complexity and its fit to the training data. When $\alpha = 0$, then the subtree T_α will simply equal T_0 . However, as α increases, there is a price to pay for having a tree with many terminal nodes, and so we end up constructing a smaller tree.

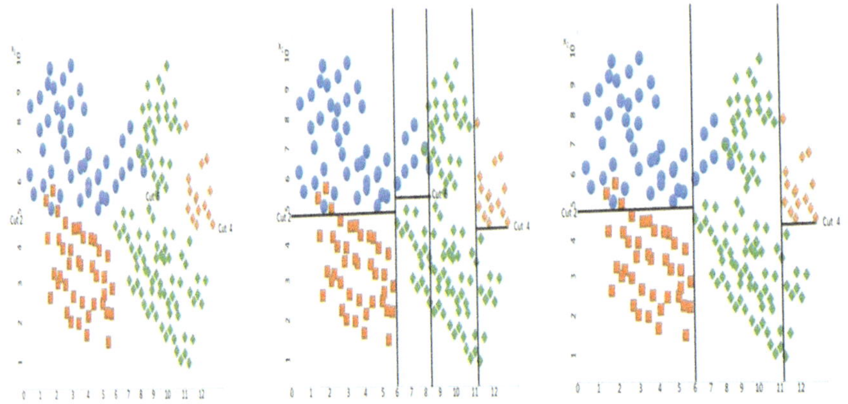

Fig. 3 **a** Collection of Points; **b** Partitions corresponding to a full Regression Tree; and **c** Partitions corresponding to a pruned Regression Tree

Example 3: Given a collection of data points in two dimensions, we will partition the input space into regions and use a representative within each region to make predictions about any data point that falls in that region. Figure 3 shows the partitions corresponding to a full regression tree and the pruned regression tree.

3.2 Classification Trees

A classification tree is similar to a regression tree, except that it is used to predict a response (output) that is qualitative (discrete) rather than quantitative (continuous). In the case of the classification tree, the prediction for input is given by the most frequently occurring response (mode) among the training data in that region. The construction of the classification tree as well its pruning is along the same lines as in the regression setting, but we use the classification error rate (the fraction of training observations that do not belong to the most common class) instead of RSS to evaluate its performance. More formally, let $\widehat{p_{mk}}$ denote the fraction of observations in the mth region of type k, then we define the classification error $CE = 1 - max\{\widehat{p_{mk}}\}$. However, Gini-index is an alternative measure that is more sensitive to changes in the tree structure and is used in many classification-tree-building methods. The *Gini-index* is defined by $Gini-index = \sum_{k=1}^{K} \widehat{p_{mk}}(1 - \widehat{p_{mk}})$. This measure the total variance across the K classes and its value is minimized when $\widehat{p_{mk}}$ is close to 0 or 1. That is node contains observations mostly of a single type. This is also referred to as node purity and helps in evaluating a quality of a split.

In addition, there are measures like *F1 score* and *model accuracy score* that are defined in terms of model sensitivity and selectivity (Rajan 2020). These two measures are very effective in problems that involve binary classification, however, they can also be extended to classification models involving multiple categories.

In the context of binary classification problems, the observations are categorized either as positive or negative. Now, for a given classification model, the input observations are categorized as (i) True Positive (TP): the observations that are correctly predicted to be positive; (ii) False Positive (FP): the observations that have been incorrectly predicted as positive; (iii) True Negative (TN): the observations that have been predicted correctly as negative; and (iv) False Negative (FN): The observations that have been incorrectly predicted as negative. Now, based on the above categorization, we define

- Sensitivity $= \frac{\text{TP}}{\text{TP+FN}}$ to be the proportion of actual positives that are correctly predicted as positives.
- Specificity $= \frac{\text{TN}}{\text{TN+FP}}$ to be the proportion of actual negatives that are correctly predicted as negatives.
- Precision $= \frac{\text{TP}}{\text{TP+FP}}$ to be the proportion of predicted positives that are correctly predicted as positives.
- $F1$ Score $= \frac{\frac{1}{\text{precision}} + \frac{1}{\text{sensitivity}}}{1/2} = 2 * \frac{\text{precision} * \text{sensitivity}}{\text{precision} + \text{sensitivity}}$ to be the harmonic mean of sensitivity and precision and ranges between [0, 1] with 0 being a poor score and 1 being a good score.
- Accuracy $= \frac{\text{TP+TN}}{\text{TP+TN+FP+FN}}$ to be the proportion of predicted positives that are correctly predicted as positives.

3.3 *Bagging and Random Forests*

The performance of the decision tree methods for constructing regression trees is evaluated based on estimating their RSS on test data using K-fold validation. That is the input data set is split randomly into K nearly equal-sized groups (or folds) and one fold is set aside for testing to estimate the RSS on test data and the remaining K − 1 fold are used for training the model. This causes these methods to yield different trees depending on the random subset of observations used for the training data (caused by random split of data for K fold validation). This is one of the primary causes of variability in its predictions. Secondly, in many instances, limited size of the training data results in models not able to accurately capture the statistical properties of the population from which it was derived. Finally, an accurate estimate of the variance associated with its predictions would be necessary for providing stricter statistical guarantees. To address these three points, we introduce Bagging and Random Trees, two generic decision tree methods, which builds on the earlier tree-building method to provide stricter statistical guarantees on its predictions. These two methods make use of bootstrapping, a widely used technique in statistics that helps in: (i) generating a large training data set from a smaller data set with statistical characteristics similar to the population from which the smaller data set was derived; and (ii) providing an accurate estimate on the variance associated with its predictions.

Bagging (James 2013): This is a method that obtains B training sets from the input data set by bootstrapping. Then it averages the predictions obtained from the B training sets to reduce the variance in its predictions. Essentially it aggregates the predictions of bootstrapped samples. In the context of regression trees, bagging involves constructing B regression trees using B bootstrapped training sets, and average the resulting predictions. These regression trees are fully grown (not pruned). Notice each regression tree constructed using bagging is likely to have high variance, but low bias. However, averaging these B trees reduces the variance in its predictions. For evaluating its performance, we estimate its error on test observations using the following straightforward way.

Each bagged tree uses around two-thirds of the input observations and about one-third of the input observations not used to train the model can be used as test data. These observations are often referred to as Out of Bag (OOB) observations. Now, we estimate the test MSE for each observation by comparing its true response with the average of the predictions from the B/3 models for which that observation was an OOB observation. Then we sum up the MSE for all the n observations to get a valid estimate of the test error for the bagged model.

Bagging improves prediction accuracy when compared to prediction using a single tree, but at the expense of interpretability. The loss of interpretability is due to averaging the predictions over a collection of B trees. However, we can obtain the importance of each predictor using the RSS (for bagging regression trees) or the *Gini-index* (for bagging classification trees) by looking at the effect on RSS due to splits over a given indicator.

Random Forests (Hastie et al. 2009): In bagging, certain splits are shared across most of the bagged trees resulting in a higher correlation among the trees. As a consequence, averaging many highly correlated quantities does not lead to a large reduction in variance as averaging many uncorrelated quantities. So, we introduce Random Forests, a statistical learning method that de-correlates the construction of bagged trees, thereby leading to a greater reduction in its variance when compared to bagging. This method like bagging builds a number of forest of decision trees on bootstrapped training samples. However, when building a tree, it only considers a random sample of m predictors instead of the full set of p predictors when considering a split. At each split, a fresh set of $m \approx \sqrt{p}$ predictors are chosen. The rationale for this is as follows: Suppose that there is one very strong predictor in the data set, along with a number of other moderately strong predictors. Then in the collection of bagged trees, most or all of the trees will share this strong predictor in the top split. Consequently, all of the bagged trees will look quite similar to each other. Hence, the predictions from the bagged trees will be highly correlated and thereby not help in reduction in the variance of its predictions. Random forests overcome this problem by forcing each split to consider only a subset of the predictors. Therefore, on average $(p - m)/p$ of the splits will not even consider the strong predictor, and so other predictors will have more of a chance. We can think of this process as de-correlating the trees, thereby making the average of the resulting trees less variable and hence more reliable. The main difference between bagging and random forests is the choice of predictor subset size m. For instance, if a random forest is built using $m = p$, then this amounts simply to bagging.

4 Clustering

Clustering refers to a very broad set of techniques in unsupervised learning for finding subgroups, or clusters in a data set. When we cluster the observations of a data set, we seek to partition them by grouping similar observations. That is, we have to not only figure out these groups but also determine what distinguishes them. This is different from classification problems where the data items are associated with groups and we need to predict them. The similarity between observations and groups is defined in terms of a distance function that is defined differently in different contexts. In this section, we present two widely used clustering approaches: K-means clustering and hierarchical clustering. In K-means clustering, we seek to partition the observations into a prespecified number of clusters. On the other hand, in hierarchical clustering (Kaufman and Rousseeuw 1990), we do not know in advance how many clusters we want. For a comprehensive treatment of unsupervised methods for prediction and inference, we refer the reader to Terzi (2016), James et al. (2013), Han et al. (2019), Breiman et al. (1984), Tibshirani (2013) and Hastie et al. (2009).

4.1 K-Means Clustering

More formally, Let $X = (x_1, x_2, \ldots, x_n)$ be the n observations, and the distance between any two observations $x_i, x_j \in R^p$ be defined as $d(i, j) = |X_i - X_j|^2$, the Euclidean distance in p-dimensional space. Let K be the number of clusters (known a priori) that we want to construct by partitioning the observations in X. A clustering $C(X)$ of X can be defined as a function $C : X \rightarrow \{1, ..., k\}$ that maps each observation $x_i \in X$ to a group labeled. That is, $C(x_i) = k$ means that x_i is mapped to group k, and n_k is the number of points in the group labeled k. Given a clustering $C(X)$, $I(C(X))$, the in-cluster distance is defined as $I(C(X)) = 1/2$ $\sum_{k=1}^{K} \left(\frac{1}{n_k}\right) \sum_{C(x_i)=k,,C(x_j)=k} d(i, j)$.

Now, we would like to design an algorithm that given a collection of observations X, and a distance function d that defines the distance between any pair of observations in X, determines a clustering $C(X)$ that minimizes the in-cluster distance $I(C(X))$. This problem is known to be NP-hard and involves considering essentially most of the n^K possible ways of partitioning the n points into K groups. So, we look at heuristics that instead of minimizing in-cluster distance $I(C(X))$ looks at minimizing $W(C(X))$, within-cluster variation, an approximation to in-cluster distance that can be approximated well by looking at only at a very small fraction of the partitions of X.

Key Idea: Initially, assign each observation to one of K clusters $C_1, C_2, ..., C_k$ uniformly at random. For $k \in [1..K]$, let center(k) denote the center of cluster C_k. Now, reassign each observation to the cluster whose center is closest to it. Now,

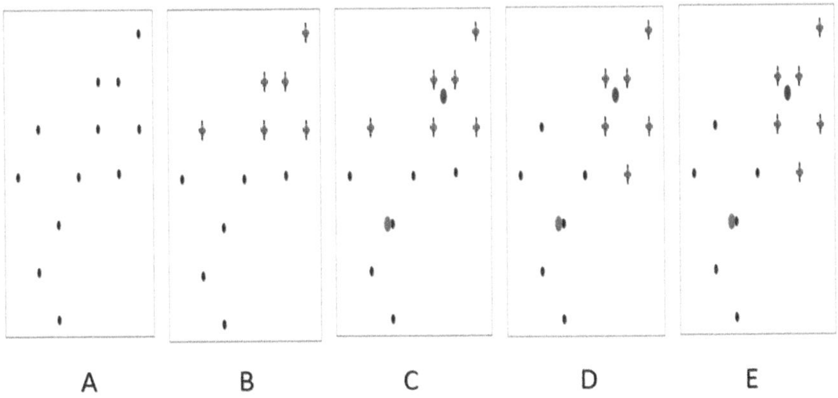

Fig. 4 Han et al. (2013): K-means clustering: **a** Initial data set, **b** Random assignment clusters, **c** Compute cluster centers, **d** Reassign clusters, **e** Re-compute cluster centers

recompute the cluster center and repeat the reassignment process until the cluster centers do not change.

K-Means Clustering (Lloyd 1957).

1. We assign each observation $X_i \in X$ uniformly at random to one of the K clusters $C_1, C_2, \ldots, C_k$.
2. Repeat until within-cluster variation doesn't change

   (a) For $i \in [1..n]$, find the cluster center $k \in [1..K]$ that is closest to X_{ii} and let $C(X_i) = k$.
   (b) For $k \in [1..K]$, re-compute center(k), the center of cluster C_k, to be the average of the observations in C_k.

The performance of this heuristic and its convergence to an approximate solution depends on the initial random assignment of cluster centers and the heuristic does not provide guarantees on its optimality. So, running this algorithm multiple times with different random starting points may help improve the algorithm's performance statistically (Fig. 4).

4.2 Hierarchical Clustering

Hierarchical clustering produces a clustering where the number of groups is not determined a priori and the distance functions used are different. The hierarchical clustering algorithms can be classified into two types: Agglomerative (bottom-up) start with groups consisting of single data points. The groups that are closest (as defined by the distance function) are repeatedly merged until we have only one group left, and divisive (top-down): start with a single cluster repeatedly split a cluster whose split results in two clusters with a maximum distance between them until each point is in a cluster by itself. We will only focus on the agglomerative approaches as they are simpler (interpretable), their analysis is simpler and their predictive power is comparable with the divisive approaches. Given pairwise distance function $d(i,j)$ between observations, hierarchical clustering produces a consistent result, without the need to choose initial starting positions (number of clusters) for the following commonly used linkages (distance functions for measuring the dissimilarity between clusters): (1) Single linkage: the dissimilarity between two clusters G and H is the smallest distance between two points in opposite groups; (2) Complete linkage: the dissimilarity between two clusters G and H is the largest distance between two points in opposite groups; (3) Average linkage: the dissimilarity between two clusters G and H is the average distance between points in opposite groups.

Agglomerative Hierarchical Clustering (Kaufman et al. 1990).

1. Let $X = (x_1, x_2, \ldots, x_n)$ be the n observations , and the distance between any two observations $x_i, x_j \in R^p$ be defined as $d(i, j) = |X_i - X_j|^2$, the Euclidean distance in p dimensional space. Let C_1, C_2, ..., C_n. be the n clusters such that $C_i = \{X_i\}$.
2. Compute the dissimilarity between each pair of clusters.
3. Repeat

   (a) Create a new cluster by merging the two closest clusters based on the dissimilarity measure used.
   (b) Update the dissimilarity distance for the existing clusters.

4. Until one single cluster remains (Fig. 5).

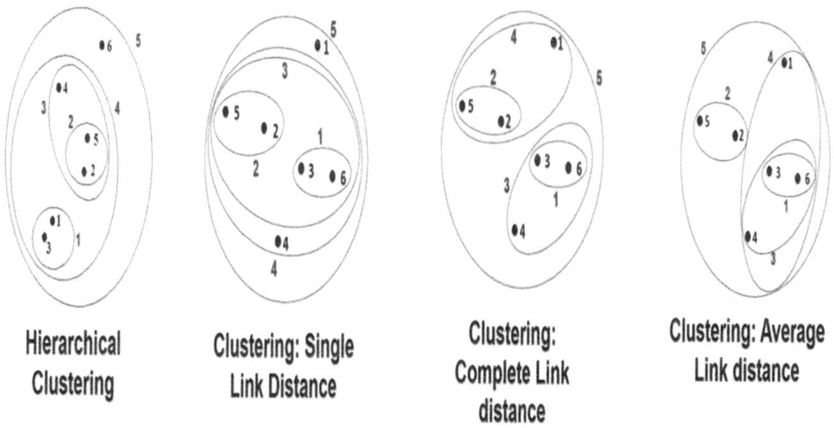

Fig. 5 Terzi (2016): Hierarchical clustering for different types of linkages

5 Data Tree Models in Precision Agriculture

Problems in Precision Agriculture (Waheed et al. 2006; Olamide et al. 2020) involve analyzing vast amounts of agricultural data like hyper-spectral data from satellites through remote sensing; data from smart devices and sensors in farms; historical climate, cropping and pest and disease, and resource utilization data over various geographies; and economic data on crops, to name a few, for allocating farm resources to maximize farm productivity and price stability while satisfying climate, environmental, ecological and quality constraints.

In recent times, climate change has led to the introduction of additional farm regulations thereby placing a greater premium on farm resources. This in conjunction with the availability of (i) smart devices and accessible technologies for gathering farm data (humidity, temperature, water and mineral content, nature of pests) and (ii) hyper-spectral data through remote sensing has provided the motivation and ideal framework for the development of computationally efficient and easy to interpret decision systems for improving farm productivity through optimally matching crops to farms. Furthermore, increased volatility in crop prices due to greater and frequent impacts of climate change on farm yields, changes in consumer preferences, and uncertain risk factors like the vulnerability of crops to diseases and pests, have made the problem of predicting the crop yield and crop price crucial for ensuring price stability. In both the problems, Classification and Regression Trees (CART) have helped in (i) determining an optimal allocation of crops to farms resulting in improved farm productivity; and (ii) accurately predict crop price and crop yields resulting in price stability to farmers.

In this section, we summarize the work of Waheed et al. (2006), where they make use of hyper-spectral data on factors like water stress, presence of weeds and nitrogen application rates to build Classification and Regression Trees (CART) for categorizing farms. Their experimental setup can be described as follows: The farm for growing corn was split into four blocks each consisting of twelve 12×12 plots that are obtained as follows. Each block is split into irrigated and non-irrigated halves. Then within each block, two weed management strategies, no weed control and full weed control, are chosen by splitting each block into two equal sub-blocks and randomly assigning one of the two weed management strategies to each sub-block. Each of these sub-blocks is further split into 3 equal plots and randomly assigned to nitrogen treatment at one of the three levels 50, 150, and 250 kg N ha1. The hyper-spectral data was recorded with a hand-held spectroradiometer at three developmental stages of corn—early growth, tasseling, and fully maturity.

Example 4: Classification of input dataset into 4 categories where non-irrigated, irrigated, weeds, and no-weeds, represent 35, 30, 25, and 10% of the data, respectively. The Classification Tree is constructed by starting with a single category consisting of all farms and perform a recursive binary partitioning until the size of each partition is greater than a chosen threshold or some stopping criteria chosen by the user. *Gini-index* is used for measuring performance. Figure 6 shows the constructed classification tree.

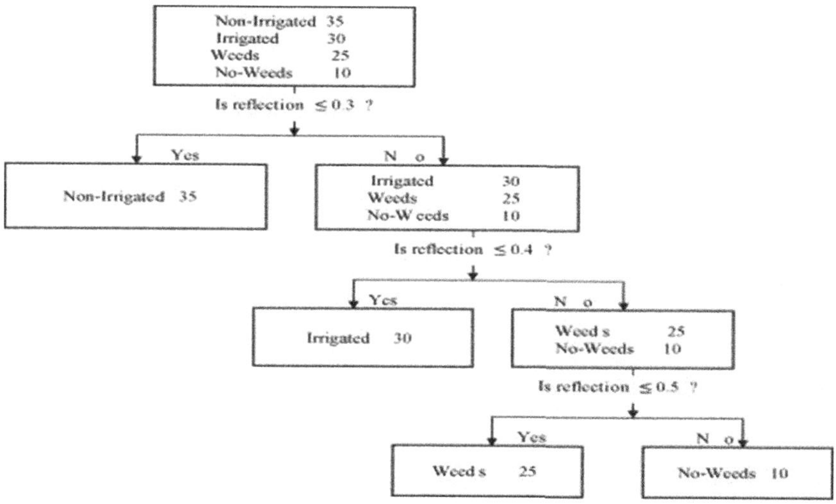

Fig. 6 Waheed et al. (2006): A classification tree with perfect classification ()

Intuitively Gini-index measures the purity or homogeneity in classification. So at the top level, the "Non-irrigated" data items are isolated into one category (left node) and the rest of the dataset into another category (right node). Then, it will further split the right node (irrigated, weeds, and no-weeds categories) further by isolating the "irrigated" into one category (left node) and the rest of the dataset into another category (right node). Then it splits the right node into weed and no weeds categories. The final tree has four pure nodes. This is an instance of perfect classification. This result is ideal but in most situations, we expect the constructed classification trees to classify some data items incorrectly.

Their CART algorithm could classify the 12 treatment combinations with 75–100% accuracy at all 3 recorded stages of development, although the best validation results were obtained at the early growth stage. Also, the classification trees generated when using one or two of the three factors (irrigation, weeds, or nitrogen), the classification accuracy was high.

6 Summary and Future Scope

In this chapter, we first motivated the study of DT techniques for solving prediction problems by highlighting their computational efficiency and higher model interpretability in comparison with other standard statistical learning techniques (Linear and Non-linear regression, SVM, and NN). Then, we have attempted to provide a self-contained overview of DT models, statistical methods, and their analysis for solving prediction problems in both supervised and unsupervised learning contexts.

We believe that this chapter will provide the necessary conceptual basis for understanding the applications of DT-based models and methods for solving various prediction problems in Precision Agriculture. In addition, this chapter can also serve as a primer for a more detailed study of DT and other statistical learning techniques in Data Science.

References

Cheesinglee (2016) Logistic regression versus decision trees. bigml BLOG. https://blog.bigml.com
Diawara N (2019) Modern statistical methods for spatial and multivariate data. Steam-H. Information Springer Nature Switzerland AG
Dietz F (2019) Automation in data science. Towards data science. https://towardsdatascience.com/
DTREG (2021) Decision trees compared to regression and neural networks. predictive modelling software. https://www.dtreg.com
Goodfellow I, Bengio Y, Courville A (2016) Deep learning. MIT Press
Gupta S (2018) Brief intro of medical image analysis and deep learning. Towards data science. https://towardsdatascience.com/
Han J, Kamber M, Pei J (2013) Data mining: concepts and techniques. Morgan Kaufmann
Hastie T, Tibshirani R, Friedman J (2009) Elements of statistical learning, vol 103. Springer Texts in Statistics Book
James G, Witten D, Hastie T, Tibshirani R (2013) An introduction to statistical learning. Springer series in statistics. Springer, New York, NY
Kaufman L, Rousseeuw PJ (1990) Finding groups in data—an introduction to cluster analysis. A Wiley-Science Publication, Wiley
Leo Breiman JH, Friedman RAO, Stone CJ (1984) Classification and regression trees. Wadsworth & Brooks/Cole Advanced Books & Software, Monterey, CA
Lianne, Justin (2020) Practical guide to cross-validation in machine learning. https://www.justintodata.com/cross-validation-machine-learning
Lloyd SP (1957) Least square quantization in PCM. Bell Telephone Laboratories Paper
Mitchell TM (1997) Machine learning. McGraw-Hill
Molinaro AM, Simon R, Pfeiffer RM (2005) Prediction error estimation: a comparison of resampling methods. Bioinformatics 21(15):3301–3307
Nielson MA (2015) Neural networks and deep learning. Determination Press.
Olamide K, Oludele A, Monday E, Shade K, Chigozirim A (2020) Evaluation of decision tree algorithms in precision agriculture. IJCAT 7(3), March 2020
Prince SJD (2012) Computer vision: models, learning, and inference. Cambridge University Press.
Rajan S (2020) Analyzing the performance of the classification models in machine learning. Towards data science. https://towardsdatascience.com/
Rosebrock A (2012) Deep learning for computer vision with Python Starter Bundle (2017) PYIMAGESEARCH
Stone M (1974) Cross-validatory choice and assessment of statistical predictions. J Royal Stat Soc Ser B (Methodol) 36(2):111–147
Terzi E (2016) Clustering, lecture notes of CS565: Data Mining Fall 2016. Boston University
Tibshirani R (2013) Data mining. Lecture notes from data mining, Spring 2013. Carnegie Mellon University
Waheed T, Bonnell RB, Prasher SO, Paulet E (2006) Measuring performance in precision agriculture: CART—a decision tree approach. Agricult Water Manag. Elsevier 84:173–185

Deep Reinforcement Learning for Agriculture: Principles and Use Cases

Ratnik Gandhi

Abstract The rapid growth of machine and deep learning solutions has led to many advancements, including self-driving cars, multi-lingual audio assistants, diseases detection and other industrial robotic solutions. Recognizing importance of these state-of-the-art solutions based on artificial intelligence, applications have been developed that can benefit the domain of agriculture and natural resources management. In this chapter, we aim to study recent advances in artificial intelligence and present applications of these advances to agriculture domain. Towards this aim, we review recent literature and introduce two foundational concepts, a network concept of deep neural networks and a learning concept of reinforcement learning. We present important use case of crop yield maximization, and through the applications of Deep Reinforcement Learning (Deep RL), showcase approaches for solving two important problems using Deep RL. Further, we showcase how these solutions can be deployed on production-ready cloud infrastructure, thus making our solutions more practical and scalable.

Keywords Crop yield maximization · Deep reinforcement leaning · Object detection · Q-learning

1 Introduction

Considering the growth of population and limited landmass, if agriculture is the only means to feed the world, it must scale well and scale rapidly. Domain of agriculture is one of the most classic domains of human knowledge that has evolved over time with many new techniques and research and advances applied to it.

Recent advances in sheer compute power, digitization of historical data and volume of new data due to economic availability of sensors have led to rise of the Big Data domain. The domain of agriculture has also benefited from these advances, where simple sensors for monitoring parameters such as temperature, moisture and advance image-based sensor for identifying pests in the plant have been used.

R. Gandhi (✉)
NTT Data Inc., Burnaby, Canada

G. P. O. Reddy et al. (eds.), *Data Science in Agriculture and Natural Resource Management*, Studies in Big Data 96, https://doi.org/10.1007/978-981-16-5847-1_4

These advances have led us to an opportunity, where we can leverage the knowledge available in these data and identify ways through which we can automate many tasks/sub tasks for sustainable farming and leading a way to scalable agriculture solutions.

Through this chapter, we make an effort to contribute at the intersection of these distinct domains. Our work is one of the early works that not only proposes but builds, deploys and make available through open sourcing solutions of state-of-the-art Deep Reinforcement Learning-based AI solutions to the problems of crop yield maximization and crop produce detection. The designed frameworks can be used for automating and controlling greenhouse-based farming processes. It can help in optimal allocation of natural resources required for plant growth—saving resources and maximizing crop yield. Further, the computer vision RL framework can be used for identifying and monitoring growth of crop produce, reducing significant amount of time for physical inspections by human monitoring. The designed frameworks are open sourced. The open sourcing allows an opportunity for other interested researchers to contribute to the important field of agriculture and crop management, and the cloud infrastructure-based deploy-ability offers solution to scale up rapidly to big data solutions.[1] In this work, we detail architecture of end-to-end, micro services-based RL solutions for the above-mentioned case studies. The solutions are deployed, on the cloud, work with Web API interface and has Tableau dashboards for visualization and analytics.[2]

1.1 Literature Review

For developing a path towards sustainable agriculture, new species of plants are being developed with the help of deep reinforcement learning. These new designs can have far reaching impacts on agricultural yields and environment. Restrictive scope of existing ML solutions for agriculture is identified in Binas et al. (2019), and they present an idea/framework for designing new plant species based on environmental conditions and given constraints on resource such as water and fertilizers. Objective of this optimization being crop yield maximization. The short note proposes a study of such experiments, first in greenhouse-controlled environment and using reinforcement learning (RL) for coming up with new plant species. The note also mentions that the policy feedback required to improve (self) learning of RL method, in case of plants, takes longer time in practice, thus motivating use of plant simulation models such as crops in silico (Marshall-Colon et al. 2017).

Use of computer vision and deep recurrent neural networks (RNNs) for understanding satellite radar images for classifying crop cover and crop production has

[1] An alternate way to handle high volume of data or designing scalable solution is by selecting most important features in the data (Gandhi and Raval 2020).

[2] Similar stateless, scalable, micro-service-based cloud-deployed ML solution is presented in Mehta et al. (2019).

been carried out in Ndikumana et al. (2018). Use of reinforcement learning on IoT edge devices sensor data for irrigation is developed in Zhou (2020). The results show that with the help of these AI capabilities, the precision of agricultural crop area mapping and irrigation control can be improved effectively.

There have been many works that present classic ML solutions to the problems in Agriculture domain, but the work that comes closest to the results presented in case studies in this chapter, where deep reinforcement learning is applied to solve the problems, is reported in Elavarasan and Vincent (2020). The work presented builds a complete crop yield prediction framework based on RNN and RL. These systems can directly comprehend environmental and agricultural parameters' raw data and map them to the crop prediction values. The study works with agriculture parameters data available through state government in India and proposes a deep Q-Network (DQN)-based solution for solving the crop yield prediction regression problem. Many alternate ML model solutions are compared against their model for accuracy. The case studies presented in this chapter differs from this work as follows. Our case studies predict the actions for changing the parameters for increasing the crop yield and thus is more active in terms of strategy. Further, the solution for identifying crop produce, with the help of computer vision-based Deep RL agent presented in this chapter is not available in Elavarasan and Vincent (2020).

An approach for balancing the quality of image resolution for object detection, to save image purchase cost and image processing compute cost, via Deep RL is presented in Uzkent et al. (2020). The solution focuses on training the Deep RL agent with dual reward. When an image with larger objects is available, an RL agent must consider lower resolution images, while for many small objects images, the agent must consider high-resolution images. The results present in Uzkent et al. (2020) for training the Deep RL agent are promising, where the agent reduces compute cost by half. Further, agent is able to reduce the requirement of high-resolution images by a factor of 30%.

Considering from the perspective of Deep RL solution from the Region Proposal Network (RPN), the work that comes closest to work presented in our case study is Pirinen and Sminchisescu (2018). The result presented a drl-RPN, a deep reinforcement learning-based visual recognition model consisting of a sequential region proposal network (RPN) and an object detector. In a typical RPN candidate object regions (Regions of Interest - RoI) are selected greedily. The approach in Pirinen and Sminchisescu (2018), on the other hand, focuses on optimizing an objective closer to the final detection task. This is achieved by a sequential attention mechanism which is trained via Deep RL. The resulting solution uses a single policy framework. Solution presented in our case study differs from Pirinen and Sminchisescu (2018) as follows. Our Deep RL agent is trained using value-based learning of Q-function and DQN network while their approach is based on policy learning. Both approaches make use of VGG16 as a feature extraction backbone. We do not need to use the detector framework, as our Tomato Dataset (Tomato 2020) is annotated.

As an important unique contribution, case studies present in our work are production-ready deployed solution over the cloud, and the solutions are available in public domain (with their respective datasets on Kaggle) (Challenge 2020; Tomato 2020; AGC 2021; RPN 2021). The solution and dataset of Elavarasan and Vincent (2020) or Pirinen and Sminchisescu (2018) is not available in public domain.

1.2 Organization

Section 2 introduces concept of reinforcement learning. Section 3 presents two case studies: crop yield maximization and crop produce detection. The section details the architecture of solution, implementation details and related results. Finally, we conclude this chapter noting certain observations in Sect. 4.

2 Reinforcement Learning

Reinforcement learning is an area that can be approached from various disciplines. From the perspective of operations research, it falls in the mathematics domain, from the bounded rationality perspective, it falls in the economics domain. Neuroscience can look at RL from reward system perspective. In this chapter, we will mainly be considering RL from machine learning perspective and understanding it from computer science domain.

Unlike classical paradigms of machine learning (ML), which falls mainly in two categories, supervised and unsupervised learning, RL forms a separate paradigms of ML. RL has supervisor but as a reward signal. The feedback to the learning agent is not necessarily instantaneous.

The data required for RL is state-action pairs and goal of the learning agent is to maximize its future rewards over many time steps. A typical RL system key concepts can be summarized in Fig. 1.

There is a learning agent $\boldsymbol{i}$ that takes an action $\boldsymbol{a_t}$, at time $\boldsymbol{t}$, from an action space $\boldsymbol{A}$. The agent performs this action in an environment $\boldsymbol{Env}$. The agent $\boldsymbol{i}$ exists and operates in the environment $\boldsymbol{Env}$. On performing action $\boldsymbol{a_t}$, the agent $\boldsymbol{i}$ receives reward $\boldsymbol{r_t}$ and an observation feedback s_{t+1} from the environment. The observation s_{t+1} is called a

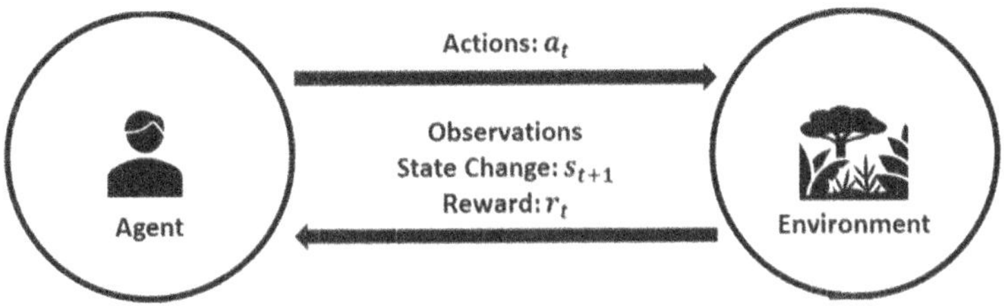

Fig. 1 Reinforcement learning concepts

new state (change) in the environment ***Env***. Reward r_t is the feedback that measures the success or a failure of agent ***i***'s action a_t. Agent ***i***'s total reward is,

$$R_t = \sum_{i=t}^{\infty} r_i \tag{1}$$

Alternatively, to promote importance of near time actions and their corresponding rewards, a reward discounting factor γ is introduced. A reward received ***j*** time steps in the future is only worth γ^{j-1} times what it would be worth if it was received immediately. In this regard, the total discounted reward of agent ***i*** is,

$$R_t = \sum_{i=t}^{\infty} \gamma^i r_i \tag{2}$$

The myopic feedback for the current action is reward. On the other hand, a strategic feedback is given by a value function. Value of a state is the total reward an agent can expect over the time, starting from current state. Given state-action pair (s_t, a_t), the value function Q is defined as,

$$Q(s_t, a_t) = E\{s_t, a_t\} \tag{3}$$

In other words, the Q-function captures the expected total future rewards agent ***i*** can receive in state s_t by taking action a_t. Q-function can be thought of as a table look up, where rows of the table are states ***s*** and columns represent actions ***a***. Ultimately, the learning agent ***i*** needs to find the best action given current state ***s***. This is called a policy $\pi(s)$. Policy captures the learning agent's behavior at any given time.

$$\pi(s) = \arg Q(s, a) \tag{4}$$

A typical RL agent can be categorized by value-based or policy-based learning agent. Where, aim of the agent is to perform necessary learning, either by optimizing Q-function or designing an appropriate policy $\pi(s)$. A subcategory that combines value and policy learning is called an actor critic model. Third category is called model-based learning in which agent attempts to learn the model of the world and then plans action using the model. The process often requires updating and replanning. In this chapter, we will be focusing on the value-based RL agent.

The RL framework is most suitable for agriculture domain problems, because a significant amount of agriculture-related data is unlabeled. Further, considering the availability of outcomes (feedback) of the choices of actions at a significantly later time and considering high cost of adverse actions, human experts tend to come up with better strategies which are conservative and incremental in nature. An RL agent simulated on an agriculture environment, on the other hand, can help identify strategies that are beyond conventional human knowledge and are aggressive in nature.

2.1 Deep Reinforcement Learning

Reinforcement learning is a Markov Decision Process (MDP) paradigm that helps model behavior of an agent, allowing us to solve complex decision-making problems. It has been used to study finite space Multi-arm Bandit problems (Burnetas and Katehakis 1997). The power of deep neural networks and deep learning allow study of the problem from continuous state spaces perspective. A Deep RL agent can be trained with value learning, policy learning or model learning. This chapter's focus is on value learning (*Q*-Learning)—Deep RL agent that uses deep *Q*-Network for learning and solving the problem.

2.1.1 Deep *Q*-Learning

The agent uses a policy for estimating $\boldsymbol{Q}$ value that maximizes the future rewards. *Q*-learning is just like a finding optimal $\boldsymbol{Q(s,a)}$ value, given a state $\boldsymbol{s}$—a look up in *Q*-table for finding actions $\boldsymbol{a}$. When the action and/or state spaces are large or continuous, the size of the *Q*-table becomes unmanageable. This necessitates need for a DNN framework, giving rise to the Deep RL paradigm. In this setting, a DNN is used for approximating the $\boldsymbol{Q}$-function values. The loss function $\boldsymbol{L}$ for approximating *Q*-function is,

$$L = E\left[\left(r + \gamma Q\left(s^{'}, a^{'}\right) - Q(s, a)\right)^{2}\right] \tag{5}$$

where the first two terms (inside the expectation $\boldsymbol{E}$) on the right hand side in the Eq. (5) are called the target $\boldsymbol{Q}$ value, and the third term is the predicted $\boldsymbol{Q}$ value for taking an action in a particular state. It is the value received when taking an action.

Next step in the training of a DQN for approximating $\boldsymbol{Q}$ values is to compute the gradient of the loss function given in (5) and update the network weights with back propagation.

It must be noted that setting parameter values for agriculture problems requires fine tuning of many continuous values. This implies the action space for the RL agent is continuous and it immediately motivates Deep RL framework.

3 Case Studies

In this section, we present case studies that addresses the problem of crop yield maximization and crop produce detection. We discuss the datasets, compute environment, solution approach and related results.

3.1 Crop Yield Maximization Using RL

This case study focuses on the problem of controlling various parameters in a Greenhouse environment designed for the crop of cherry tomatoes. The RL agent is trained using value function learning. For simulating the greenhouse environment, states and actions, we use Autonomous Greenhouse Control challenge dataset (Challenge 2020). The learning of RL agent offers insight into plantation knowledge of tomato crops and how it can be encoded into AI algorithms. The solution that we developed also showcases how to automate decisions—mimicking human expertise. Further, the study highlights the importance and impact of change in parameters like weather conditions and other contributing factors on the tomato crop growing patterns.

3.2 Problem Statement

The objective of this case study is to design a deep learning-based RL agent that can automatically control and change different parameters like temperature, air, water supply and light in given greenhouse environment, so that the production quality of the cherry tomatoes crop can be maximized. The expected outcome of the RL agent learning is to find the optimal strategy for setting the greenhouse parameters.

3.3 Dataset

For this case study, we have used Autonomous Greenhouse Challenge (AGC) 2nd Edition dataset available on Kaggle (Challenge 2020). It contains data on outdoor and indoor greenhouse climate, irrigation, status of actuators, requested and realized climate set points, resource consumption, harvest, crop-related parameters, tomato quality, analysis of irrigation and drain samples and root-zone/slab information. Data were collected over a period of 6 months. The crop considered was of cherry tomatoes being grown inside six high-tech glasshouse compartments, located at the Wageningen Research Center in Bleiswijk, Netherlands. Our interest is to develop an RL agent that can determine the set points for climate, irrigation and a number of cultivation-related parameters and control the production of cherry tomato crops.

Table 1 Details of the parameters available in the AGC dataset (Challenge 2020)

Sr. No.	Description
1	Weather data: – Weather information defined by ten parameters/columns is available from this data
2	Greenhouse climate data: – This provides information about the indoor greenhouse climate, status of irrigation and actuators – Each datum is defined across 18 features
3	Climate and irrigation set points: – This data defines the set points set for 16 features inside the green house
4	VIP (Realized set points): – Contains the achieved values of the above-mentioned 16 set points
5	Production: – Information about the produced tomatoes and fruits belonging to two classes (A & B) – The dataset is defined by nine parameters
6	Crop parameters: – Data about each crop under observation—collected each week – Six parameters were measured for each crop
7	Resources: – Set of nine resources that are used during the experiment
8	Tomato quality and dry matter content: – Information about each tomato under observation measured across eight parameters
9	Lab analysis: – Lab analysis for concentration of around 17 minerals along with pH and EC of irrigation and drainage water
10	Root zone data: – Values of six parameters observed by the sensors

The objective is to maximize net profit, by minimizing use of resources (e.g., water, nutrients, energy-heating and electricity, CO_2), while optimizing income as a function of production and fruit quality. The dataset contains raw and processed data. Raw data were collected via climate measuring boxes and sensors, climate and irrigation process computer, weather station, manual registrations (performed by the greenhouse staff). Table 1 describes in detail the parameters available in the AGC dataset.

3.3.1 Approach

Our aim is to determine the set points for parameters (such as temperature, amount of natural and artificial light, heating, CO_2 concentration and fertigation). To achieve this by an RL agent, we consider use of sensors data to self-learn and assume that the greenhouse is accessible via a digital interface remotely. The provided data in the AGC dataset (Challenge 2020) simulates actions carried out by five different teams on these parameters and reflects their outcomes. Thus, creating a complete simulation

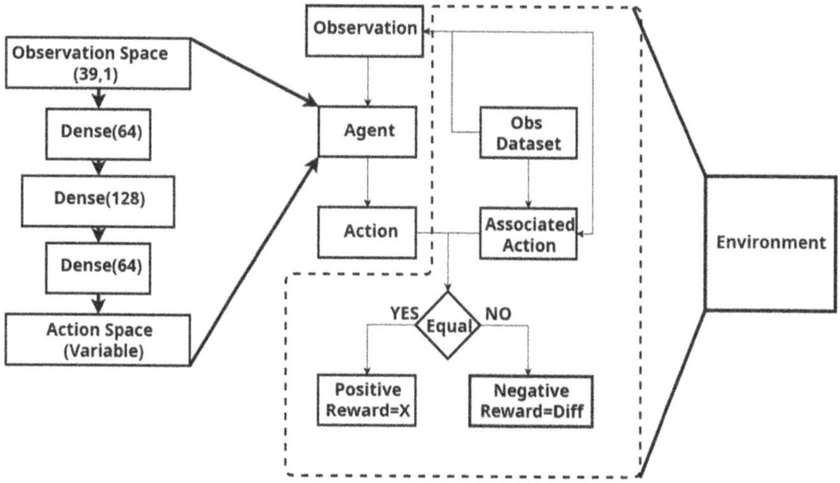

Fig. 2 The image represents the approach used to solve the autonomous greenhouse problem

environment for an RL agent to train and operate in. This makes the dataset complete with a built-in environment—required to formulate it as a Deep RL problem. Figure 2 summarizes the overall approach for solving the Autonomous Greenhouse Control and Monitoring using Deep RL framework.

With the help of data (in the form of actions taken by the five teams and their responses), and more aggregated data from the greenhouse sensors data (such as amount of water provided to the plants, intensity of white light under which plants were kept), a dataset of observation and corresponding correct actions were created to formulate the problem as an RL problem.

Our interest is to design an RL agent that mimics setpoint strategy for changing parameters of the best team (of the five teams) under given circumstance. The agent is designed to mimic a team's action. This means we assume that the settings used by teams are optimal or close to optimal human knowledge. This assumption is founded on the fact that these teams were competing to find best strategy solution to win the competition.[3] A learning agent in RL framework deals with exploring an environment and learning from rewards/penalties it receives.

The AGC dataset provides 47,000 rows of observation data for each team—collected every 5 min over the period of 12 weeks. This data was used in the following way:

- When an environment is initialized, it starts with row 1 data of team 1. This observation is fed to the agent and it predicts an action.

[3] It has been shown that RL agents are capable of finding better than human optimal strategy for solving given problems. See Deep RL agent for Atari Breakout game for more details (Mnih et al.2015).

- If the predicted action is equal to the action from the dataset, the agent receives a positive reward, else a penalty. The penalty and reward are calculated based on the difference between the prediction and actual values. The penalties allow us to increase the speed of learning—providing with faster convergence.
- Every team's data is an episode of learning the environment. That is,
  - An episode contains 47,000 steps (one episode is basically training the data on one team's actions).
  - There are five teams, so our deep RL agent can train for a total of five episodes.

The Deep RL agent that we train is a DQN-based that attempts to learn value functions. The agent uses a DNN with three hidden dense layers. See Fig. 2 for the DNN specifications. We discretize our action space for training and using our Deep RL agent.

We have used Python libraries such as Pandas, Numpy, Tensorflow for implementing and training our model. Next, we discuss details of our production-ready solution deployed over cloud infrastructure.

3.3.2 Web Service and Solution Architecture

While designing services for managing a greenhouse, the most usable service is a greenhouse monitoring and controlling service that can provide visualization of various sensor readings and offers and interface for taking actions. In our case, an RL agent is controlling the greenhouse. In this regard, the built web service provides with the following:

- Lets the user start/stop the greenhouse simulation.
- Lets the user visualize the environment in the greenhouse.
- Lets the user visualize actions and rewards of an agent with a random policy vs a trained agent's policy.
- Lets the user send the simulation data to a database that can be visualized using third-party tools such as Tableau.

Overall architecture, given in Fig. 3, uses the following sub-solutions:

- A backend API (using FastAPI) that lets the frontend dashboard control the simulations and fetch observations and actions from the RL model.
- A frontend dashboard offering visualization of the data and control over the data (using DASH Python).
- A Postgresql database that maintains simulation data so that it can be visualized on Tableau.

Working of the solution is as follows:

(1) A user is given simulation controls on the frontend dashboard.
(2) Once the simulation is started, it looks for data from the backend API in the following order.

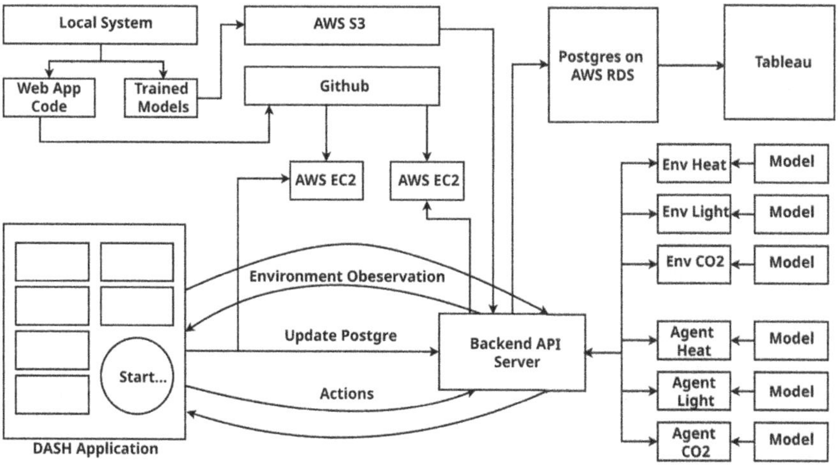

Fig. 3 The figure shows the architecture of the solution where, for simulation visualization we use DASH application, for managing the backend API server for environment and agent models we use AWS S3. AWS EC2 instance for deploying the frontend and AWS RDS for deploying the PostgreSQL-based persistent storage is used

(a) On receiving the first environment observation data, as input from the dashboard, obtained through a POST request on the endpoint of the API, the deep RL agent returns a predicted action.

(b) Frontend again sends a POST request of this predicted action to the endpoint. This is termed as "step" in the environment. The backend then returns the next environment observation and the corresponding rewards for this action. Visualization of the actions, observations and rewards is completed in this step. See Fig. 5 for details.

(c) Go back to step 2(b)—send a POST request with the previously fetched observation to get the next action and repeat the process.

(3) Once the user is done analyzing the simulation, they can press the update Tableau button and all the collected simulation data is sent to a database deployed over the cloud. This database is then connected to a Tableau worksheet that can further be used to analyze data in detail. It must be noted that every time the backend API is started, it fetches the trained model that has been uploaded to AWS S3 instance.

In terms of data flow, the environment observation consists of 39 columns which includes greenhouse indoor/outdoor climate, status actuators and other parameters. The action by an agent is to predict parameter values. An agent can be given a random action, or we can give the agent an action executed by one of the five teams. A random action fetches a random reward, while a predicted action reward, termed as agent reward, is also computed. In terms of RL environment creation, our solution includes estimating closest action space, steps taken, index setting for five teams and reward definition. Our deep RL framework is based on deep Q-network learning. Using this DQN framework, agent's training, actions and reward were derived.

3.3.3 Results

In this subsection, we discuss the results related to training and results of applications of the trained Deep RL agent on the problem of crop yield maximization. Note that the agent's goal is to mimic actions of the five teams that maximizes the overall crop yield.

Figure 4 shows the results of the average training reward for our Deep RL agent. The figure plots average rewards every 500 steps. When the agent is trained with Team 0's action data (episode 1), it initially performs similar to a random agent taking actions (bottom curve is random agent's reward while central curve is reward collected by the agent learning on Team 0's data). However, with time, it learns strategy and the average rewards gradually increases with time. Once the episode gets over and a new learning episode starts (agent is already trained on the Team 0's policy/strategy), the model initially makes mistakes in choosing an action, as the strategy for controlling parameters (such as light), used by the Team 1, is different from what Team 0 uses.

However, the agent quickly catches up to the new strategy and the average reward gets to the value it was before and then the model performs well for the rest of the data.

Tabs given at the top of Fig. 5 allow user to switch between simulation status and action parameter choice. It also allows the refreshing visualization on Tableau frontend. The top left sub-plot contains visual information of the actions taken by the agent, original actions and random actions. Similarly, the top right sub-plot shows

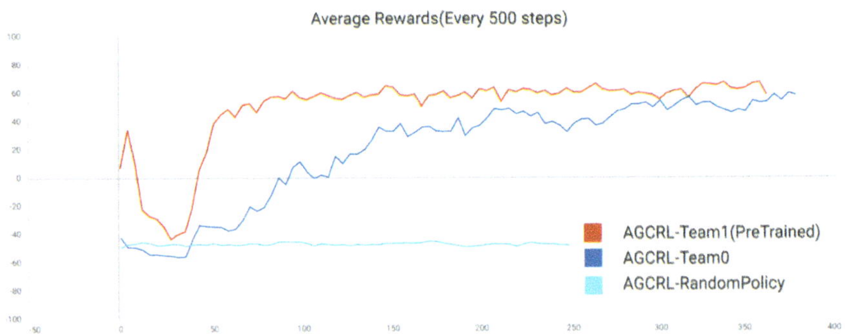

Fig. 4 Average training reward of Deep RL agent. Baseline average reward of random action policy is at the very bottom. The central reward curve is of the agent when it starts training afresh. The top curve is the same agent starting its training after the first episode training

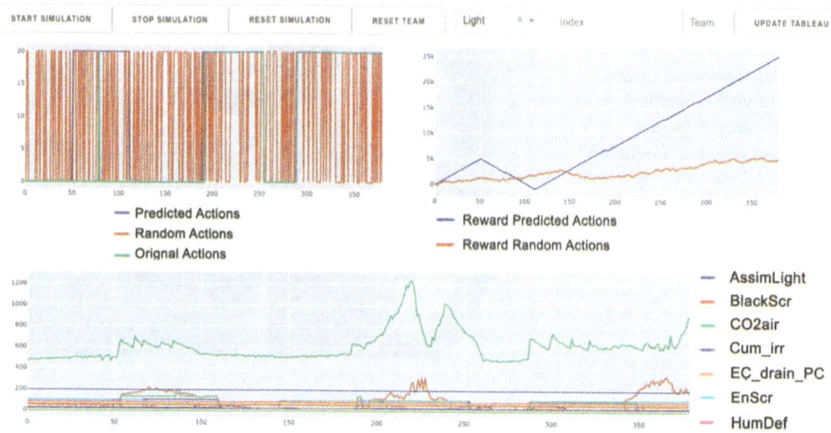

Fig. 5 The DASH app-based dashboard for monitoring environment activities. Image in the top left monitors predicted, random and actual actions by the agent. Corresponding rewards are captured on top right plot. The bottom plot monitors the observational parameters in the environment

an interactive visualization of the accumulated rewards for predicted actions vs. the rewards for random actions.

The bottom sub-plot in Fig. 5 contains visualization of all the observable parameters in the greenhouse simulation. These interactive plots were created using Plotly library. These results clearly show that the Deep RL agent trains faster and is able to mimic reward maximizing actions for changing the set points for maximizing the crop yield.

3.4 Crop Produce Detection Using RL

This case study focuses on building a deep reinforcement learning-based agent for detecting specific crop objects like cherry tomatoes in an input image. In terms of classification problem, the process of object detection is to identify whether the object of interest belong to the same semantic class of images or not. An RL agent will perform this classification work by stepwise learning to form a bounding box around the object of interest in the given input image. Overall objective would be to create a bounding box that fits on the object as tightly as possible.

The solution uses a sequential attention mechanism that has been trained using deep reinforcement learning. VGG16 (Simonyan and Zisserman 2014) network is used for learning the latent feature of input images. The images are fed to the visual backbone in the Pascal VOC (Visual Object Classes) format for feature extraction. These latent features of the input image are given to the RL-based deep Q-network as a state in the environment. The RL agent, based on the earlier rewards and state information, identifies an action for updating the bounding box inside the image.

Fig. 6 Typical tomato images in the dataset. In some images, tomatoes are explicitly visible and red. While in other images, tomatoes are small, occluded or green in color (Tomato 2020)

As the dataset already has labeled bounding boxes around the object of interest, a reward for the update bounding box action is computed on the basis of Intersection over Union (IoU is the metric that gives information about the extent of overlap of two bounding boxes) so as to help the agent follow the right search trajectory. An image with improved bounding boxes is generated as an output. The process is regressing on the bounding box eventually defining a tight bounding box around the object by appropriately processing the rewards.

3.4.1 Dataset

For this case study, we use the Tomato Dataset (Tomato 2020) available on Kaggle, first reported in Li et al. (2020). This dataset consists of images of cherry tomatoes crop production. The images are taken from the glass greenhouse of the Chongming Base of the National Facility Agriculture Engineering Technology Research Center. The dataset consists of two sub datasets: images data and annotations of images data.

The images data contains tomato plant images that are of variable size (most are 1560×2080 pixels) and may or may not contain a tomato. There are a total of 895 JPEG format images. Annotation data contains 895 XML format files that provide annotation information for the input tomato plant images. The method of annotation used here is the bounding box, where we are given coordinates information $x_{\min}$, $y_{\min}$, $x_{\max}$ and $y_{\max}$ in raw pixel value of a bounding box where a tomato is present. An image can have multiple bounding boxes. Figure 6 shows typical images in the input Tomato Dataset.

This dataset is also suitable for conventional supervised deep learning architectures like YOLO (Redmon et al. 2016).

3.4.2 Problem Statement

Tomatoes are an important Horticulture crop and there has been considerable amount of research done in improving its yield. Manual measurement of the phenotypic information is tough and hence computer vision provides a faster and scalable means for phenotyping using imaging. In this case study, our aim is to make use of a RL-based agent for detection of tomatoes in the tomato yield images. The objective is that the RL agent will adaptively select the spatial resolution of the tomatoes in each image, that is provided to the detector, and label it with bounding box information, thereby facilitating tomatoes detection.

3.4.3 Proposed Work

In this subsection, we present the overall approach for agent training and how the solution can be deployed using cloud framework.

As shown in Fig. 7, the Kaggle Tomato Dataset (Tomato 2020) is flown through Roboflow to obtain the data in Pascal VOC format for the downstream consumption. Subsequently, the VGG16 backbone allows us to extract latent features which are then sent to a Deep Q-Network for defining actions. In our case, the actions are to move the bounding box's top, left, right or bottom edges inwards by a constant factor. These actions allow the bounding box to shrink—eventually latching onto the object of interest (a tomato). The reward calculator calculates a reward for each state the system reaches in. The reward is calculated on the basis of IoU operation on the predicted bounding box and the original provided in the annotation dataset. The output at the end of these state transitions, on the basis of the rewards and penalties,

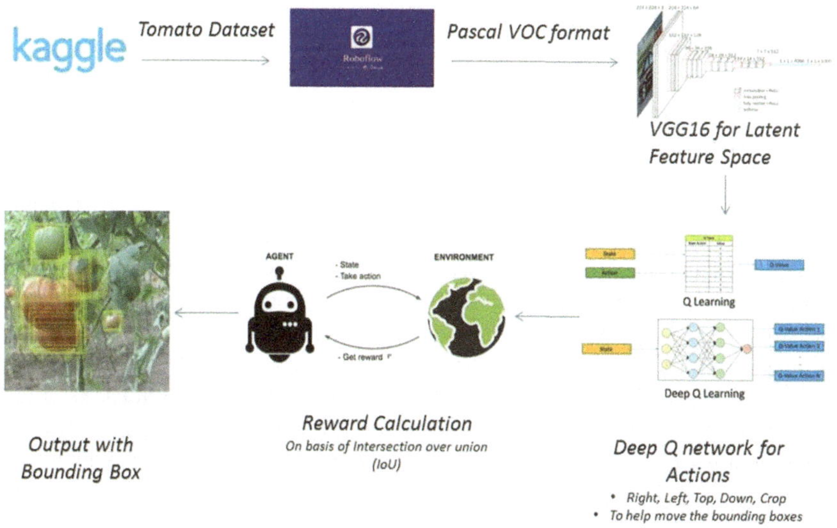

Fig. 7 This image shows the Deep RL-based RPN solution training pipeline. The data is captured from Kaggle and sent through VGG16 backbone for feature extraction. Subsequently, a DQN framework helps identify the Deep RL agent what actions the model can take. A corresponding reward is calculated, and an output image with bounding boxes is generated

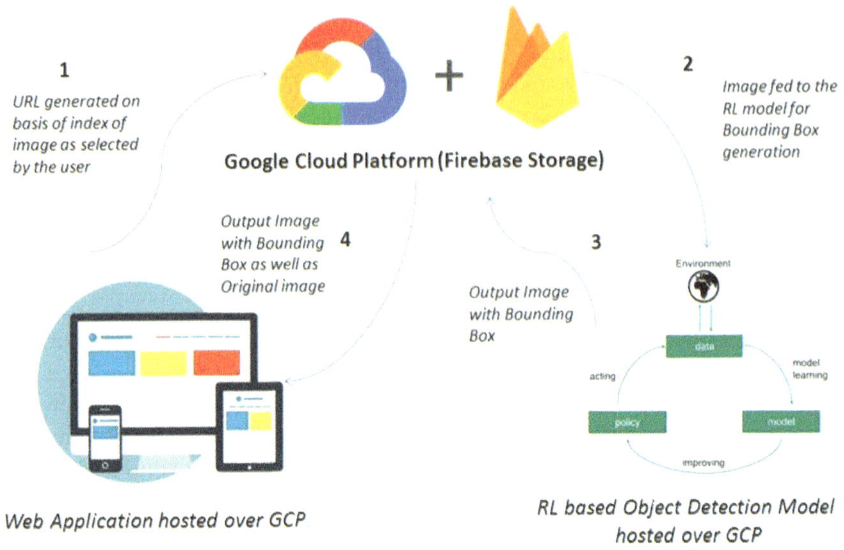

Fig. 8 The architecture of the solution of Deep RL-based RPN for tomatoes detection. Once the Deep RL agent is trained, the model is deployed as a micro-service over cloud with API interface. Through web interface, appropriate web-function is called, which in turn interacts with the model and receives an output

is an image with tight bounding box around a tomato. In our implementation, the IoU operation threshold has been set to 70%. This means that the predicted bounding boxes by the Deep RL agent, around a tomato, are accurate by 70% or more.

3.4.4 Architecture

We give details of the production-ready architecture for deploying the Deep RL-based crop produce framework over cloud infrastructure.

Figure 8 details the overall solution presented in this case study. The model as well as the dataset and a web application for users is hosted over Google Cloud Platform. The storage used is Firebase. Through the user interface, a user is allowed to select an image from train or test dataset or a custom image locally. In case of selecting train or test image, the URL for image is generated to index it over Firebase storage, and then that URL is provided to the agent for generating bounding box. The RL model in turn returns the image with bounding box which is generated as an output.

3.4.5 Implementation Details

We follow a multi-step approach in the implementation. First step is to check if the bounding boxes are in a standardized Pascal VOC XML format. The model is format-agnostic, and a standardized bounding box format is used for convenience. Once the dataset is loaded, the image paths along with their respective XML paths are logged into a CSV file for providing access for training the model. The log is used for loading an image, extracting bounding box information for the same image and preparing the image for the training. Before feeding the image to VGG16, the image is resized to 224 × 224 pixels, color channels are validated to be RGB and finally the image tensor format is verified. Output feature vectors from the VGG16 backbone encodes the important characteristics of the input image. Deep RL agent takes five different actions—move the edges of the bounding box on the image in four conventional directions—up, down, right and left or do not move (terminal action). The movement actions and the terminal action constitute the predefined actions that our RL agent can choose from.

The next step is to use the extracted feature vectors as a state for the deep Q-network. The Q-network outputs expected total future reward values for the given state-action pairs. The deep Q-network has two hidden layers of size 1024 each. The input to the Q-network is the feature vector we extracted from the VGG16 network, while the output is probability distribution (Q-values) on the actions. The RL agent can either take an action with the highest q-value or explore other actions using an exploration–exploitation strategy. The RL agent is initialized with a bounding box of the size of the entire image, and we give the agent freedom to take up to 60 actions on the given image. After each action, we evaluate whether the action increased the IoU with the actual bounding box and the region proposed by our RL agent. The agent is rewarded on an increase in IoU and penalized otherwise. We refer to a single pass over all the training images as one episode.

Once the Q-network is trained, the weights are saved. The next step in the flow is the web user interface. The test dataset is hosted on Google's Firebase storage platform, and a user can choose an existing or a random image and visualize how our agent performs on it. Overall system has a lot of moving parts working in tandem to achieve our goal in creating a Deep RL-based RPN framework for detecting tomato produce from given images.

3.4.6 Results

Array of images in Fig. 9 shows the output of the designed RPN solution. The left column consists of all the images that are input to the trained Deep RL RPN agent, and images in the column on the right are the output images. The current solution captures one of the objects of interest from given images.

We presented use of RL agent with VGG16-based convolutional neural network for object detection. The solution automatically determines where to find the object of interest (tomatoes). The model outputs the images and category-dependent search trajectories.

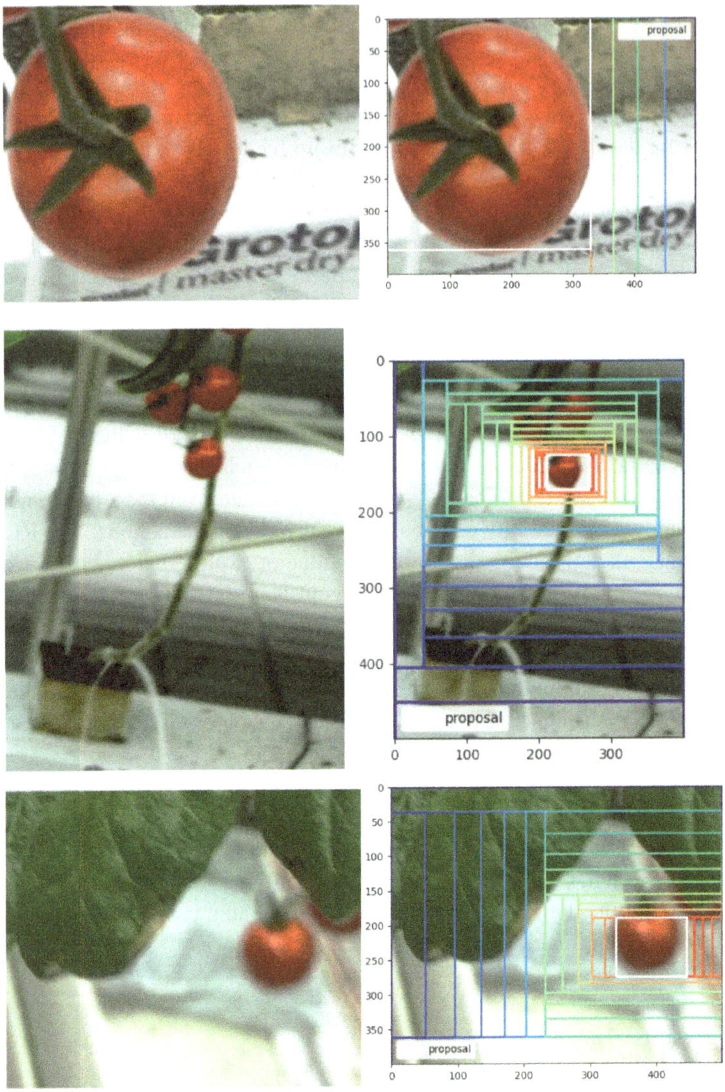

Fig. 9 Input and output of the Deep RL-based RPN for tomatoes detection

4 Conclusion and Future Directions

In this chapter, we presented state-of-the-art AI concepts of deep reinforcement learning. We motivated two important use cases, namely, crop yield maximization and crop produce detection from an image. Further, we presented robust and scalable production-ready Deep RL-based solution for handling these problems. The contributions consolidate applicability of Deep RL-based solutions to classical problem of agriculture and farming. Further, it lays foundation for developing big data-based AI solutions by contributing open source implementations over cloud infrastructure. The works presented in the case studies are generic, and designed models can be trained on alternate datasets for leveraging the designed RL agents.

Besides improving accuracies of the presented deep RL agent, following are some interesting future research directions and extensions of the works presented in this chapter. The agent designed in the crop yield maximization case study is currently trained against five teams' strategy data. This was done to get immediate rewards. It would be interesting to identify new approaches for training the agent based on actual observable data with long-term rewards. This research direction has possibility of generating policy actions that might be better than human optimal strategies. This approach would need a more detailed environment simulation or a real greenhouse. On the computer vision front, the current deep RL RPN solution finds a single object of interest in the given image. An interesting extension of the work would be to train the agent to find multiple objects in the given images.

Acknowledgements The author would like to thank Kaushal Patil, Supan Shah, Meet Akbari, Dhaval Deshkar, Vidish Joshi, Pathik Patel, Vyoma Patel, Shrishti Sharma and Arpitsinh Vaghela for their contributions to the case studies.

References

AGC G (2021). Autonomous greenhouse control and monitoring using deep reinforcement learning. https://github.com/Kaushal1011/AutonomousGreenhouseControl

Binas J, Luginbuehl L, Bengio Y (2019) Reinforcement learning for sustainable agriculture. In: ICML workshop on climate change AI.

Burnetas AN, Katehakis MN (1997) Optimal adaptive policies for Markov decision processes. Math Oper Res 22(1):222–255

Elavarasan D, Vincent PM (2020) Crop yield prediction using deep reinforcement learning model for sustainable agrarian applications. IEEE Access 8:86886–86901

Gandhi R, Raval MS (2020) Algorithms for optimization and machine learning over cloud. In: Nature-inspired computation in data mining and machine learning. Springer, pp 23–46

Li Z, Xu L, Zhu S (2020) Pruning of network filters for small dataset. IEEE Access 8:4522–4533

Marshall-Colon A, Long SP, Allen DK, Allen G, Beard DA, Benes B, von Caemmerer S, Christensen AJ, Cox DJ, Hart JC, Hirst PM, Kannan K, Katz DS, Lynch JP, Millar AJ, Panneerselvam B, Price ND, Prusinkiewicz P, Raila D, Shekar RG, Shrivastava S, Shukla D, Srinivasan V, Stitt M, Turk MJ, Voit EO, Wang Y, Yin X, Zhu X-G (2017) Crops in silico: Generating virtual crops using an integrative and multi-scale modeling platform. Front Plant Sci 8:786

Mehta M, Derasari R, Patel S, Kakadiya A, Gandhi R, Chaudhary S, Goswami R (2019) A service-oriented human capital management recommendation platform. In: 2019 IEEE international systems conference (SysCon). IEEE, pp 1–8

Mnih V, Kavukcuoglu K, Silver D, Rusu AA, Veness J, Bellemare MG, Graves A, Riedmiller M, Fidjeland AK, Ostrovski G et al (2015) Human-level control through deep reinforcement learning. Nature 518(7540):529–533

Ndikumana E, Dinh HTM, Baghdadi N, Courault D, Hossard L (2018) Deep recurrent neural network for agricultural classification using multitemporal SAR Sentinel-1 for Camargue, France. Remote Sens 10:1217

Pirinen A, Sminchisescu C (2018) Deep reinforcement learning of region proposal networks for object detection. In: proceedings of the IEEE conference on computer vision and pattern recognition, pp 6945–6954

Research WU (2020) Autonomous greenhouse challenge, 2nd ed. https://www.wur.nl/en/project/autonomous-greenhouses2nd-edition.htm

Redmon J, Divvala S, Girshick R, Farhadi A (2016) You only look once: unified, real-time object detection. In Proceedings of the IEEE conference on computer vision and pattern recognition, pp 779–788

RPN G (2021). Region proposal network for tomatoes detection using deep reinforcement learning. https://github.com/Supan14/RPNDeep-RL

Simonyan K, Zisserman A (2014) Very deep convolutional networks for large-scale image recognition. arXiv:1409.1556

Tomato (2020) Tomato dataset. https://www.kaggle.com/kangdali/tomato

Uzkent B, Yeh C, Ermon S (2020) Efficient object detection in large images using deep reinforcement learning. In: Proceedings of the IEEE/CVF winter conference on applications of computer vision, pp 1824–1833

Zhou N (2020) Intelligent control of agricultural irrigation based on reinforcement learning. In: Journal of physics: conference series, vol 1601. IOP Publishing, p 052031

Part II
Data Science Applications in Agriculture

Computer Vision and Machine Learning in Agriculture

Mehul S. Raval, Sanjay Chaudhary, and J. Adinarayana

Abstract The productivity and efficiency of a traditional agriculture crop production cycle can be improved by computer vision and machine learning. Innovative Information and Communication Technology (ICT) solutions in sensing, processing, and learning have attracted significant attention in precision agriculture. The power of visualizing the real world allows spatial or spatiotemporal information gathering and its representation. On the other hand, learning allows processing the information for planning, reasoning, and inference. The combination of computer vision and machine learning facilitates the automation of various tasks in the crop cycle. Compared to conventional agricultural practices, tasks are performed automatically and with greater fidelity. Computer vision has improved due to the following: availability of solid-state photo-detectors, increased sensor size, better spatial and spectral resolution, increased frame rate, and shutter speed. The development of remote sensing platforms like Unmanned Aerial Vehicles, terrestrial robots, and satellites allows capturing images at different scales. All the above factors increase the quality of the visual information. At the same time, information processing has become more accessible due to the growth of parallel computing (Graphics Processing Unit), efficient open-source machine learning libraries, large-scale annotated datasets, and reproducible research. Enrichment in information and increased processing power have seen innovations in precision agriculture like crop health and growth monitoring, prevention and control of pest and crop disease, automatic harvesting of crops, automated crop quality testing, and automated farm management. Computer vision and machine learning go hand in hand with precision agriculture, and therefore, they are seamlessly integrated into this chapter. It showcases publicly available large datasets

M. S. Raval (✉) · S. Chaudhary
Ahmedabad University, Ahmedabad, India
e-mail: mehul.raval@ahduni.edu.in

S. Chaudhary
e-mail: sanjay.chaudhary@ahduni.edu.in

J. Adinarayana
Indian Institute of Technology Bombay (IITB), Mumbai, India
e-mail: adi@iitb.ac.in

G. P. O. Reddy et al. (eds.), *Data Science in Agriculture and Natural Resource Management*, Studies in Big Data 96, https://doi.org/10.1007/978-981-16-5847-1_5

and state-of-the-art algorithms developed while using them. After thorough analysis, it is evident that computer vision combined with intelligent techniques allows processing in more complex environments. It, in turn, will produce generalized and robust automation systems for precision agriculture.

Keywords Artificial intelligence · Automation · Computer vision · Machine learning · Precision agriculture

1 Introduction

The advent of mechanization and electricity has altered agricultural practices forever. Today, information and automation are further accelerating positive change in this domain. However, the sector still faces challenges due to the rise of population and urbanization. The world's population will cross the 10 billion mark by 2057 (United Nations 2019). By 2050, 70% of the world's population will be living in urban areas compared to 50% today (FAO 2009). It will fuel the rise in income level, which will push up demand for food, which is estimated to rise by 70% in comparison to the present level (FAO 2009).

On the other hand, the growth rate of cereal is dropping from 3.2% (1960) to 1.5% (2000) (FAO 2009). Climate change is another threat to agriculture. Regression studies predict that climate change may decrease the African continent's agricultural output during the 2080–2100 period by 15 to 30% (FAO 2009). The probability of increased crop production due to the rise in arable land is small due to rapid urbanization. Due to the reasons given above, it is necessary to increase food production and build policies for equitable distribution. The consequence of rapid urbanization and climate change leads us to the hypothesis—*use technology for increasing crop production*. It is necessary to understand the crop production cycle for technology-based interventions. A typical cycle during crop production (Wikipedia—Agricultural Cycle 2021) is shown in Fig. 1.

It is a regular and seasonal crop production cycle meant for growing and harvesting of crops. It is essential to understand the properties of the soil and select the seed for the crop. It is also important to maintain vital conditions such as water, air, temperature, energy, and enzymes for proper germination (Wikipedia—Agricultural Cycle 2021). The agriculture industry is still facing challenges like food safety, food security, poverty alleviation, sustainability, and management of natural resources.

Precision Agriculture (PA) (Pierce and Nowak 1999; Stafford 2000; Zhang et al. 2002; Gebbers and Adamchuk 2010) or digital agriculture is a possible solution to overcome many prevailing problems of conventional agriculture. The fusion of multiple domains like Industry 4.0, Internet of Things (IoT), Artificial Intelligence (AI), Cloud resources and Computing, Blockchain, Big Data, Robotics, and Computer Vision (CV) empowers PA. There is no universal definition of PA. It is interesting to study the development through the lenses of definition as it evolves. Table 1 captures the definitions of PA as it evolved through the 1990s.

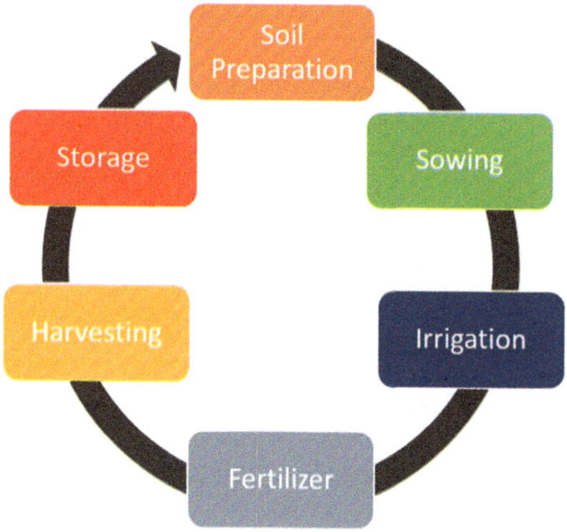

Fig. 1 Typical crop production cycle (*Source* Wikipedia—Agricultural Cycle 2021)

Based on the above definitions, it can be concluded that PA is a technology that can help to maximize the yield while conserving resources. A typical PA cycle is shown in Fig. 2. The decision to opt for PA can be taken based on the answers to the following questions:

- How much improvement is possible in the yield due to PA?
- Does the yield variability justify investments?
- What is the size of the farm?
- What is the cost of PA per unit of measurement?
- Environmental sustainability?

Economic studies indicate that large farm size is beneficial for profitable PA (Zhang et al. 2002).

The goal of PA is to maximize crop production, in turn, maximizing income while minimizing inputs and environmental impacts. Therefore, the PA cycle performs the correct operation at the right time and place, using the exact amount of material with precision. Based on Fig. 2, the essential blocks for PA are: observations through sensors, data processing, and visualization, deriving inference from processed data, and applying decisions to initiate agricultural-related actions. Vision is one of the vital sensing modalities in PA. The role of vision-based automation is discussed in the next section.

1.1 Computer Vision Preliminaries

Like other domains, CV also has many exciting definitions. Haralick and Shapiro (1991), define it as "*Computer Vision, image understanding or scene analysis is that combination of image processing, pattern recognition, and Artificial Intelligence*

Table 1 Definitions of precision agriculture

Year	Definition	Authors
1994	The intent of precision agriculture is to match agricultural inputs and practices to localized conditions within a field to do the right thing, in the right place, at the right time, and in the right way	Pierce et al. (1994)
1996	The targeting of inputs to arable crop production according to crop requirement on a localized basis	Stafford (1996)
1997	Precision agriculture is a management strategy that uses information technologies to bring data from multiple sources to bear on decisions associated with crop production	National Research Council (1997)
1999	Precision agriculture is the application of technologies and principles to manage spatial and temporal variability associated with all aspects of agricultural production for the purpose of improving crop performance and environmental quality	Pierce and Nowak (1999)
2004	Precision farming is a systems approach to managing soils and crops to reduce decision uncertainty through better understanding and management of spatial and temporal variability	Dobermann et al. (2004)
2010	Precision agriculture, or information-based management of agricultural production systems	Gebbers and Adamchuk (2010)
2016	Precision Agriculture (PA) can be defined as the management of spatial and temporal variability in the fields using Information and Communications Technologies (ICT)	Fountas et al. (2016))
2019	Precision Agriculture is a management strategy that gathers, processes, and analyzes temporal, spatial, and individual data and combines it with other information to support management decisions according to estimated variability for improved resource use efficiency, productivity, quality, profitability, and sustainability of agricultural production	ÏSPA (2019)

technologies which focuses on computer analysis of one or more images taken with a single/multiband sensor or taken in time sequence. The analysis recognizes, locates the position and orientation and provides sufficiently detailed symbolic description." Huang (1996), suggests, "*Computer Vision has a dual goal. From the biological science point of view, CV aims to come up with computational models of the human visual system. From the engineering point of view, computer vision aims to build autonomous systems which could perform some of the tasks which the human visual*

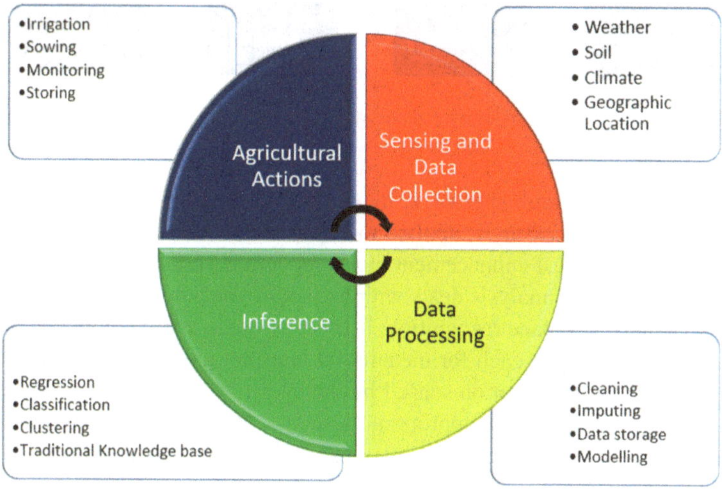

Fig. 2 Precision agriculture cycle

system can perform." CV is an interdisciplinary domain that makes computers understand visual data—images or videos (Wikipedia Computer vision 2021). It understands and automates the task performed by the Human Visual System (HVS). The domains empowering CV are shown in Fig. 3.

A classic CV task involves: image acquisition, processing, analyzing, and understanding. A typical processing chain of CV tasks is shown in Fig. 4. Image acquisition captures signals through sensors like cameras, tomographic devices, and radar. It produces a 2D image or multidimensional volume as output. The sensors may also use

Fig. 3 Domains of Computer Vision (CV); AI: Artificial Intelligence; BS: Biological Sciences; CS: Cognitive Sciences; MS: Mathematical Sciences; PS: Probability and Statistics; PY: Physics; SP: Signal Processing

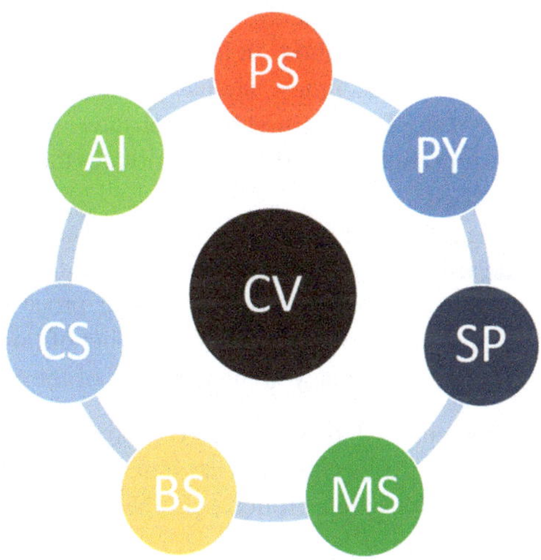

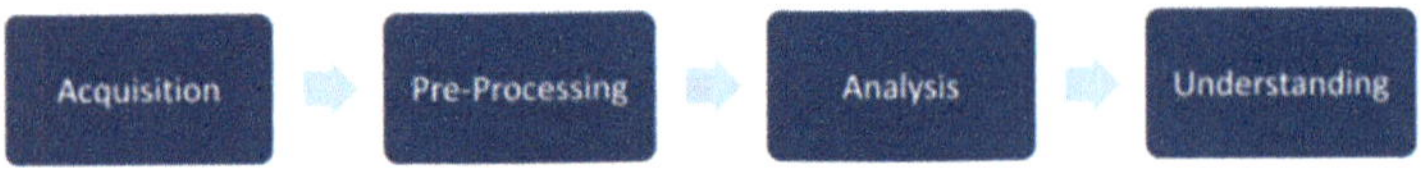

Fig. 4 Important blocks in the computer vision system

Near-Infra Red (NIR) multispectral or hyper-spectral imaging for capturing biological processes. Preprocessing is applied to improve visual signal quality for further processing, e.g., contrast enhancement, noise reduction, resampling, or scale-space transformation. Image analysis deals with extracting meaningful information from the visual signals, e.g., face recognition. It needs to break an image into constituent parts and further analyze each for meaningful interpretation. Image understanding involves converting information sensed by the eyes into a world's description that makes sense and generates an appropriate response. It works with segmented parts to find out what is happening in the image or video. It may further initiate actions based on the derived inference.

1.2 Computer Vision in Precision Agriculture

CV has been extensively used in the agriculture domain due to the following reasons (Mahajan et al. 2015):

- Technology has improved the sensing quality with reduced cost of the equipment.
- Computational power on the edge and centralized locations has grown tremendously over the years.
- There is an increasing trend of non-destructive testing in the agro and food industry. Such a method fits food safety criteria and is also economically efficient (Zareiforoush et al. 2015).
- The manual methods in farming are time-consuming, and they have low repeatability and accuracy. Human resources also have to be specially trained for the task. There is a substantial inter- and intra-rater variability during quality testing (Barbedo 2013). CV facilitates the development of automated and efficient methods.
- The collection of samples from the field and then testing them in the lab can be destructive. A CV-based system can do on-field testing, minimizing the losses (Barbedo 2013) during the testing process. Overall, it is non-destructive and automated. Robust on-field capturing of information, processing, and testing can reduce redundancy at many phases of the crop production cycle.

Though CV has offered several advantages, it is also essential to understand its limitations. The challenges are both intrinsic and extrinsic to the methods. Some of the limitations are as follows (Barbedo 2016):

- The merging of background and foreground (Region of Interest (RoI)) often causes poor segmentation. It may cause a problem while incorrectly selecting RoI and cause inaccuracy.
- Capturing visual signals in the field environment is always challenging. Illumination depends on the time of the day, season, angle of the sun, and capturing visuals. Therefore, variation in image characteristics due to illumination also degrades performance.
- The indistinct boundaries of RoI and non-RoI regions result in incorrect segmentation; e.g., merging healthy and diseased tissue makes the detection of diseased portions difficult.
- The occlusion of the RoI with other objects will also cause false segmentation, false analysis, and understanding.
- The edge devices have limited computing power and memory; therefore, real-time field processing is complicated.
- Usually, the inputs from CV are processed by Machine Learning algorithms. The inherent problem of differences in the training and testing data distribution can also impact results.

Modern CV or agro vision is used for crop monitoring and phenotyping, harvesting, weed control, pest and disease control (Parikh et al. 2016), vehicle guidance, and yield mapping (Lu and Young 2020). CV depends on Machine Learning (ML) techniques for automation and decision support systems. Most computer vision tasks combine with ML intending to detect the object of interest and this helps in quantification, classification, and predicting the phenomenon. Commonly applied ML or Deep Learning (DL) techniques in agriculture are: regression, dimensionality reduction, and clustering. A successful combination of CV and ML requires robust hardware systems and a data-driven analysis pipeline. Data-driven approaches require high quality, large scale, and annotated datasets. Therefore, the following section reviews available public datasets for critical computer vision applications.

2 Important Datasets and Their Characteristics

The success of a data-driven PA pipeline depends on the availability of datasets. The challenging aspect is preparing the dataset. It involves the cost of data acquisition, pre-processing, annotation, classification, and many times measuring physiochemical processes. The growth of reproducible research has paved the way for data sharing, leading to improved data-driven approaches. It can save costs and facilitate benchmarking of machine learning algorithms (Lu and Young 2020; Lobet 2017). The CV community has benefited from the availability of large-scale natural image datasets (Kuznetsova et al. 2020). It has boosted the accuracies for classifiers and allowed the creation and testing of novel architectures. The section discusses a large (>1000 images) public dataset for weed control, fruit detection, and plant phenotyping.

2.1 Public Datasets for Weed Control

2.1.1 Grass-Broad Leaf Dataset (Dos Santos Ferreira 2017)

This dataset was captured by an Unmanned Aerial Vehicle (UAV) flying at a height of 4 meters from ground level in a soybean field. All images with weed occurrence were selected from a set of images captured by the UAV, resulting in 400 images. The RGB images were segmented using a superpixel algorithm (Achanta et al. 2012). It was further manually annotated into four classes–soybean, grass, broadleaf, and soil. The segments constitute the dataset with 15336 segments3249 of soil, 7376 of soybean, 3520 of grass, and 1191 of broadleaf weeds. The dataset is stored in the TIFF format and has different sizes of segments. Representative images from the four classes are shown in Fig. 5.

Fig. 5 Sample images of **a** Weed-Broad Leaf; **b** Grass; **c** Soil; **d** Soybean (*Source* dos Santos Ferreira 2017a)

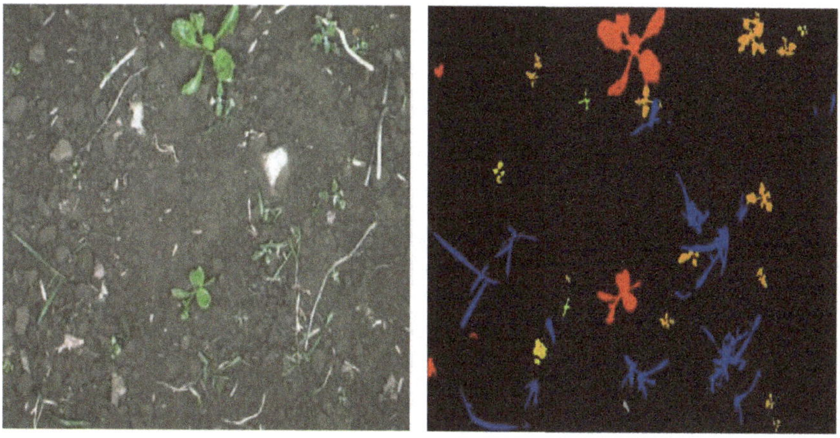

(a) (b)

Fig. 6 Representative image and corresponding ground annotations (*Source* Chebrolu et al. 2017)

2.1.2 Sugar Beets Dataset (Chebrolu et al. 2017)

This dataset is a large-scale agriculture dataset that is useful for plant classification. It provides localization and mapping covering growth stages for weed control. An available agricultural robot was used in a sugar beet farm in Germany over three months in 2016. The data was recorded thrice in a week, starting from seedlings to stopping when the robot would damage the crop. The robot was equipped with multiple sensors—four multispectral channel cameras and a Red-Green-Blue-Depth (RGB-D) camera that captured detailed information about the plantation. Under controlled lighting conditions, they generated images of size 1296 × 966 pixels in the PNG format. LIDAR and GPS sensors provided localization, navigation, and mapping. The dataset has 283 multiclass images annotated at the pixel level for sugar beet and nine different types of weeds. Another 12340 images have three-pixel level annotations—weed, crop, and background. Figure 6 shows the representative images along with the corresponding ground annotations.

2.1.3 Synthetic Sugar Beets Dataset (Di Cicco et al. 2017)

Agriculture robotics have a massive potential for weeding applications. The robot must detect the weed and classify the weed and the crop with great accuracy. The data-driven approach needs a large annotated dataset that can be substantiated by using synthetically generated data. The synthetic dataset is useful in training the model and augment the natural image set. Such effort reduces human efforts in data collection and annotation. The synthetic sugar beets dataset (Di Cicco et al. 2017) is an effort to generate a large-scale dataset for weed control using robots. It

(a) (b)

Fig. 7 **a** Synthetic image with a plant. **b** Corresponding ground truth image. (*Source* Di Cicco et al. 2017)

is algorithmically generated using a computer graphics technique, and it modeled plants and real-world textures. The dataset has 8518 synthetic RGB images of size 480 × 360 pixels in PNG file format with four classes of sugar beets and three types of weeds. Each image is annotated pixel-wise for crops, soil, and weeds. Figure 7 represents a synthetic image of a plant and also a ground truth image.

2.1.4 Leaf Counting (Teimouri 2018)

The dataset is meant for estimating weed growth by counting the number of leaves. It helps estimate the amount of herbicide spraying needed for weed removal. The dataset is useful for classification based on the number of leaves only; it cannot be used for weed classification and semantic segmentation. It has 9372 RGB images in PNG format with several leaves counted. The images are captured using a mobile phone camera, a consumer camera, and a gray industrial camera across Denmark. The images are in various resolutions for 18 types of weeds at different stages of growth. The dataset is classified into 9 classes based on the number of leaves, ranging from 1 to more than 9. The number of images for each class varies from 160 to 3292. The dataset has varying soil types and lighting conditions.

2.1.5 Deep Weed (Olsen et al. 2019)

This is a dataset that contains more than 10000 images and is valuable for training deep neural networks. It contains a large set of weed images collected from nine different species. It has 17509 images captured in Australia and in situ with other flora. The RGB images are captured in JPEG format and have a size of 256 × 256 pixels. A ground weed-controlled robot without any light control in the field captured the images. Each weed species has more than 1000 images, e.g., Chinee apple—1125, Lantana—1064, Parkinsonia—1031. The dataset also has 9106 negative images making it a reasonably balanced dataset.

2.2 *Public Datasets for Fruit Detection*

2.2.1 Orchard Fruits (Bargoti and Underwood 2017)

The dataset contains images and annotations for apples (1120 images), mangoes (1964 images), and almonds (620 images). They are available in the PNG format with 8-bit and 16-bit RGB color resolution. The images of apples and almonds have 308 × 202 pixels, and those for mangoes have 500 × 500 pixels. The images are captured at three different locations in Australia using an automated ground vehicle and handheld camera. The images are cropped to a lower resolution, and they have circular and rectangular bounding box annotations. The apple dataset has additional pixel-level annotations. Figure 8 shows the annotated images.

The images are captured during the daytime at two locations in Victoria and Queensland, Australia. The apple and mango data was captured by traversing rows of an orchard on a general-purpose ground vehicle. The apples have been captured from a near distance, and mango trees are located further away and are captured using a high-resolution camera. The almond trees had large canopies, and a tree can host far more fruits than apples or mangoes. Therefore, a high-resolution handheld DSLR camera was used to capture the images. The data varied from 2 MP to 17 MP. Each image for mangoes and apples contains about 100 fruits, and the almond images have up to 1000 fruits.

Fig. 8 Annotated images from orchard dataset (*Source* Bargoti and Underwood 2017)

2.2.2 Date Fruit Dataset (Altaheri et al. 2019)

The date fruit dataset fulfills the requirements of the applications during the pre-harvesting and harvesting stages. The dataset contains 8079 images of date bunches captured from 29 palms. Five varieties of dates are captured using a color camera in six sessions. The sessions are planned to cover four different maturity stages of the date bunches. The dataset has the following variations: angle of capture, illumination conditions, scale, and occlusion. It contains classes for type, maturity, and harvesting decision.

2.2.3 MangoYOLO Dataset (Koirala et al. 2019)

This dataset is useful for deep learning and to benchmark the mangoYOLO architecture. You-Only-Look-Once (YOLO) (Redmon et al. 2016) is a fast object detector. Its variants—YOLOv2 and YOLOv3 were used for fruit detection and orchard load estimation. Interestingly the dataset is collected at night under artificial lighting, and it resulted in better contrast images. RGB images are captured using a camera mounted on a ground vehicle. The 1730 images of size 612×512 pixels are available in JPEG format. The bounding box annotations are stored in an XML file. The data set divides 1730 images as follows: 1300 for training, 130 for validation, and 300 for testing.

2.2.4 WSU Apple Dataset (Karkee et al. 2019)

This dataset has been created by the Washington State University Agriculture robotics laboratory. The dataset has 2298 images of apple trees with fruits captured by robotic vehicles during harvesting. The images are captured over multiple years, at different times of the day, and in multiple varieties. The bounding box provides the fruit annotations. The lab also released additional 1600 images captured by an RGB-D camera.

2.3 Public Datasets for Plant Phenotyping

2.3.1 Global Wheat Head Detection Challenge (David et al. 2020)

Detecting a wheat head reveals different characteristics—health, size, maturity, and population density. Therefore, it is essential to detect wheat heads from plant images. Many methods exist, but they have reported results on a tiny dataset captured under well-controlled conditions. The machine learning algorithms, when applied to such datasets, yields excellent results. However, they have very little applicability when

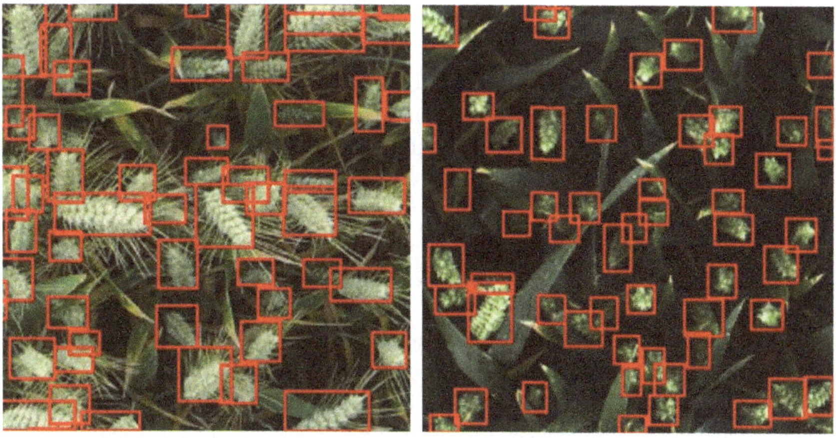

Fig. 9 Annotated wheat head images from GWHD (*Source* David et al. 2020)

applied in field scenarios. The challenges in this problem are: extremely high variability in environmental conditions, development stages, differences in genotype, head orientations, motion artifacts, and occlusion (David et al. 2021). The challenging problems make it a challenging CV problem.

The Global Wheat Head Detection (GWHD) dataset is large, diverse, and well-annotated to represent the challenging conditions. It has been collected from many countries, covering different growth stages and phenotypes. It has 4700 high-resolution RGB images, and 1.9 lac annotated wheat heads. It is publicly available for benchmarking. A web-based tool is used for evaluating and labeling the shared sub dataset. It has all the tools for annotations. Only clearly visible wheat heads have been labeled by omitting the partially or entirely occluded wheat heads. The GWHD dataset has 4698 square patches extracted from 2219 high-resolution images. There are 40–60 labeled heads per image on average. A few images have 100 heads, and a maximum of 120 heads are present in the images. The difficulty for benchmarking is increased by adding 100 images with no heads and representing the field conditions. Figure 9 shows the representative images from the GWHD dataset.

2.3.2 Sorghum Biomass Prediction Dataset (Sorghum Biomass Challenge 2021)

Sorghum is an agricultural feed substitute and a source of biofuel. An increase in the demand for sorghum motivated different plant breeding strategies. The ARPA, or Energy-funded Transportation Energy Resources from Renewable Agriculture Phenotyping Reference Platform, or TERRA-REF monitored the whole growth cycle over an acre of crop using an RGB-D camera, thermal, hyperspectral, and laser 3D-scanner sensors. The dataset allows us to predict the biomass at the end of the season. The training dataset has 277,327 images from 176 sorghum species in 349 unique plots. The metadata contains sorghum species, the date on which the image was captured, the amount of sunlight accumulated at the field site, the height of the camera, and the weight of the plant cut down at the end of the season. The testing dataset has 19442 images from 110 unique plots. 33% plots included images from the early growth cycle, 33% from the mid-growing season, and 33% from the late growing season.

2.3.3 Open Plant Phenotype Database (OPPD) (Leminen et al. 2020)

The OPPD is a collection of plant seedlings for 47 weed species. The intra-species variations were created by cultivating the plant in a controlled environment for ideal, drought, and natural growth conditions. The data collection took place over four seasons, and plants were tracked from seedling emergence to the stage with the growth of leaves. The resulting dataset has 7590 images stored in the JPEG format, which covered 64292 plants. Each image is annotated with a label of the weed species, and bounding boxes are given using the annotation tool. The dataset can be used for plant classification and instance segmentation.

2.3.4 Root Dataset (Arabidopsis Thaliana Root Segmentation Challenge 2021; Gaggion et al. 2020)

The growth of the root helps to assess the response of plants to soil properties and environmental challenges. The developmental assessment can also be linked to their genetic background. Root segmentation allows extracting information like length, density, and lateral roots. The analysis over a large dataset allows capturing the dynamics related to growth and real-time parameters. Arabidopsis thaliana root segmentation (Gaggion et al. 2020) captures video sequences under controlled conditions capturing plant root growth. The challenge is to segment the roots. The plants are grown in Petri boxes for 2–4 weeks, and each video has one frame every 12 h. There are five images with annotations per sequence. The frame has a resolution of 3280 × 2464 and is captured with an infrared camera. The binary ground truth is manually generated from 34 videos with 5–7 frames per video. The ground truth depicts the root without the background. An expert has carried out three-class labeling for 26 videos, and the labels are—main root, lateral root, and background.

2.3.5 Leaf Dataset (Leaf Counting Challenge 2021; Minervini et al. 2016)

The challenge is to count the number of leaves. The dataset has been derived in labs or collected from a public dataset. It contains the top view of rosette plants—Arabidopsis and tobacco. A camera captures tobacco images with a single plant in its field of view. In contrast, the Arabidopsis was captured, having many plants in the field of view. The images are taken over several days and stored in the TIFF format. The images are annotated by hand, and a ground truth label for each leaf in the scene is generated. They are stored in the PNG format with the leaf center denoted as a single pixel.

Table 2 summarizes the characteristics of the datasets discussed in this section.

Table 2 Dataset summary

Type	Name	Platform/Sensor	# images/size/classes	Dataset URL	Source
Weed	Grass- broad leaf dataset	UAV/RGB	15336/variable/4	https://data.mendeley.com/datasets/3fmjm7ncc6/2	dos Santos Ferreira (2017b)
	Sugar beets dataset	Robot/RGB-D and Multispectral	12340/1296 × 966/3	https://www.ipb.uni-bonn.de/data/sugarbeets2016/	Chebrolu et al. (2017)
	Synthetic sugar beets dataset	Synthetic	8518/480 × 360/3	http://www.diag.uniroma1.it//~labrococo/fds/syntheticdatasets.html	Di Cicco et al. (2017)
	Leaf counting	Cameras/RGB	9372/Variable/9	https://vision.eng.au.dk/leaf-counting-dataset/	Teimouri (2018)
	Deep weed	Robot/RGB	17509/256 × 256/9	https://www.kaggle.com/coreylammie/deepweedsx	Olsen et al. (2019)

(continued)

Table 2 (continued)

Type	Name	Platform/Sensor	# images/size/classes	Dataset URL	Source
Fruit	Orchard fruits	Robot/RGB	3704/Variable/4	https://data.acfr.usyd.edu.au/ag/treecrops/2016-multifruit/	Bargoti and Underwood (2017)
	Date fruit dataset	Camera/RGB	8079/Variable/5 and 4	https://dx.doi.org/10.21227/x46j-sk98	Altaheri et al. (2019)
	MangoYOLO dataset	Robot/RGB	1730/612 × 512/2	https://figshare.com/articles/dataset/MangoYOLO_data_set/13450661/2	Koirala et al. (2019)
	WSU apple dataset	Robot/RGB-D	2298/Variable/Variable	https://research.libraries.wsu.edu:8443/xmlui/handle/2376/17721	Karkee et al. (2019)
Pheno-typing	Global wheat head detection challenge	Camera/RGB	2219/Variable/NA	https://doi.org/10.5281/zenodo.4298502	David et al. (2020)

(continued)

Table 2 (continued)

Type	Name	Platform/Sensor	# images/size/classes	Dataset URL	Source
	Sorghum biomass prediction dataset	RGB-D camera, thermal, Hyperspectral, Laser 3D scanner	277,327/Variable/NA	https://www.kaggle.com/c/sorghum-biomass-prediction/overview/submission-and-evaluation	Sorghum Biomass challenge (2021)
	Open Plant Phenotype Database (OPPD)	Camera/RGB	7590/Variable/47	https://vision.eng.au.dk/open-plant-phenotyping-database/	Leminen et al. (2020)
	Root dataset	IR Camera/IR	34 Videos/3280 × 2464/NA	https://sites.google.com/sinc.unl.edu.ar/root-segmentation-challenge/home#h.x8m73yix931s	Gaggion et al. (2020)

(continued)

3 Machine Learning in Precision Agriculture

ML is at the heart of processing the datasets discussed in the previous section. ML allows computers to interact with the environment by sensing the signals from the Internet of Things (IoT) devices, learning without being explicitly programmed (Samuel 1959), and activating the farm actuators. Integration of sensing by CV and processing by ML has facilitated a revolution—Agriculture-4.0 (Yang et al. 2020). ML is one of the critical blocks in the domain of AI. The prime goal of ML is to understand the nuances of learning and then build this capability in computers. The

Table 2 (continued)

Type	Name	Platform/Sensor	# images/size/classes	Dataset URL	Source
	Leaf dataset	Camera/RGB	10000/Variable/NA	https://data-challenges.fz-juelich.de/web/challenges/challenge-page/85/overview	Minervini et al. (2016)

general impact is to improve the problem-solving capability of machines. It can be done by understanding the computational aspect of learning and building algorithms based on these models to solve future problems. Figure 10 depicts various hierarchies in AI and ML.

ML is partitioned in a supervised form, that is, when corresponding to each input, ground truth labels are available. Only a few ground truth labels are available in the semi-supervised form. While in unsupervised learning, there are no labels. The goal is to learn the functional mapping between input and output from the training data in the latter two. The goal for unsupervised learning is to find the patterns from the data. In PA, supervised learning algorithms are used very heavily to solve classification

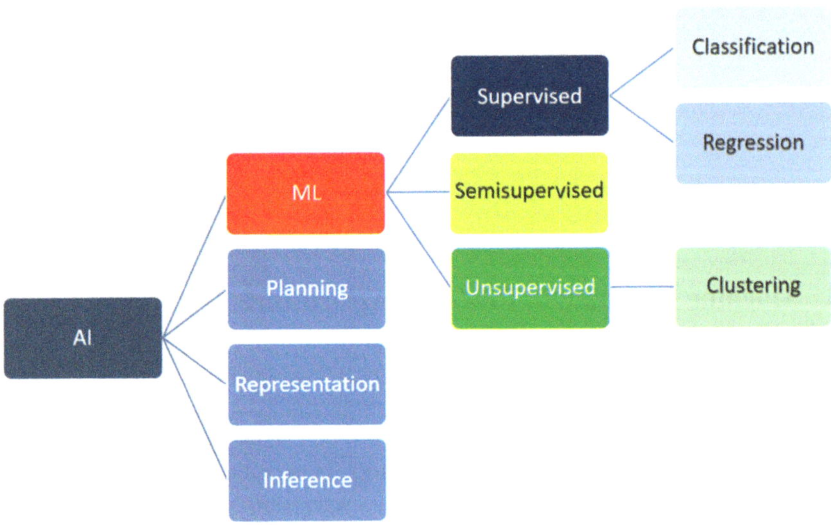

Fig. 10 Various hierarchies in AI and ML

or regression problems. In classification, the goal of the supervised algorithm is to predict the categorical label, e.g., the fruit quality {Good, Bad, Ripe, Unripe}. In the case of regression, the output prediction is a number which is a continuous variable on a number line, e.g., yield at the end of each season in Metric Tonnes.

3.1 *Shallow Neural Networks and Deep Neural Networks*

A Neural Network (NN) is a collection of basic computing elements such as artificial neurons which mimic some computational functions of a biological neuron. The neurons are arranged in layers that are cascaded together. A neuron in the layer is connected to every other neuron in the next layer. Each connection has a parameter weight signifying the importance of the connection. The learning of NN would mean tuning weights to obtain the correct prediction by the network. The neurons also use a nonlinear activation function before generating an output. It helps in realizing an extremely high degree of non-linearity between the input and output. A nicely engineered feature suitable for the field environment is the leading cause of limiting results generated from a shallow NN.

Large annotated datasets and parallel computing availability through the Graphical Processing Unit (GPU) have fueled DL using Deep Neural Network (DNN). The *deep* connotation means the NN has more layers to learn the hierarchy. They can realize very complex and intelligent tasks beyond the reach of Shallow Neural Networks (SNN). The stacking of layers allows hierarchical feature learning, which better resembles the biological processes of the brain. Table 3 summarizes an SNN and a DNN.

Table 3 Summary of SNN and DNN

Shallow neural network	Deep neural network
It uses 1 to 3 layers of neurons	It will use many layers of neurons
All the neurons are fully connected	Neurons share weights, e.g., Convolutional Neural Network (CNN)
The input is usually converted to a vector and then fed to the NN	It allows 1D, 2D, and 3D processing of the input data
Features are extracted from the object of interest and then fed to the NN	The DNN automatically extracts features from the data and predicts the output. Features are learned hierarchically, i.e., initial layers learn simple features like edges, colors, and deeper layers learn more complex features
It is possible to analyze the network and explain its functioning	It is not easy to analyze the network due to depth and non-linearity
It is suitable for small datasets and can be trained on CPUs	It uses GPUs and trains over vast datasets. The richness of learning surpasses the performance of SNN by large margins

CNN is a subclass of NN that allows the use of spatial data. They have proven extremely powerful in CV applications such as object detection, classification, and regression. It follows a standard structure and follows an alternating convolutional layer and pooling layer. In classical architecture, the last few layers are fully connected, and the last layer is a softmax layer. CNN transforms an input volume to an output volume through neuron activation.

The following section describes the techniques which combine CV and ML in PA. The section covers techniques applied to the dataset discussed in Sect. 3 for: 1. Weed control; 2. Fruit detection; 3. Plant phenotyping.

4 Applications of Computer Vision in Precision Agriculture

4.1 Weed Control

Weeds designate the plants which grow with desirable plants. They compete for resources such as sunlight, water, fertilizers, and soil nutrients. A weed can cause stunted growth in the desired plants and reduce their yield. Therefore, it is crucial to identify the weed using CV and ML and remove it from the field. Selective weeding is the key to the successful growth of the crop of interest.

A work using CNN was developed to detect weeds (dos Santos et al. 2017a) in soybean crops. The images of soybean plants and weeds made up of broadleaf and grass are segregated so that specific herbicides could be used to eliminate the weeds. A soybean plantation in Brazil was photographed using a drone. The image database (cf. Grass-broadleaf dataset, dos Santos et al. 2017a) with 15000 images containing soybean, soil, grass, and broadleaf was created. AlexNet—a CNN architecture and a superpixel algorithm is used for segmentation. The network achieved an accuracy of 98% to detect weeds—grass and broadleaf without soil and soybean. The results of the CNN classification are also compared with a Support Vector Machine (SVM), AdaBoost, and Random Forest (RF). The shape, color, and texture features extracted from the images are used for the classification. However, CNN outperformed all the other classifiers.

An exciting work developed a synthetic weed dataset (Chebrolu et al. 2017) intending to minimize the time, effort, and cost required to annotate the actual dataset. A large amount of photorealistic data was possible by tuning the procedural generative model parameters and using natural world texture. The generated artificial set can be directly used to tune DL algorithms. The paper compares the performance using natural and artificial datasets to compare the efficacy of the approaches. The procedural generative method created four datasets, each with 1300 images. The variation in the datasets was created by using a combination of sugar beet plants and weeds. The performance and quality of the synthetic dataset were evaluated using semantic segmentation performed by SegNet (Badrinarayanan 2017). It is an encoder–decoder architecture with a final softmax layer to perform pixel-wise classification. The semantic segmentation provides robust segmentation as it can classify in the case of occlusion as well. The basic SegNet (8-layered network) and RGB-SegNet (26 layered network) generated exciting results when applied to natural and synthetic datasets. The in-class average accuracy for RGB-SegNet is 59% to 67% over the actual dataset and 61% to 66% on the synthetic dataset. The basic SegNet had 82% to 88.9% accuracy on a real dataset and 70% to 80% over synthetic datasets. It shows that the performance of SegNet is comparable on natural and synthetic datasets.

This method (Teimouri 2018) used CNN to automatically estimate the species of the weeds and their stages of growth. The images are captured in situ, covering 18 weed species from cotyledon until eight leaves are visible. The CNN algorithm was trained on 9000 images for automatically dividing the weeds into 9 growth stages. A test set of 2500 images with variation in soil type, crop, image resolution, and light condition is used to test CNN. It achieved a maximum accuracy of 78% on identifying *Polygonum spp* and an accuracy of 48% in black-grass. It showed an accuracy of 96% in identifying deviation among the leaves and 70% in estimating the leaves.

The deep wood (Olsen et al. 2019) proposed deep learning for the automatic classification of weeds in Australia. It used the Inception-V3 (Szegedy 2016) and ResNet-50 (He 2016) as the CNN models. The networks were fine-tuned using transfer learning (1000 classes of ImageNet to 9 weed classes). The 17509 images in the dataset are partitioned in 60% (Train)—20% (Validation)—20% (Test) sets. It used stratified k = 5 cross-validations, and standard binary cross-entropy loss has been used for training. Several augmentations such as rotation, scale, color, illumination, and perspective have been used for robust training. The average top – 1 accuracy percentage for Inception V3 varied from 85.3% (Chinee Apple) to 97.6% (Siam Seed). The top-1 accuracy (%) for ResNet-50 varied from 88.5% (Chinee Apple) to 97.2% (Parkinsonia). Tensorflow took 180 ms for inference with a 5 Frame per Second (FPS) rate.

4.2 Fruit Detection

CV and ML help estimate the yield of fruits and perform their mapping on the field. The farmer must perform optimal resource utilization, maximize the return of investment and minimize the time budget. Precise fruit locations can also help with automation through robotic harvesting systems.

The classical approach of feature extraction and fruit detection has been overtaken by deep learning based approaches (Bargoti and Underwood 2017). Outdoor image collection also poses challenging conditions for computer vision-based fruit detection. Since the fruit size is relatively small compared to trees, it is essential to use a high-resolution camera covering a large field of view. For example,18 megapixels (MP) images may have up to 1500 fruits (Bargoti and Underwood 2017). The paper experimented with Faster RCNN (Ren et al. 2016). The input image is passed through either five convolutional layered ZF net or the deeper VGG16 net for region proposal generation. The work also used transfer learning and standard augmentation techniques. The multiclass classifier to detect apples, almonds, and mangoes is evaluated. The best F1 score for VGG16 net for apples, mangoes, and almonds is 0.9, 0.9, 0.77, respectively.

Altaheri et al. (2019a) developed a machine vision system for date fruit classification in an orchard environment. The framework used deep learning to develop three classification models based on type, maturity, and the harvesting decision. The approach again relied on transfer learning and fine-tuning on a pretrained model. The dataset (Altaheri et al. 2019) has more than 8000 images for 5 types of dates and with pre-maturity and maturity stages. The VGG16 based network provided an accuracy of 99.01% for type, 97.25% on maturity, and 98.59% on harvesting decisions. The average classification decision is 35.9 ms per image on GPU and 1298 ms per image on CPU.

The mangoYOLO (Koirala 2019) compares 6 DL architectures to detect mangoes in tree canopies. About 1500 images from 3 orchards were collected using a 5 MP RGB camera and a floodlight mounted on a vehicle moving at 6 Kmph. The Faster RCNN (Ren et al. 2016) with VGG16 (Simonyan and Zisserman 2014), ZF net (Zeiler and Fergus 2014), YOLO v1 (Redmon et al. 2016), YOLO v2 (Redmon and Farhadi 2017), and YOLO v3 (Redmon and Farhadi 2018) along with Single Shot Detector (SSD) (Liu et al. 2016) were used in the experimentation. A new architecture based on YOLO v3 and YOLO v2 (tiny) model has been proposed in the paper with the name mangoYOLO. The model variant achieved the best F1 score of 0.968 and an average precision of 0.983 on the test set. It can process a 2048 $\times$ 2048 pixels frame in 70 ms, achieving a 14 FPS rate.

4.3 Plant Phenotyping

Plant phenotyping is one of the fastest growing areas of PA with the application of CV and ML. It plays an essential part during the study of attributes related to plants. Plant phenotyping refers to a quantitative description of its anatomical, ontogenetical, physiological, and biochemical properties (Walter et al. 2015). It can provide accurate numbers on properties and help predict the behavior of plants under different conditions, e.g., different soil or environmental conditions. This understanding helps during effective crop management.

Image analysis empowers the non-destructive testing of plants on a massive scale. The domain has matured from controlled conditions in labs to phenotyping under uncontrolled conditions in the fields. Modern era plant techniques depend on terrestrial and remotely sensed imaging for generating a quantitative description of plant performance during their life span. CV-based research in this dimension has grown significantly over the past few years. The availability of low-cost hardware and robust machine learning pipelines have fueled domains like wheat head detection, yield estimation, weed, and root analysis. State-of-the-art divides plant phenotyping into four subtypes (Walter et al. 2015): spectral reflectance and absorption; temperature and measures of water status; size, morphology, or plant growth; and root system analysis. DL is ideally suited for carrying many of the tasks, and the following discussions will showcase the applications.

Crowdsourcing is an exciting way to collate challenging and annotated datasets. The popularity of the web platforms for hosting challenges and the availability of tools for data processing have made accessibility easier for participants. The Global Wheat Head Challenge (GWHD) (David et al. 2020) is created to check if a solution to detect wheat heads by ML approaches works well in different geographic regions. The GWHD (David et al. 2020) is a challenge with a bias (based on geography) between training and test sets. Europe and North American samples form the training dataset, and Asian and Australian samples form the test dataset. The source code of the three winners is available as open-source, and winners used existing DNNs like EfficientDet (Tan et al. 2020), Faster RCNN, and YOLO v3. Surprisingly there was no domain adaptation done in the solutions. The dataset generalizes well over different architectures, and it is not a determining factor in deciding the winner. It seems that data augmentation techniques, such as mix up (Zhang et al. 2017), mosaic (Bochkovskiy et al. 2020), cut mix (Yun et al. 2019), and pseudo labeling (Lee 2013) are the differentiating factors. The top method using EfficientDet and Faster RCNN along with mix up and mosaic augmentation had an accuracy of 0.68.

Sorghum biomass can be predicted using high resolution in spatial and temporal domains using UAV sensed images (Masjedi et al. 2018). The two methods are adopted in the paper for biomass prediction. The first method used non-linear regression over data obtained using LiDAR and hyperspectral sensors. The second method used a biophysical crop model—Agricultural Production Systems sIMulator (APSIM) (Hammer et al. 2010). The UAV data is parameterized, and then biomass is simulated. Eighteen genotypes of sorghum were planted in rows in a field. Ground truth biomass was acquired using a harvester for all the plots. The method used two nonlinear regressors—Support Vector Regression (SVR) and Multilayered Perceptron (MLP). After threefold cross-validation, the R^2 was 0.48 for SVR and 0.55 for MLP after two months of sowing. The APSIM crop model was estimated using leaf growth rate, which provided $R^2 = 0.67$ for the images captured by the RGB camera.

The OPPD dataset (Leminen et al. 2020) is a public dataset for plant detection and classification. The dataset covers variability in visual appearance as each of the 47 plant species is cultivated under different growth conditions. Baseline experiments for future comparisons are carried out to detect plant instances and then plant classification. The goal for plant instance detection is to locate all the plants in the image. A detection model using Faster RCNN (Ren et al. 2016) with ResNet50-v1 (He et al. 2016) reported that the average precision (AP) computed with Intersection over Union (IoU) variation from 0.5 to 0.95 is 37.01. The plant species identification used ResNet50- v1 (He et al. 2016) and reported an average accuracy of 77.06. The low number indicates a need to build better models over this dataset.

One of the vital tasks in plant phenotyping is counting leaves. The paper (Aich and Stavness 2017) used a leaf counting challenge (Minervini et al. 2016). The challenge was to find several leaves of the rosette plant from the RGB image. The data-driven technique used deconvolution for segmentation and CNN for leaf counting. Thus, the architecture for segmentation and counting is trained separately. The deconvolution network constructed a succinct feature map and then generated class probability maps. The counting network used both the original image and the corresponding binary segmented image for counting leaves. The approach achieved 1.62 as the mean and 2.30 as the standard deviation in an absolute count difference, averaged over all five test datasets.

5 Summary, Challenges, and a Way Forward

The CV and ML pipeline is as follows: Image acquisition—Pre-processing–Segmentatio–nObject localization—Classification chain. Each of the components needs fine-tuning depending on the external conditions. Robust image acquisition techniques have allowed CV to capture data across visible, NIR, IR, or hyperspectral modalities. Appropriate algorithms have to be carefully developed and tuned according to the situation for a robust performance of the above pipeline. Improved signal-to-noise ratio due to better imaging sensors and pre-processing techniques (e.g., noise removal) allows better image analysis. Segmentation is the most crucial step for down-the-chain analysis. Image segmentation allows the breaking up of the image into its constituent parts. Segmentation for an image captured under laboratory conditions is straightforward to obtain. However, when images are captured in open field conditions, it brings in a very high variability (cf. Sect. 2.1). The use of a "Big data" processing pipeline along with appropriate domain-specific augmentation improves results. A summary of ML algorithms using CV is shown in Table 4.

As seen in plant phenotyping, there is significant variability in the physiological attributes of plants. A dataset like GWHD (David et al. 2020) increases the diversity even further when crops from different geographic locations are combined. It may introduce a bias specific to the location in the dataset, which can impact the generalization of the ML methods. The data-driven approaches analyzed in this chapter use deep neural networks. Such networks are susceptible to the amount of data and its quality. It means that sensors have to capture enough images for the deep learning architecture not to overfit. It is significant for the CV algorithm used in PA to generalize well to a multitude of scenarios. The ideas of stratified k-fold cross-validation, domain transfer, and augmentation help to generalize the algorithm. Though CNN has gained prominence in processing agri-data, many applications still use Shallow Neural Networks. They need feature engineering, which is the selection of appropriate features for making decisions. It is important to select features based on the conditions of the dataset. If selected features are highly correlated, then their dimensionality must be reduced using an appropriate transform. It reduces the computational and execution requirements. Support vector machine (SVM), ANN,

Table 4 Summary of ML algorithms on datasets

Problem	Task	Network	Results	Source
Weed detection in Soybean	Segmentation, Classification	CNN- AlexNet	Accuracy 98% (Average)	dos Santos et al. (2017a)
Weed detection in Sugar beets (Synthetic dataset)	Segmentation, Classification	CNN- SegNet	Accuracy – 60%–80%	Chebrolu et al. (2017)
Weed species identification and growth estimation	Segmentation, Classification	CNN	Accuracy 48%–78%	Teimouri (2018)
Weed classification	Segmentation, Classification	CNNs Inception-V3 and ResNet-50	Top—1 Accuracy (Inception-v3) 85.3%–97.6% ResNet—50 88.5%–97.2%	Olsen et al. (2019)
Fruit detection Apple, Mango, Almond	Segmentation, Classification	CNN-Faster RCNN	F1 scores Apple/Mango–0.9, Almond –0.77	Bargoti and Underwood (2017)
Date Fruit classification: Type, Maturity, Harvesting decision	Segmentation, Classification	CNN-VGG 16	Accuracy 99.01% type, 97.25% maturity, 98.59% harvesting decision	Altaheri et al. (2019a)
Mango	Segmentation	CNN YOLOv2 and v3	F1 score 0.968 14 FPS processing	mangoYOLO Koirala (2019)
Wheat head detection	Segmentation	CNN EfficientDet/Faster RCNN and Augmentation	Accuracy 0.68	GWHD David et al. (2020)
Sorghum biomass prediction	Regression	Support Vector Regressor/ANN	R^2 0.48-SVR/0.55 ANN	Masjedi et al. (2018)
Plant detection and classification	Segmentation, Classification	Faster RCNN/ResNet50-v1	Detection Average Precision—37.01 Classification Accuracy—77.06	Leminen et al. (2020)
Rosette plant leaf counting	Segmentation	CNN	1.62/2.30 Mean/SD on count difference	Aich and Stavness (2017)

and RF are good examples of classifiers or regressors, which are helpful on a smaller dataset.

This chapter showed the use of CV in PA by presenting state-of-the-art datasets and their allied techniques. The review of CV systems shows that they have widespread use in PA applications like weed control, fruit detection, phenotyping, grading, harvesting, disease control, and many more. One of the significant advantages of using them is automating the task and performing robust non-destructive analysis. It will significantly improvise the efficiency of many existing farm practices. A CV and ML combination has shown tremendous potential for precision agriculture development. The systems to be developed must meet the need of all stakeholders. As per the PA cycle, the algorithm drives the decision and actual farm practices; therefore, it must be robust to satisfy subsequent handling, transportation, and storage demands. It has been shown that several holes need to be filled for complete automation in PA and its integration with the physical on-farm systems. One may note that CV with AI-driven solutions will improve the robustness of PA automation systems. It will push the development further in using multimodal intelligence within the PA architecture. Such attempts will surely improve food security and quality. The content present in the chapter will motivate more researchers into the domain and apply these concepts holistically while solving open problems. Some of the future challenges to be addressed for realizing the full potential of CV-ML in precision agriculture are as follows.

Domain review suggests that using the existing technology PA has potential for both small and large fields. It can take advantage of low cost to achieve high precision. However, the level of complexity involved in agriculture processes is overwhelming. More so, the automation is limited to only a part of the chain. End-to-end automation is yet a distant dream. CV-ML technology itself is not the solution to all the problems in agriculture, nor is it popular across the board. Vision-based systems are still in infancy in the PA domain. Data collection efforts by individuals will not be able to scale and generalize. The existing datasets which have started appearing in the public domain are also restricted to a single crop or disease. The vision technique still needs a lot of workforce support. Many areas are to be explored by CV-ML technologies for going to a completely automated farm management system. Therefore, it is necessary to build a large-scale dataset and share it in the public domain. It is crucial as the field expands and more challenges will be observed. Open-source reproducible research will help to overcome these challenges quickly.

It is also necessary to focus on developing manpower with a CV-ML in-depth domain knowledge and understanding the finer aspects of agricultural demands. Producing such manpower is going to be a very challenging problem. The emerging domain will need programming, statistical analysis, machine learning, CV, parallel computing, automation, farm practices, and other agriculture aspects. The penetration of PA will push the demands for such skilled manpower (Dong et al. 2019).

The domain of agriculture is full of uncertainty. The sheer scale and range of crops and practices make it even more complex. For example, the practice of growing the same crop will be very different across geography. The different practices will yield very different results. Therefore, when CV-ML is applied to such a complex domain, their robustness and generalization must be severely tested. As newer challenges are added to the environment, the robust solution of today will also need to be tuned to keep pace with the change, e.g., impact due to climate change. It seems almost impossible to create a default workflow for all situations as the lab experiments, when ported to an external farm environment, showcase very different results. Real-time, accurate, and robust solutions will be in demand.

As one of the emerging areas, the use of Augmented Reality (AR) and Virtual Reality (VR) is also penetrating PA (de Oliveira and Correa 2020). The goal for AR is to expand users' perception of their surroundings. On the other hand, VR aims to replace the perception of the real world with the perception of the artificial world. AR heavily depends on the sensitivity of the virtual world when the user interacts with the surroundings of the real world. A stimulus to a user at the right place and time can significantly augment the experience. CV will aid AR by providing an understanding and interpretation of the scene. CV will help VR by creating and simulating intelligent objects in the virtual world. Movement interactions are enabled by cameras that recognize motions and model them.

AR-VR can improve user experience through data visualization and help them to make decisions. Commonly used devices for AR-VR agri-applications are smartphones, glasses, computers, and heads-up displays. AR-VR is used extensively for pest identification, managing machinery used in agriculture, driving autonomous vehicles in or above the farms, assembly and maintenance of machinery, collection of samples, yield estimation, education, and 3D terrain mapping. AR-VR can help in sustainable agriculture development through educational programs and training. They have shown promising results for PA development and management, but they are under-explored technologies. It can open up research directions for new researchers.

Acknowledgements The authors wholeheartedly thank all reviewers for super quick responses and insightful comments. We believe that it has significantly improved the quality of the chapter. The authors would like to thank Ms. Shailaja Sampat for the initiation of the work and interactions. It is imperative to thank all researchers contributing to creating datasets, open-source software, and their contribution to reproducible research.

References

Achanta R, Shaji A, Smith K, Lucchi A, Fua P, Süsstrunk S (2012) SLIC superpixels compared to state-of-the-art superpixel methods. IEEE PAMI 34(11):2274–2282

Aich S, Stavness I (2017) Leaf counting with deep convolutional and deconvolutional networks. In: Proceedings of the IEEE international conference on computer vision workshops, pp 2080–2089

Altaheri H, Alsulaiman M, Muhammad G (2019a) Date fruit classification for robotic harvesting in a natural environment using deep learning. IEEE Access (7):117115–117133

Altaheri H, Alsulaiman M, Muhammad G, Amin SU, Bencherif M, Mekhtiche M (2019) Date fruit dataset for intelligent harvesting. Data brief (26):104514

Arabidopsis Thaliana Root segmentation (2021) https://sites.google.com/sinc.unl.edu.ar/root-segmentation-challenge/home#h.x8m73yix931s. Accessed 10 June 2021

Badrinarayanan V, Kendall A, Cipolla R (2017) Segnet A deep convolutional encoder-decoder architecture for image segmentation. IEEE PAMI 39(12):2481–2495

Barbedo JGA (2013) Digital image processing techniques for detecting, quantifying and classifying plant diseases. SpringerPlus 2(1):1–12

Barbedo JGA (2016) A review on the main challenges in automatic plant disease identification based on visible range images. Biosyst Eng (144):52–60

Bargoti S, Underwood J (2017) Deep fruit detection in orchards. In: Proceedings IEEE international conference on robotics and automation, pp 3626–3633

Bochkovskiy A, Wang CY, Liao HYM (2020) Yolov4: Optimal speed and accuracy of object detection. arXiv preprint arXiv:2004.10934

Chebrolu N, Lottes P, Schaefer A, Winterhalter W, Burgard W, Stachniss C (2017) Agricultural robot dataset for plant classification, localization and mapping on sugar beet fields. Int J Rob Res 36(10):1045–1052

David E, Madec S, Sadeghi-Tehran P, Aasen H, Zhen B, Liu S Guo W (2020) Global Wheat Head Detection (GWHD) dataset: a large and diverse dataset of high-resolution RGB-labelled images to develop and benchmark wheat head detection methods. Plant Phenomics

David E, Ogidi F, Guo W, Baret F, Stavness, I (2021) Global wheat challenge 2020: Analysis of the competition design and winning models. arXiv preprint arXiv:2105.06182

de Oliveira ME, Corrêa CG (2020) Virtual reality and augmented reality applications in agriculture: a literature review. In: Proceedings 22nd symposium on virtual and augmented reality, pp 1–9

Di Cicco M, Potena C, Grisetti G, Pretto A (2017) Automatic model based dataset generation for fast and accurate crop and weeds detection. In: Proceedings IEEE/RSJ international conference on intelligent robots and systems pp 5188–5195

Dobermann A, Blackmore BS, Cook S, Adamchuk VI (2004) Precision farming: challenges and future directions. In New directions for a diverse planet. in: proceeding 4th international crop science congress, pp 1–19

Dong W, Yuanquan C, Daoliang L, Wanbin Z, Weiming T, Taisheng D Shaozhong K (2019) Foresight of disruptive technologies in agricultural engineering. Stra Stu of Chin Acad Eng 20(6):57–63

dos Santos Ferreira A, Freitas DM, da Silva GG, Pistori H, Folhes MT (2017a) Weed detection in soybean crops using ConvNets. Comput Electron Agric (143)314–324

dos Santos Ferreira A, Pistori H, Matte FD, Gonçalves da S, Gercina (2017b) Data for: Weed Detection in Soybean Crops Using ConvNets, Mendeley Data, v2, https://doi.org/10.17632/3fmjm7ncc6.2. Accessed 24 June 2021

FAO (2009) How to Feed the World in 2050. http://www.fao.org/fileadmin/templates/wsfs/docs/expert_paper/How_to_Feed_the_World_in_2050.pdf. Accessed 8 June 2021

Fountas S, Aggelopoulou K, Gemtos TA (2016) Precision Agriculture. In: Supply chain management for sustainable food networks. Wiley, Chichester, UK, pp 41–65

Gaggion N, Ariel F, Daric V, Lambert É, Legendre S, Roulé T, Ferrante E (2020) ChronoRoot: high-throughput phenotyping by deep segmentation networks reveals novel temporal parameters of plant root system architecture. https://doi.org/10.1101/2020.10.27.350553

Gebbers R, Adamchuk VI (2010) Precision agriculture and food security. Science 327(5967):828–831

Hammer GL, van Oosterom E, McLean G, Chapman S C, Broad I, Harland P, Muchow R C (2010) Adapting APSIM to model the physiology and genetics of complex adaptive traits in field crops. J Exp Bot 61(8):2185–2202

Haralick RM, Shapiro LG (1991) Glossary of computer vision terms. Pattern Recognit 24(1):69–93

He K, Zhang X, Ren S, Sun J (2016) Deep residual learning for image recognition. In: Proceedings of the IEEE international conference on pattern recognition, pp 770–778

Huang T (1996) Computer vision: evolution and promise. https://cds.cern.ch/record/400313/files/p21.pdf. Accessed 16 June 2021

ISPA (2019) Definition of precision. https://www.springer.com/journal/11119/updates/17240272. Accessed 8 June 2021

Karkee M, Bhusal S, Zhang Q (2019) Apple dataset benchmark from orchard environment in modern fruiting wall. https://research.libraries.wsu.edu:8443/xmlui/handle/2376/17721. Accessed 16 June 2021

Koirala A, Walsh KB, Wang Z, McCarthy C (2019) Deep learning for real-time fruit detection and orchard fruit load estimation: Benchmarking of 'MangoYOLO.' Precis Agric 20(6):1107–1135

Kuznetsova A, Rom H, Alldrin N, Uijlings J, Krasin I, Pont-Tuset J, Ferrari V (2020) The open images dataset v4. Int J Comput Vis 1–26

Leaf counting challenge (2021) https://data-challenges.fz-juelich.de/web/challenges/challenge-page/85/overview. Accessed 10 June 2021

Lee DH (2013) Pseudo-label: The simple and efficient semi-supervised learning method for deep neural networks. In Workshop on challenges in representation learning, international conference on machine learning 3(2)

Leminen MS, Mathiassen SK, Dyrmann M, Laursen MS, Paz LC, Jørgensen, RN (2020). Open plant phenotype database of common weeds in denmark. Remote Sens 12(8):1246

Liu W, Anguelov D, Erhan D, Szegedy C, Reed S, Fu CY, Berg AC (2016) SSD: Single shot multibox detector. In European conference on computer vision, pp 21–37

Lobet G(2017) Image analysis in plant sciences: publish then perish. Trends Plant Sci 22(7):559–566

Lu Y, Young S (2020) A survey of public datasets for computer vision tasks in precision agriculture. Comput Electron Agric (178):105760

Mahajan S, Das A, Sardana, HK (2015) Image acquisition techniques for assessment of legume quality. Trends Food Sci Technol 42(2):116–133

Masjedi A, Zhao J, Thompson AM, Yang KW, Flatt JE, Crawford MM, Chapman S (2018) Sorghum biomass prediction using UAV-based remote sensing data and crop model simulation. In: International geoscience and remote sensing symposium 2018, pp 7719–7722

Minervini M, Fischbach A, Scharr H, Tsaftaris SA(2016) Finely-grained annotated datasets for image-based plant phenotyping. Pattern Recognit Lett (81):80–89

National Research Council (1997) Precision agriculture in the 21st century: geospatial and information technologies in crop management. Washington, DC, USA: National Academy Press

Olsen A, Konovalov DA, Philippa B, Ridd P, Wood JC, Johns J, White RD (2019) DeepWeeds: A multi-class weed species image dataset for deep learning. Sci Rep 9(1):1–12

Parikh A, Raval MS, Parmar C, Chaudhary S (2016) Disease detection and severity estimation in cotton plant from unconstrained images. In: IEEE international conference on data science and advanced analytics, pp 594–601

Pierce FJ, Nowak, P (1999) Aspects of precision agriculture. Adv Agron (67):1–85

Pierce FJ, Robert PC Mangold G (1994) Site-specific management: The pros, the cons, and the realities. In: Proceedings of the integrated crop management conference, Iowa State University, p 17–21

Redmon J, Farhadi A (2017) YOLO9000: better, faster, stronger. In: Proceedings of the IEEE International Conference on Pattern Recognition, pp 7263–7271

Redmon J, Farhadi A (2018) Yolov3: An incremental improvement. arXiv preprint arXiv:1804. 02767

Redmon J, Divvala S, Girshick R, Farhadi A (2016) You only look once: Unified, real-time object detection. In: Proceedings of the IEEE international conference on pattern recognition, p 779–788

Ren S, He K, Girshick R, Sun J (2016) Faster R-CNN: towards real-time object detection with region proposal networks. IEEE PAMI 39(6):1137–1149

Samuel AL (1959) Some studies in machine learning using the game of checkers. IBM J Res Dev.3(3):210–229

Simonyan K, Zisserman A (2014) Very deep convolutional networks for large-scale image recognition. arXiv preprint arXiv:1409.1556

Sorghum Biomass challenge (2021) https://www.kaggle.com/c/sorghum-biomassprediction/overview/description Accessed 9 June 2021

Stafford JV (1996) Essential technology for precision agriculture. In: Robert PC, Rust RH, Larson WE (eds), precision agriculture. proceedings third international conference on precision agriculture, pp 595–604

Stafford JV (2000) Implementing precision agriculture in the 21st century. J Agric Eng Res 76(3):267–275

Szegedy C, Vanhoucke V, Ioffe S, Shlens J, Wojna Z (2016) Rethinking the inception architecture for computer vision. In: IEEE international conference on pattern recognition, pp 2818–2826

Tan M, Pang R, Le QV (2020) Efficientdet: Scalable and efficient object detection. In: Proceedings of the international conference on pattern recognition, pp 10781–10790

Teimouri N, Dyrmann M, Nielsen PR, Mathiassen, SK Somerville GJ, Jørgensen RN (2018) Weed growth stage estimator using deep convolutional neural networks. Sensors 18(5):1580

United Nations (2019) World Population Prospects 2019. https://population.un.org/wpp/Download/Standard/CSV/. Accessed 8 June 2021

Walter A, Liebisch F, Hund A (2015) Plant phenotyping: from bean weighing to image analysis. Plant methods 11(1): 1–11

Wikipedia—Computer Vision. (2021) https://en.wikipedia.org/wiki/Computer_vision. Accessed 8 June 2021

Yang X, Shu L, Chen J, Ferrag MA, Wu J, Nurellari E, Huang K (2020) A survey on smart agriculture: Development modes, technologies, and security and privacy challenges. J Auto Sinica 8(2): 273–302

Yun S, Han D, Oh S J, Chun S, Choe J, Yoo Y (2019) Cutmix: Regularization strategy to train strong classifiers with localizable features. In: Proceedings of the IEEE/CVF international conference on computer vision, pp 6023–6032

Zareiforoush H, Minaei S, Alizadeh MR, Banakar A (2015) Potential applications of computer vision in quality inspection of rice: a review. Food Eng Rev 7(3):321–345

Zeiler MD, Fergus R (2014) Visualizing and understanding convolutional networks. In European conference on computer vision, pp 818–833

Zhang H, Cisse M, Dauphin YN, Lopez-Paz D (2017). Mixup: Beyond empirical risk minimization. arXiv preprint arXiv:1710.09412

Zhang N, Wang M, Wang N (2002) Precision agriculture—a worldwide overview. Comput Electron Agric 36(2–3):113-132

An Architecture for Quality Centric Crop Production System

Vikas Kumar, Deepak Hiremath, and Sanjay Chaudhary

Abstract Artificial Intelligence (AI) can be applied to find the patterns in historical data and help finding real-time predictions for making data-driven decisions. Digital transformation and automation are required for industries for better growth. AI-predictive analytics software solution that at its core, delivers the ability to recommend or select farm fields, practices, timing and inputs that have a high probability of delivering crops that meet an optimized set of quality attributes. These attributes allow the client to produce food with specific characteristics for competitive advantage and/or financial benefit. The system will allow growers to benefit by allowing "what-if" modeling of farming practices to anticipate desirable crop attributes. In this chapter, we proposed an architecture for developing a recommendation system for helping farmers produce crops with specific quality attributes which are required by the specific consumer. The recommendations can be generated by using the endogenous and exogenous data captured through IoT sensors and AI modeling. The architecture is generic and solutions can be designed to work for a range of foods/crops. The architecture is designed to develop solutions that deliver food producer's a focused value proposition. It uses an Agcentric methodology in creating the capability to recommend the fields for crop production.

Keywords Data science · Classification · Prediction · Crop quality attributes · Analytics framework · Data-driven · Supply chain management

V. Kumar (✉) · D. Hiremath · S. Chaudhary
School of Engineering and Applied Sciences, Ahmedabad University, Ahmedabad, India
e-mail: vikas.tyagi@ahduni.edu.in

D. Hiremath
e-mail: deepak.hiremath@iet.ahduni.edu.in

S. Chaudhary
e-mail: sanjay.chaudhary@ahduni.edu.in

G. P. O. Reddy et al. (eds.), *Data Science in Agriculture and Natural Resource Management*, Studies in Big Data 96, https://doi.org/10.1007/978-981-16-5847-1_6

1 Introduction

The farmers in India are conventionally taking decisions to grow crops based on traditional market requirements, which yield much lesser returns than expected. Lately, few farmers have been slowly shifting to client-based contract farming in many parts of the country. For example, McCain Foods (India), a potato supplier company, has tied up with the farmers across Banaskantha District in Gujarat state to procure potatoes based on specific quality attributes they desire. ITC Ltd., another company is following a similar model with the farmers in the state of Madhya Pradesh. Land quality indicators and the effect of environmental parameters on quality of the product is investigated in (Dutta and Sengupta 2016; Dumanski et al. 2000). KRBL Limited, one of the largest Basmati rice exporters of India procures rice from across various states, i.e. Uttar Pradesh, Uttaranchal, Punjab and Haryana. They undergo strict quality parameter checks at the procurement stage, which are based on moisture content, head rice yield, percentage of broken, admixture, length and width of grain. Other parameters such as soil quality, aflatoxin, microbial and residual tests are also used (KRBL 2021). Reliance Retail in Karnataka state for the procurement of cauliflower, estimates the seasonal demands well in advance before the cropping season and then contacts farmers while advising them on the quality parameters needed by them during the procurement. A complete set of advisory services are designed for them to achieve their intended quality attributes (Velasco et al. 2002).

This chapter proposes a solution architecture for helping agro-consumer clients and farmers to take data-driven decisions based on their requirements. The solution focuses on managing raw materials services, which are provided by the client to the farmers which require a certain specification of a crop produce. The proposed research uses an AI-predictive analytics software solution that is a combination of a decision support system as well as a recommendation system, which helps the clients select farm fields and associated farmers that can deliver crop produce with an optimized set of attributes. The system also guides the farmers to grow such crops with the target of achieving maximum profitability using various specific crop practices according to the client's requirements. The practices can be related to the selection of a crop variety, timing of various practices such as inputs and water schedules during the growing crop cycle, etc. This will help in meeting the optimized set of attributes that allow the clients to produce food with specific characters. The solution is designed to work for a range of foods/crops. While it is clear that the solution delivers a food producer-focused value proposition, it uses a very Ag-centric methodology in creating the capability to recommend the fields.

Information and Communication Technology (ICT) is changing the working culture dramatically and high availability of Internet, mobile devices and intelligent applications are the key enablers. Due to the dearth of information about various new technologies the farmers and other actors across the value chain are handicapped. ICT supporting the development and delivery of timely, targeted information and services can make farming more sustainable. It can also deliver safe, nutritious and affordable food, facilitate market integration and access to finance to make agriculture attractive and profitable (Arshad et al. 2016). The data collected during the different cultivation stages can be harnessed for taking decisions to ensure the quality product.

1.1 Related Work

ICT intervention in the Agriculture domain is becoming a driving force for improving the production and fulfilling the user demands. Research communities are working collaboratively to boost crop yield. High availability of sensors and computing resources has triggered the utilization of AI, Machine Learning (ML) and Data Science (DS) for data harnessing. Data-driven decisions backed by advanced technologies can minimize the risk and losses incurred during cultivation practices. An approach for using Near Infrared (NIR) spectroscopy is proposed to greatly increase the demand for consumers to procure a high-quality product. Accordingly, NIR spectroscopy was used to monitor and quantify the protein, gluten, moisture content and starch in some wheat varieties treated by different nitrogen fertilization rates. Partial least-squares (PLS) analysis was implemented for calibration models. The precision of the prediction model provided by the NIR equations confirms that NIR technology could be a rapid, safe and precision method for wheat quality monitoring and quantification (Sihariya et al. 2013). The use of continuous monitoring using ICT can help finding interesting results or conclusions for improving the product quality. The use of organic fertilizers is essential to attain sustainability in agricultural production. To enhance physicochemical properties of poor-quality soils, compost and charcoal could prove very useful. Compost has started to gather attention in the Indian subcontinent due to the steady increase in prices in fertilizers and also degradation of sold parameters due to excessive use. Therefore, various studies were conducted to gauge the influence of compost and charcoal on wheat yield and soil quality parameters (ICRISAT 2021). Sensors and IoT can be used for finding the results for different physical, chemical, rheological and sensory parameters revealed significant ($p < 0.05$) differences between the cultivars studied. Based on rheological and sensory parameters, it was found to be the most suitable variety for production of chapati (Sharon 1989; Amir et al. 2020). Thus, sensor data can be used to decide regarding crop quality yield to match with the required end users' needs. The effect of terrain properties on quality-related attributes is being studied in (Aguirrezábal 2015). In farming, for sustainable agriculture and landscape management one of the major attributes that need to be considered are soil properties and terrain attributes. The objectives of the study were to predict the grain yield, grain protein, and biomass in wheat grown in hilly regions. The study used an artificial neural network (ANN) model to identify the most important topographic attributes and soil properties affecting the variability of the wheat-yield components under rained conditions in the semiarid and hilly regions.

Various industries with diverse backgrounds require high and good quality of grains for specific uses, seed breeders have brought out new cultivators which are tailored to these specific demands (Velasco et al. 2002). For this, real-time crop management and monitoring is required with a dual objective of increasing high quality and quantity of produce. Plant breeding for improving crop yield is always done at the crop population level, improving quality attributes are mostly concentrated at individual plant level, like grains, flowers, stalks, etc. Most current studies about crop physiology in grains are obtained in wheat, especially used for bread and chapati, and hence illustration in this chapter, wheat and their breeds are being used for quality and quantity-based studies (Aguirrezábal 2015). Examples have been presented in the chapter how various environmental and chemical inputs affect given quality traits as a result affect the overall production and acceptability of the produce. Mathematical crop models using Big Data and ML are illustrated showing how they can be incorporated in quality modules that allow simultaneous simulation of grain yield and quality. For example, a crop model has been illustrated to show the relationship between environmental and topographic factors with that of quantity and quality. Tweaking of resources, with the production and changing agricultural practices is necessary to improve efficiency in agriculture. A better way to understand the relationship between the plant response to the environment can help improve desirable quality as well as quantity of the produce while improving resource-use efficiency. Dealing with variety by environmental interactions is one of the major concerns for producers. The chapter tries to find methods on how Big Data, AI, ML can be incorporated.

In the literature survey, it was found that continuous real-time monitoring of different environmental parameters not only helps derive new results but also is very useful to make the right decisions at the right time. The success in research efforts provided the motivation to design an ICT-enabled architecture to handle the Big Data captured and facilitate the crop yield quality estimates.

1.2 Problem Statement

The quality attributes of any Agro-products are highly dependent on field, weather, seed variety and cultivation practices. Real-time continuous monitoring of crops and right decisions at the right time ensures the desired quality in the product. The relationship between the quality attributes and different environmental parameters is too complex to derive using analytical approaches. The complex relationships and their effect can be learned by using ML techniques. AI and data science can help make data-driven decisions and root cause analysis for growing Agro-products with desired attributes. Currently, there is a huge gap between farmers and clients (i.e. food processing companies), thus a platform is required to bridge the gap between farmer's produce and desired market demand. Using the best cultivation practices and real-time monitoring of different growth stages of crops the quality of product can be ensured. The quality product will be profitable to both the farmers as well as the consumer. Farmers will get better quality yields and consumers will get the product to match their requirements.

Mathematically, we can represent the quality of a crop product Qc as the function of different parameters x_i having complex dependencies and causal effects. Our objective will be to find the optimal parameters for getting the desired quality products. Using ML approaches, we can find the relationships between the parameters and their effect on quality yield.

$$Q^c = f(x_1, x_2, x_3, \ldots, x_n) \quad (1)$$

Since there exists non-linearity between the predictors and target quality function therefore no analytic method can be used for ensuring the product quality and controlling the environmental parameters. In such a situation, only ML algorithms can help to facilitate pattern recognition and help making data-driven decisions to control the variables. In order to apply mathematical modeling for yield estimation, we need to have a platform where data can be captured through IoT-enabled sensors in real time. Our objective is to propose an ICT architecture for real-time monitoring and surveillance of crops so that the system can control the variables in such a way to maximize the quality of the product and provide benefits to the farmers as well as the end users. Following challenges can be addressed using the proposed architecture:

- Field selection based on the requirement of the quality of end product.
- Real-time crop environment monitoring to control the adverse effect.
- Predictive modeling for expected quality of product based on the current state of crop in the field.

1.3 *Objective*

The objective is to develop an IoT, Big Data Analytics and AI-based solution for helping the farmers and other agricultural stakeholders to make data-driven decisions. The solution minimizes the risk, cost and ensures the quality of the crop produced. The main objectives can be listed as follows:

- Development of an integrated platform for the real-time crop monitoring and visualizing the Key Performance Indicators (KPI) for sustainable quality yield production.
- Sensor data will facilitate real-time crop monitoring and different variables affecting the crop can be visualized on Interactive dashboards.
- Effect of dynamic environmental conditions on product qualities can be learned by using state-of-the-art classification algorithms. If classes are modelled as quality grades then the problem becomes a multi-class classification problem.
- Cultivation field selection for expected quality of agro product.
- Use of image processing for disease classification and alert generation.

2 Proposed Architecture

We are addressing the problem of ensuring the quality of agro-produce based on the real-time monitoring of cultivation practices starting from field selection to harvesting. The proposed solution facilitates data-driven decisions for increasing the profits of involved stakeholders and provides a dashboard for KPIs visualization. In this chapter, we are proposing an IoT, Big Data Analytics and AI-based solution enabling the data-driven decisions for ensuring the quality of agro-produce and hence will increase the profit of farmers and consumers. Different components of the architecture are shown in Fig. 1. The platform facilitates data integration, storage processing and analytics. The architecture is having loosely coupled microservices, which can be deployed and tested independently. The different components of architecture are explained in the following subsections:

2.1 Data Sources

The quality attributes in agro-product depend on field characteristics, real-time data captured through IoT sensors, weather and many exogenous parameters. We propose a data acquisition pipeline for data integration and prepossessing (Chaudhary 2018). For data acquisition, we need to install different kinds of sensors, i.e. temperature, soil moisture, nitrate, etc., in the field and these sensors will send the data at a certain interval. If precision agriculture is in place, then we will already have a data

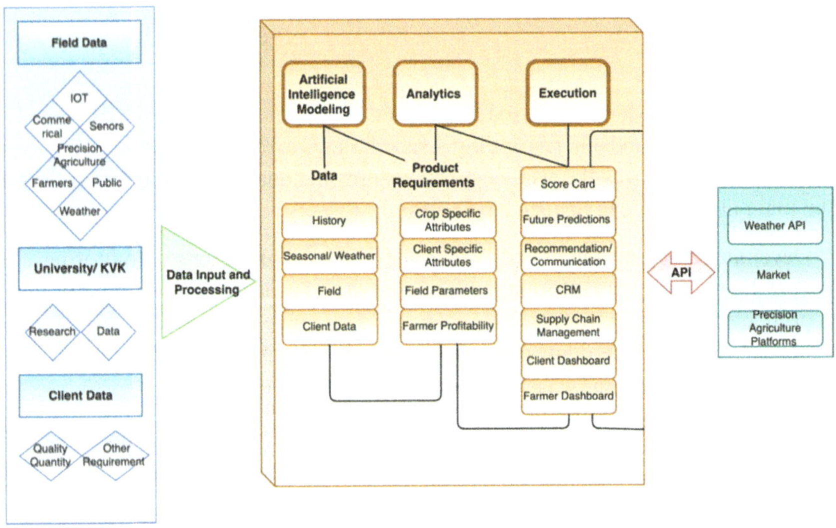

Fig. 1 Proposed Architecture for solution development

farm in place and that data can be sent to the server. The other data sources may be weather API, drone or satellite images. This data can be used as the input for the analytic engine. Different components of the architecture are described in the following subsections.

2.2 *Data Availability, Association and Exploitation*

Nowadays, data is being produced at a large scale and precision agriculture is a rich source. We can integrate the data collected from the fields with the weather APIs data and knowledge bases of agricultural universities. Available data sources, which can be used for data collection and Industry 4.0 intervention in agriculture are Lidar, Soil Texture Data, Location Data, Clustering Data, Satellite Data, Crop Rotation Data, Field Shape and Size Data, Border Data, Headland Data and Slope Data.

2.3 *AI, ML, Data Science and Analytics Engine*

The real-time data captured through IoT sensors and from other sources can be harnessed for finding hidden patterns and generating recommendations helping farmers and clients to mitigate risks. The analytic engine will ingest the streaming data and forecast the status of crop periodically using which a farmer will be able to make decisions for minimizing the losses due to quality mismatch of product. AI/ML models will help farmers grow the agricultural produce with desired quality by guiding them during the entire life cycle of the crop and thus will enhance their profit earning and reduce risk.

Crop cultivation-related data can be captured using sensors in real time and we will have huge amounts of data in different formats for analytics. Tremendous advancement in data handling and analytics technologies during the last decade has increased the innovative applications across a wide range of domains. AI can play a major role in decision-making regarding control of different parameters affecting the quality attributes of a farm product. In our proposed analytic engine, we use machine learning techniques to achieve AI-driven decision making. In the following subsections, the basic understanding of technologies proposed in the solution architecture were explained.

2.3.1 Artificial Intelligence (AI)

AI is defined as "the theory and development of computer systems able to perform tasks normally requiring human intelligence, such as visual perception, speech recognition, decision-making, and translation between languages" (ICRISAT 2021; Sharon 1989). The high availability of data captured through sensors and the Internet can be exploited for demand analysis of an agro-product. Artificial intelligence can play a very important role in quality-centric agriculture for making decisions at every stage of cultivation practices. Implementation of AI can be done on top of huge amounts of data captured through different IoT devices deployed in fields. The data from heterogeneous sources can be integrated into the developed platform. Once the data is integrated then different kinds of ML algorithms can be used for classification, segmentation, prediction, forecasting and root cause diagnosis.

2.3.2 Machine Learning (ML)

Machine Learning is the use and development of computer systems that can learn and adapt without following explicit instructions, by using algorithms and statistical models to analyse and draw inferences from patterns in data. Different kinds of classifications and regressions problems can be solved using ML algorithms. An example use case of ML is to classify the field as 'suitable' or 'not-suitable' for growing wheat crops based on the environmental and chemical variables. In this use case, supervised machine learning algorithms can be applied. A supervised algorithm works with historical labeled data in which we have labeled class associated with the variable vectors. In supervised algorithms, we split the whole data into two development and test sets and develop the pattern recognition algorithms for classification. Such techniques help to take data-driven decisions using historical data.

ML algorithms are data-hungry algorithms and proved to work accurately in learning complex patterns which cannot be identified using analytical techniques.

2.3.3 Data Science (DS)

Data science is related to statistical analysis and pattern finding by using correlation and variance analysis. These techniques are very helpful for feature engineering and to make decisions regarding parameters selection for solving real-world problems. Data science methodologies include statistical, mathematical and parameter estimations. These techniques help a user to understand the effect of a variable on the target variable. These techniques are the foundation for implementation of ML algorithms and are part of AI implementation.

2.3.4 Big Data Analytics

Data captured through IoT sensors in real time is streaming data and can be captured at a very high frequency. Thus, data will have Volume, Variety and Velocity (3 V) properties and hence will need Big Data Analytics (BDA) Infrastructure to store and process the data. In our architecture, we store and process the data on an open-source BDA framework and will perform AI/ML-based processing and will produce the KPIs using a user-friendly dashboard.

2.4 Application Programming Interface (API) Integration

API is a computing interface that defines interactions between multiple software or mixed hardware-software intermediaries and essentially have the following properties:

- It defines the kinds of calls or requests that can be made,
- How to make them,
- The data formats that should be used,
- The conventions to follow, etc.
- It can also provide extension mechanisms so that users can extend existing functionality in various ways and to varying degrees (Velasco et al. 2002).
- An API can be entirely custom, specific to a component, or designed based on an industry standard to ensure interoperability.
- Through information hiding, APIs enable modular programming, allowing users to use the interface independently of the implementation.

In our proposed solution, we can make use of existing agriculture domain-related data APIs for data integration and will build a data lake for storing big data. The decision regarding cultivation practices depends not only on field parameters but also on weather, climate and market demand of products. Therefore, besides the field characteristics data, different APIs for collecting weather, precision agriculture and market demand APIs will be integrated in the Analytics engine.

2.5 Implementation Strategy and Use Cases

We are preparing the data collection and quantifying the association in order to find the probability of the extent to the particular variety is suitable for the selected field. Field specifications data and advantaged raw material attributes are proved to be highly correlated. For example, wheat varieties cultivated for the purpose of Chapati or bread (GW 503, GW 190, GW 173, GW 496 and Lok 1, etc.) are very different from those cultivated for making Rava (Semolina) or Maida (Refined flour) (GW 1 and A 206). Their growing methods, package of practices and quality control mechanisms are also very different. Our implementation strategy will include the following steps:

1. Data collection, integration and curation

    a. Lidar, Soils Textures, Soil Quality, Location Data, Clustering Data, Satellite Data, Crop Rotation Data, Field Shape and Size Data, Border Data, Headland Data, Slope Data for generating the recommendation regarding field selection for crops having specific quality.

2. Exploratory Analysis
   a. Data Integration, correlation analysis, variance and distribution analysis
3. Association rule mining
   a. Inferencing and root cause analysis
4. Prediction and forecasting KPI development
5. Surveying the qualities of Wheat
6. Generalization of system for any crop
7. Prototype development using open source technology stacks.

2.6 Use Case for Quantity Yield Prediction and Variety Recommendation (Sample Use Case for Wheat Seed Yield Prediction)

The use case is taken from the contribution made by the author Anubhav Rana on a GitHub repository. In this example, we showcase the use of ML for the prediction of optimal seed quantity required. The problem is a regression problem and helping to make seed quantity-related decisions based on integrated data from different sources (Rana and Anubhav 2018).

Improving the procedure for estimating the total amount of seed that needs to be produced can save farmers and producers millions of dollars. By building a predictive model for the dry yield (how much useful product was harvested/extracted) of crops in India, a procedure has been developed to forecast the dried yield per acre of a farm. Recommending specific high yield varieties to locations within the region was another method that could be used to improve the efficiency, productivity and quality of the produce in the farm. The ideas behind this methodology can be applied to other crops and can be scaled to bigger regions.

2.6.1 Business Understanding

$1330 million is the value of the total seed produced in the country. So, a variability of 1.5% in the projected production can result in a loss of about $19.95 million for a surplus. Minimizing this uncertainty can be vital especially when scaling up to bigger regions.

The problem is to estimate the amount of seed that will be required for regions in India and by building a predictive model for the dry yield per acre, the total seed production can be better estimated. Recommending specific high yield varieties for the different locations and providing recommendations for each variety are few of other ideas that can be incorporated into the solution.

2.6.2 Data Understanding

Crop Data

Anonymized crop data from seed companies and research institutes, which contains the yield for different varieties with various crop attributes at different locations.

Weather and Location Data

Location data which includes datasets that are not anonymized was used to obtain weather data based on the date when the variety was sown. Other parameters such as latitude and longitude data as well as temperature data over the period of the cropping season can be used. Further terrain and soil-based parameters like altitude soil composition and soil data can also be used as additional parameters.

Data Preparation

The Gross Yield was removed by converting the Dried Yield Per Acre (Dried Yield/Standing Area). The Sown and Harvest dates were combined to create the Days till Harvest. This method is used to find the total production in a perceived time period. Other quality attributes can be combined for quality like moisture content which should be around 10–12% ideally.

2.6.3 Modeling

Following inference and recommendations can be generated:

- Regression for a better understanding of the coefficients to recommend the varieties.
- Matrix Factorization—Singular Value Decomposition (SVD): Matrix factorization was attempted to try to recommend varieties to particular locations for which data was not available. Using the dry yield as a 'rating', varieties were recommended and compared to the regression recommendations. This was not entirely successful as domain knowledge along with other factors is very influential in this area. There are biological and agricultural reasons for specific varieties being planted at particular locations and this matrix factorization technique would struggle to incorporate that (Figs. 2, 3 and 4).

Prediction models:

After performing the feature engineering, we performed model selection by applying different regression techniques and found that the following two ensemble techniques are performing best in such cases:

- Random Forest Regressor.
- Gradient Boosted Regressor.

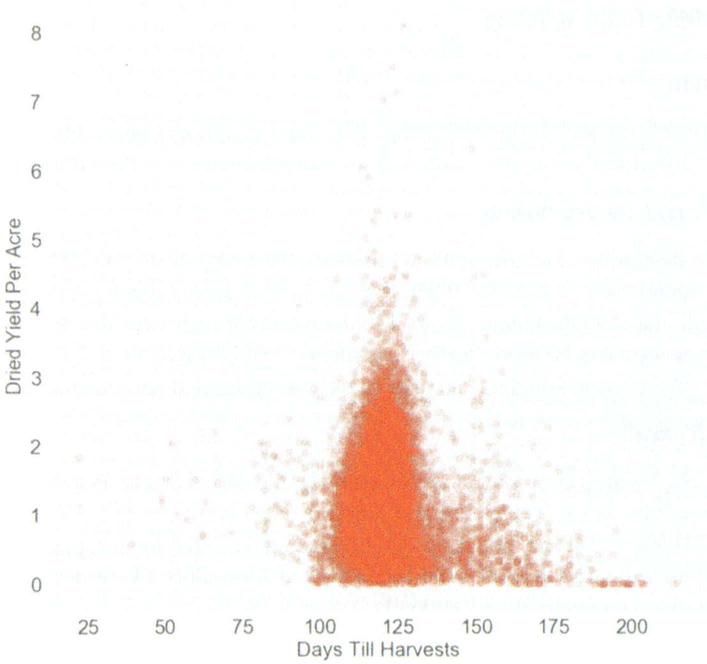

Fig. 2 Days till Harvest vs Yield (Rana and Anubhav 2018)

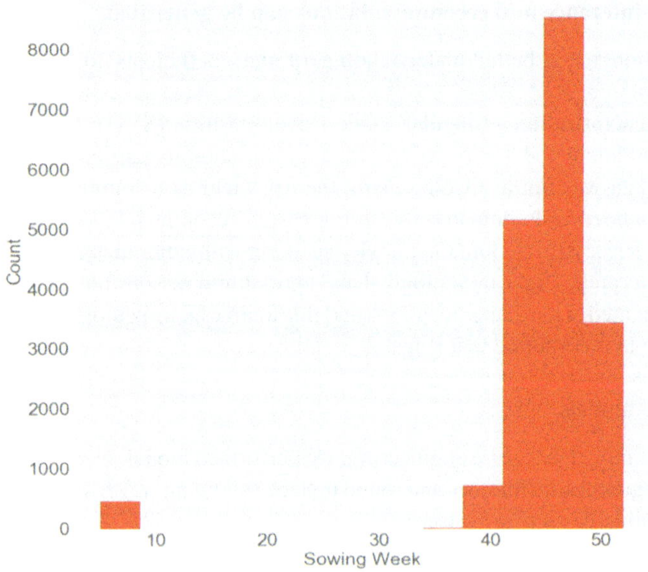

Fig. 3 Sowing Week of Year Histogram (Rana and Anubhav 2018)

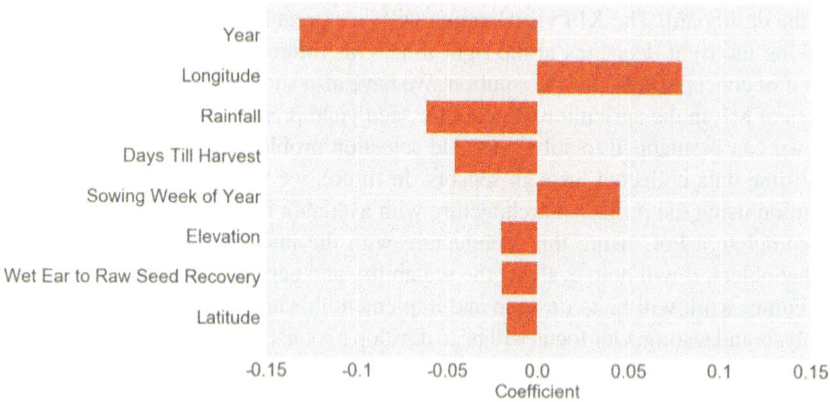

Fig. 4 Top 8 Features (Rana and Anubhav 2018)

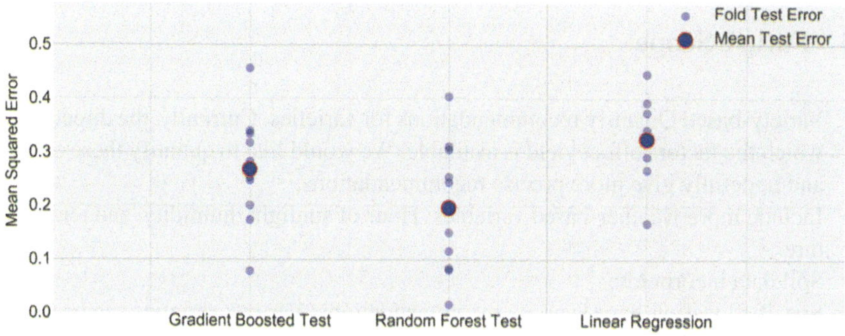

Fig. 5 Testing Errors (Rana and Anubhav 2018)

Evaluation:

The performance evaluation for different models is done using 10-fold cross-validation and we found that polynomial models did not perform very well. The mean error for different models is shown in Fig. 5.

3 Summary

In this chapter, an architecture was proposed for development of a platform for data integration, processing and analytics. The platform having all the components as explained in different sections can help storing and processing real-time data. A real time huge amount of data can be utilized for mining key insights. The insights derived from data through AI, ML and data science algorithms can be visualized on the dashboard. The KPI visualization will be a great help to the stakeholders for making the right decisions at the right time. Our future work will be to develop a proof of concept (PoC) for the solution. We have also shown one of the potential use cases of ML in the agriculture domain for seed yield prediction. The initial use case shown can be mapped to solve the field selection problem based on historical and real-time data collected through sensors. In future, we will develop an end-to-end solution using the proposed architecture with available input data. We are planning to complete a PoC using this architecture with the integrated data from multiple stakeholders, it will help evaluate the scalability and performance.

Future work will be to develop and implement this architecture for performance analysis and testing. Our focus will be to develop a robust architecture for the agriculture domain facilitating data integration, processing, analytics and harnessing using state-of-the-art technologies. Currently, the architecture is in ideation stage and our goal is to make a prototype and then a product.

4 Future Scope

- Variety-based Quantify recommendations for varieties: Currently, the direction in which the factors affect yield is available. We would like to quantify these effects and hopefully give more precise recommendations.
- Include more weather-based variables: Hour of sunlight, humidity, and temperature.
- Soil data incorporate.
- Specific location-based variety recommendations: Various varieties can be tested in different locations, various recommendations can be made for these varieties in those locations where data was unavailable before.
- Attempt to use mixed models (treating some predictors as random variables).

Acknowledgements We acknowledge the contribution made by Anubhav Rana on GitHub repository, we also thank the editors for encouraging us to present our proposed idea for AI-enabled agriculture practices.

References

Aguirrezábal L et al (2015) Improving grain quality: ecophysiological and modeling tools to develop management and breeding strategies. Crop physiology, 423–465

Amir RM et al (2020) Comprehensive assessment and evaluation of selected wheat varieties for their relationship to chapatti quality attributes. Food Sci Technol 40, 436–443

Arshad W et al (2018) Fatehjang-2016: a high yielding and rust resistant wheat (Triticum aestivum L.) variety for rainfed areas of Punjab. J Agric Res (03681157) 56(3)

Chaudhary S, Kumar V (2018) Service oriented architecture and spatial data integration for agro advisory systems. Geospatial infrastructure, applications and technologies: india case studies. Springer, Singapore, pp 85–199

Dumanski J, Pieri C (2000) Land quality indicators: research plan. Agric Ecosyst Environ 81(2):93–102

Dutta A, Sengupta S (2016) A case study of Pepsico contract farming for potatoes. IOSR J Bus Manag, 75–85

Elevation data, https://www.whatismyelevation.com. Accessed. 01 June 2021

Fisher, Sharon. OS/2 EE to Get 3270 interface early. Google Books (1989)

Latitude and Longitude, https://www.findlatitudeandlongitude.com/. Accessed 01 June 2021

ICRISAT. ICT/Digital agriculture, https://www.icrisat.org/ict-innovations. Accessed 01 June 2021

Imran NJ, Ashfaq A, Asad AK (2017) Grain yield, yield attributes of wheat and soil physico chemical characteristics influenced by biochar, compost and inorganic fertilizer application. Agric Res Tech Open Access J. https://doi.org/10.19080/artoaj555795

India Meteorological Department, Ministry of Earth Sciences. Customized Rainfall Information System (CRIS) Hydromet Division, http://hydro.imd.gov.in/hydrometweb/(S(trppmyyq14aajr45iopnqsmi))/DistrictRaifall.aspx. Accessed 01 June 2021

KRBL. KRBL's role in the procurement of paddy, http://krblrice.com/procurement.html. Accessed 01 June 2021

Norouzi M et al (2010) Predicting rainfed wheat quality and quantity by artificial neural network using terrain and soil characteristics. Acta Agriculturae Scandinavica Section B–Soil Plant Sci 60(4):341–352

Rana A (2018) Yield prediction and variety recommender, https://github.com/anubhavrana/Capstone-Yield-prediction-and-Variety-recommender. Accessed 01 June 2021

Sihariya G, Hatmode VB, Nagadevara V (2013) Supply chain management of fruits and vegetables in India. Proceedings of the Tenth AIMS international conference on management, Bangalore, Indian

Somers DJ et al (2007) Genome-wide linkage disequilibrium analysis in bread wheat and durum wheat. Genome 50(6):557–567

Velasco L, Fernández-Martínez JM (2002) Breeding oilseed crops for improved oil quality. J Crop Product 5(1–2):309–344

Velasco, L., et al. "Genetic and environmental variation for tocopherol content and composition in sunflower commercial hybrids." The Journal of Agricultural Science 139.4, 425 (2002)

Crop Classification for Precision Farming Using Machine Learning Algorithms and Sentinel-2 Data

Jay Prakash Kumar, Dimple Singhania, Sanket Nayankumar Patel, and Mihir Dakwala

Abstract Accurate and timely monitoring of crops can help in better agriculture management. Remote sensing-based crop monitoring helps to track the real-time crop condition. Small farm holding, intercropping, multi-cropping, inadequate irrigation infrastructure, and other challenges are quite common to Indian farms. Satellite-based crop monitoring uses pixel-based analysis poses many challenges as the majority of Indian farms are smallholdings. Hence, to overcome these challenges a novel method has been proposed based on image processing, and machine learning (ML) algorithms in this chapter. In the study, Morbi or Morvi district of Gujarat state, India has been selected, where majority of holdings are small and marginal with cultivation of cotton, cumin, wheat, bajra, sesamum, and groundnut as major crops. In the study, optical satellite data from Sentinel-2 was used to delineate farm boundaries using standard deviation and perform crop classification by using Random Forest (RF) algorithm with approximately 85% accuracy and an F1-score of 0.84. Open-source Geographic Information System (GIS) platform QGIS was used to perform post-processing operations to handle raster, and vector data analysis. The proposed method can be applied to other areas where different farm holding sizes and multiple crops are grown.

Keywords Classification · Farm boundary · Object-based precision farming · Random forest · Sentinel-2 data

1 Introduction

Agricultural activities of a region need to be monitored for planning and decision-making to meet the market demand. These activities can be carried out either manually by inspecting and keeping records of information collected from farmers for each farm field or using remote sensing as a tool to monitor phenological changes, crop health, land use, and land cover details of each farm field without manual efforts

J. P. Kumar (✉) · D. Singhania · S. N. Patel · M. Dakwala
Amnex Infotechnologies Pvt. Ltd., Ahmedabad, Gujarat 380054, India
e-mail: jay4@amnex.com

G. P. O. Reddy et al. (eds.), *Data Science in Agriculture and Natural Resource Management*, Studies in Big Data 96, https://doi.org/10.1007/978-981-16-5847-1_7

in a shorter time. Precision Agriculture (PA) is farming management based on information and technology that helps to detect, understand, and manage spatio-temporal differences within farms for optimizing productivity and reducing resources (Hakkim et al. 2016). Remote sensing satellite images can give detailed, up-to-date information regarding land use and agricultural patterns through their high-resolution sensor capabilities that can be difficult to gather manually with much precision. Since agricultural policies are carried out at field level, the precise boundaries of farms assume a greater importance in PA. However, surveying fields and farmers would be time-consuming and will not be accurate to a certain extent in the case of large areas. Manual interpretation of remote sensing images is a time-consuming and labor-intensive task (North et al. 2018). This necessitates developing an innovative, automated, accurate, and practical method that can work on large areas like the whole district in effectively generating farm boundaries by using high resolution satellite data to keep farm boundary data up-to-date (Maxwell 2010). Once the farm boundary is obtained, each farm can be classified or labeled with the help of ground control points by using suitable ML techniques.

Many researchers are working in this field of research to overcome the time and manual efforts that were needed to generate crop boundaries and identify crops at the farm level by using satellite images. Different methods were formulated to generate field boundaries from satellite imagery automatically. Different methodologies were used to delineate farm boundaries such as Crop Field Edge Probability Map generation, Candidate Agriculture Field Extraction and Linear Dimensionality Reduction (DR). Missing data and hyperspectral data non-availability were the main limitations to these methods (Yan and Roy 2015). In another study, farm boundaries were delineated by using Unmanned Aerial Vehicle (UAV) by applying feature extraction technique. This method has limitations such as over-estimation without preprocessing and other challenges (Crommelinck et al. 2016).

Pre-processing techniques to create BOA-corrected (Bottom-of-Atmosphere) reflectance images, and classification were performed using Time-weighted Dynamic Time Warping (TWDTW), and for assessing the classification accuracies obtained by Pixel-based TWDTW (PB-TWDTW) and Object-based TWDTW (OB-TWDTW). High intra-class variability and phenological cycle changes were major challenges in this method (Belgiu and Csillik 2018). Delineation of farm boundaries using edge detection and image segmentation was performed. This approach also had limitations as single-time stamp data was used (Watkins 2019). Wang and Boesch (2007) delineated forest boundaries by using the JSEG (J-measure-based SEGmentation) algorithm. The method has major drawbacks in finding thresholds on distinct illumination distributions. In another study, farm boundaries were generated by using Multi-Resolution Segmentation (MRS), Contour detection (CD) by using the globalized probability of boundary (gPb) but comes with over-segmentation and under-segmentation limitation (Marshall et al. 2019).

Delineation of farm boundaries was carried out by using Principal Component Analysis (PCA), Segmentation, Edge Detection filters, Normalized Difference Vegetation Index (NDVI) or Soil Adjusted Vegetation Index (SAVI). This method also fails where the landscape is heterogeneous (Davidse 2015). Research performed by using cadastral boundary feature extraction on high-resolution satellite images but poses a challenge with invisible farm boundaries (Wassie et al. 2018). Agriculture parcels delineation is carried out by using a texture-based algorithm but results depend on the input parameter selected (Alemu 2016). In another method, delineation of agriculture parcels by using super-pixel processing and RUSBoost algorithm was implemented. This also has the challenge to determine the optimal features for training (Garcia-Pedrero et al. 2017). Standard deviation methods automatically delineate farm boundaries from multi-temporal satellite (SPOT satellite) data that covers varying cropping patterns and phenological cycles, which gives better results. This method faces challenges while working with small farm holding (North et al. 2018). Object-based crop classification by using different ML models was performed but model selection totally depends on the balance between the improvement in the accuracy and computational resources or expertise required for model interpretation (Peña et al. 2014). Farm parcel or parcel-based approach of crop classification using Random Forest (RF) has increased the classification accuracy rather than pixel-based approach (Ok et al. 2012). A detailed overview of the work carried out by different authors is summarized in Table 1.

The present study aimed to create an end-to-end pipeline to generate farm boundaries and perform crop classification by using multi-temporal optical satellite data for any agricultural cropping situation. Overall objectives of the study are firstly to develop a method that can delineate farm boundaries or farm parcels and cover farms of all shapes and sizes irrespective of cropping pattern and the phenological cycle of the crop, secondly, perform crop classification to label each farm boundaries by using ground control points and develop end to end pipeline for farm boundary delineation and crop classification on commercial aspects.

2 Study Area

The study area is located at 22° 82′ N latitude and 70° 83′ E longitude in Morbi (or Morvi) District of Kathiawar peninsula of Gujarat state (Fig. 1). Morbi district is primarily an agricultural district with cotton, cumin, and wheat as the predominant crops. The other major crops cultivated are bajra, sesamum, groundnut, etc. The elevation of Morbi lies 48 m above mean sea level (MSL). The prevailing climate in Morbi is known as a local steppe climate. The average annual temperature of the district is 27.0 °C. The mean annual rainfall of Morbi is 521 mm. The farming patterns are irregular, the majority are small in size and frequently change cropping patterns.

Table 1 Overview of different methods implemented and results obtained

Methods implemented	Results achieved	Summary	References
Optimal parameter setting, Field Boundary Delineation, Crop Row detection	The ratio of missing is 0.78 and the ratio of false detection is 0.73	Achieved a reasonable result in identifying crop row detection and can further be improved by fast Fourier transform This algorithm finds linear features and rectangular shapes. For other types of shapes and features, algorithms can be modified accordingly	Alemu (2016)
The method applied for classification on pixel-based and object-based were Time-weighted Dynamic Time Warping	In three study areas, the object-based TWDTW method performed better than the pixel-based TWDTW method and overall accuracy was between 78.05 and 96.19%	The pre-processing techniques were used to create BOA-corrected (Bottom-of-Atmosphere) reflectance images. The classification was done using Time-weighted Dynamic Time Warping then the accuracies were derived for Pixel-based TWDTW (PB-TWDTW) and Object-based TWDTW (OB-TWDTW). High intra-class variability and phenological cycle changes were major challenges in this method	Belgiu and Csillik (2018)
The semi-automatic method applied by walking the boundary with handheld GPS & imaging using UAV platform, and manual on-screen digitizing	The output of the method was confirmed using other data sources and processes. The results give a useful objective to already existing practices	The image segmentation process is not accurate, due to the heterogeneity of the landscape	Davidse (2015)
This methodology makes use of superpixels and supervised classification to transform segmentation issues in machine learning problem statements	The Visual representation of two high-resolution images shows promising results	A novel methodology has been developed to generate superpixels and determine which adjacent superpixels should be merged using supervised classification, showing better results	Garcia-Pedrero et al. (2017)

(continued)

Table 1 (continued)

Methods implemented	Results achieved	Summary	References
Farm boundaries were decided using crowd-driven manual digitization tools (CropLandExtent), contour detection, and multi-resolution segmentation	Quantitative completeness and spatial correctness were used to assess the used methodology	Manual farm digitization using a large number of people is promising but it does involve inaccuracy, high-quality control, and training	Marshall et al. (2019)
Extended directional edge filters were convolved on the stacked images generated using the local standard deviation of time series data	One-to-one, under segmentation, over-segmentation, many-to-many have been visualized for evaluating results	This process is applicable to large areas and can be automated	North et al. (2018)
Classification done using single classifiers and combined in hierarchical classifiers, the C4.5 decision tree, logistic regression (LR), support vector machine (SVM) and multilayer perceptron (MLP) neural network methods were assessed	Slight variations were found between the single and hierarchical classifications while calculating overall efficiency	The classification of summer crops was outperformed by SVM and MLP in contrast to the C4.5 decision tree but a higher computational time and model complexity were observed	Peña et al. (2014)
The forest delineation was suggested using a combination of image segmentation and wavelet-based texture features	The efficiency of the model was found to be 0.93 and 0.81	Automatic forest boundary delineation was performed using the JSEG method and experimentally found threshold handles different illumination conditions successfully. The method needed to optimize further for distinct lightness distributions conditions	Wang and Boesch (2007)

(continued)

3 Methodology

3.1 Dataset Used

In the study, Level-2A atmospherically corrected cloud-free imageries of Sentinel-2 satellite was used, which was obtained from Copernicus open access hub, an open

Table 1 (continued)

Methods implemented	Results achieved	Summary	References
WorldView-2 satellite images were used to extract cadastral boundaries in rural areas using the Mean-shift segmentation plug-in in QGIS	Buffer overlay methods were used to quantitatively analyze the extracted boundaries	Cadastral features that are based in a nearly non-vegetated terrain and having land covers of different reflectance near boundaries, it is observed that the mean-shift segmentation approach is more helpful in extracting boundaries. Also, some boundaries were still not extracted due to identical reflectance on adjacent parcels	Wassie et al. (2018)
Multi-temporal Sentinel-2 satellite images were used to delineate farm boundaries using edge detection, image segmentation, and knowledge-based crop classification	The results show that the two highest overall accuracies (82.9–83.6%) of the canny edge with watershed segmentation methodology are recorded. The lowest Overall Accuracy 61.3% was obtained when the per-pixel approach was applied	Watershed workflow exceeds other methods used in this research	Watkins (2019)
Missing data for multi-temporal satellite data were filtered using Laplacian Eigenmaps (LE) nonlinear Dimensionality Reduction (DR) algorithm and a series of random forest classifications were performed using these data	It has been observed that the accuracies obtained using the Laplacian Eigenvalues Spectral angle mapper Refined (LE-SAM-R) DR band were better than the conventional metrics-based approach	DR algorithms cannot handle missing data	Yan and Roy (2015)

and free data hub for Sentinel-1, Sentinel-2, Sentinel-3, and Sentinel-5P images. Five scenes of Level-2A data of Sentinel-2 during *rabi* season (November–February) were chosen in order to cover different phenological growth stages of crops. Differences in the gap of scenes collection were maintained to cover the phenological changes of the crop. 10-m-resolution red (R), green (G), blue (B), and NIR (NIR) bands of Sentinel-2 were used in the study to capture the detail changes. The dimensions of each tile were 2399 m wide and 1523 m in height. Details of Sentinel-2 satellite data

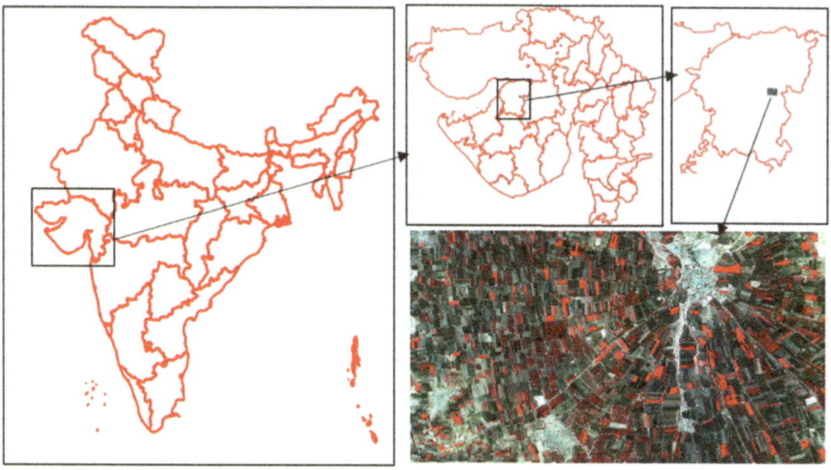

Fig.1 Study area depicted on Sentinel-2 False color composite image

used in the study are shown in Table 2. Each band (R, G, B, and NIR) for a single time is shown in Fig. 2.

After the generation of farm boundaries, in order to label each farm, an adequate number of ground truth samples were required. In the study, ground truth points were collected between November 2018 to March 2019 by using a handheld Global Positioning System (GPS) with positioning accuracy of ±2 m. Normally distributed sample points for jowar fodder, maize fodder, coriander, cumin, onion, vegetable, wheat, and gram were collected during the field survey.

3.2 Satellite Data Analysis

The acquired temporal Sentinel-2 data was processed in order to get meaningful information on farm boundaries by applying object-based crop classification techniques. The workflow followed in the study is described in Fig. 3.

3.2.1 Pre-processing

Images of a particular region, at varying time periods, were considered to capture phenological changes of crop farms. This helps in capturing multiple crop fields at any given time and identifies farms of long-duration crops. The images of each time are stacked as a single image of dimension (row, column, channels), where channels comprise all the four bands (R, G, B, NIR) of five time periods in this case.

Table 2 The characteristics of Level-2A data of Sentinel-2 Satellite used in the study. (Band wavelength values based on data shown in ESA Sentinel-2 https://sentinel.esa.int/web/sentinel/missions/sentinel-2/instrument-payload/resolution-and-swath)

Date	Bands	Dimension (Width, Height) (m)	Central wavelength (nm)
25 Nov 2018	Red, Green, Blue, and Near-Infrared	2399, 1523	Red–664.6 Green–559.8 Blue–492.4 NIR–832.8
24 Dec 2018	Red, Green, Blue, and Near-Infrared	2399, 1523	Red–664.6 Green–559.8 Blue–492.4 NIR–832.8
18 Jan 2019	Red, Green, Blue, and Near-Infrared	2399, 1523	Red–664.6 Green–559.8 Blue–492.4 NIR–832.8
28 Jan 2019	Red, Green, Blue, and Near-Infrared	2399, 1523	Red–664.6 Green–559.8 Blue–492.4 NIR–832.8
17 Feb 2019	Red, Green, Blue, and Near-Infrared	2399, 1523	Red - 664.6 Green - 559.8 Blue - 492.4 NIR - 832.8

3.2.2 Standard Deviation

Each crop has its own reflectance and spectral signature values. The digital number of satellite images helps to identify different crops. Temporal data captured during different time periods helps to identify the long-duration and short-duration crops. The fields can also have multiple cropping patterns. To identify the difference between a crop field with another, variation between farm fields needs to be identified. This variation is calculated by using a sliding window filter that calculates the standard deviation of the image locally. Standard Deviation (SD) was calculated at each point by utilizing the pixels in a neighborhood range around that point for each satellite image. Different shapes of the farm were extracted using a round kernel filter of diameter 5 pixels (i.e., 21 pixels). SD images of each band from previous steps were stacked. Stacked SD images were then averaged at each point, giving us the single-channel band containing all the information related to spatial variability. The single-channel band image is shown in Fig. 4.

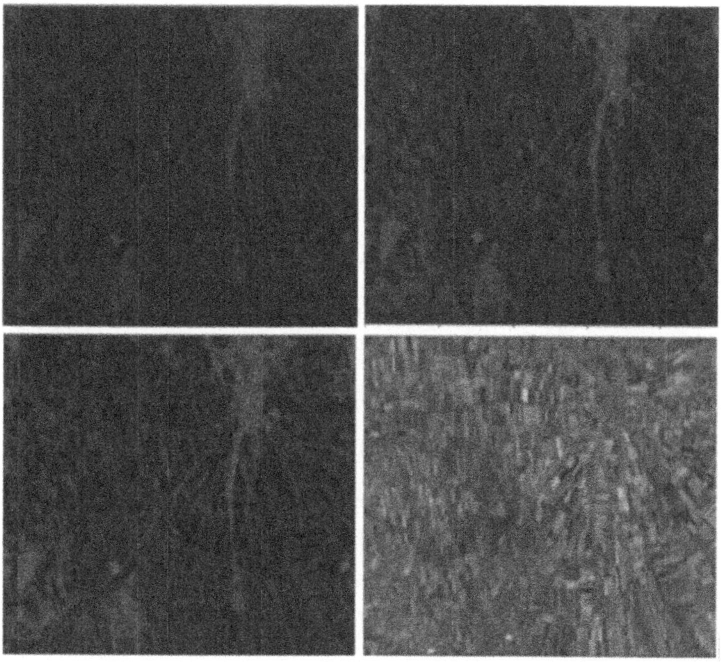

Fig. 2 Four spectral bands (Red, Green, Blue, and NIR) are mentioned in the image from captured tiles of the study area

3.2.3 Edge Detection Using Filters

The SD image is a single channel boundary image consisting of clustered boundaries. To obtain clear and true farm boundaries, we use the filters at different angles at a regular interval of 11.25° (Fig. 5). These filters fetch the line features at different angles. Linear features in the images were calculated by using 16 pairs of directional operators (right and left edges). These filters calculate edges at an angle of 11.25° summing together to 180°.

All the 16 images generated using a directional filter were combined by using a logical OR operator. Skeletonization of the image was done to get the boundary of 1-pixel thickness. Post Processing such as thinning, raster to vector, simplifying, cleaning was done by using QGIS (3.6.3-Noosa) software, which is an open-source desktop geo-processing software that helps in performing raster and vector operations (Team et al. 2016). This desktop application helps to deal with simple to complex GIS operations such as analyzing and handling geographical information by visualizing geographical statistics, through layers such as building maps, etc. Following operations were performed to create farm boundaries as vector polygons and store them in shapefile format by performing raster and vector operations. Thinning operation thins non-null cells of input raster that denote linear features, into single-cell

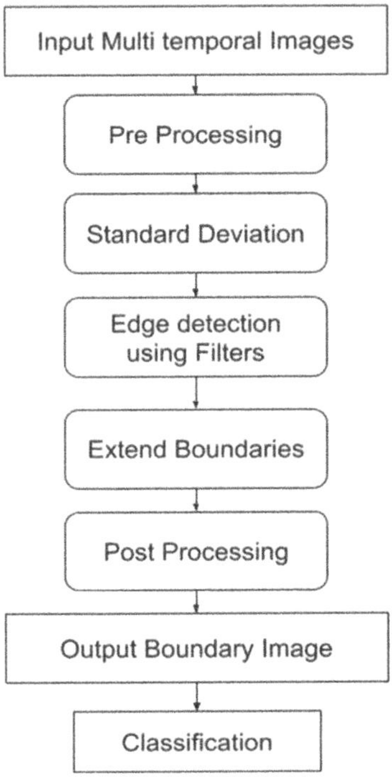

Fig. 3 Methodology followed in the analysis of satellite data for farm boundary generation and crop classification

Fig. 4 Average SD image generated using time-series satellite images

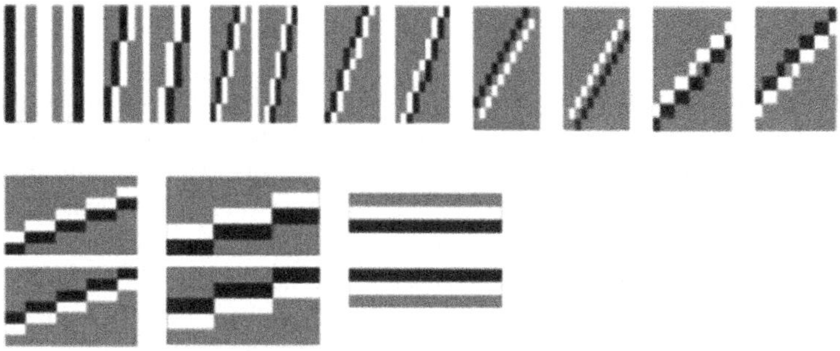

Fig. 5 Different filters are used to capture true farm boundaries at different angles

width. Lines are often thinned to single-pixel width before they are transformed to vector data.

The Raster to Vector process helps to convert the discrete raster data into continuous vector data for better analysis in the study. Simplify operation generalizes lines or polygons vector file with a modification by removing extraneous bend while preserving the essential shape. By cleaning operation, the dangles are removed so that small unwanted extensions are removed. The threshold for this dangle removal was set to 70 m after several trials.

3.2.4 Extend Boundary

The farm parcels generated are accurate to a certain level. Few farms remain unclosed due to changes in the phenological cycle of crops at different temporal sets. These farms also need to be considered. To make the farms completely bound an algorithm was devised. A sliding window was passed over the binary raster image obtained after simplifying, cleaning, and the endpoints of the left-out dangle were identified. The new binary image containing only the endpoints of the dangles is created of the same resolution as the input image. Keeping the standard deviation image as a reference, compare the SD of each endpoint in the new binary image. The endpoints whose SD is above the threshold are kept as sure boundary endpoints and these are extended.

3.2.5 Post Processing

The completed boundary raster image is processed to convert into polygons. This raster image was simplified, cleaned, and converted to a polygon feature and small areas that are not part of farm boundaries are removed by considering the threshold of 1000 m^2 after experimenting with different values.

3.2.6 Classification

The farm boundary polygon shapefile was then classified by using training data points that are generated from randomly selected farms during ground truth collection. These points are used for training RF classifier. While training, class imbalance was reduced to improve accuracy. Bias, variance, and noise are important factors for creating errors while training. ML methods such as ensemble techniques (Random Forest) help to reduce these factors (Verbaeten and Van Assche 2003). For labeling unknown farm parcels using the trained model, we generated random points in each farm and classified these points with the help of the training model. If the crop class assignment of each farm is more than 80% then we assign those farm parcels with the crop class else the farm parcel is revisited. This method helps to improve the stability and accuracy of ML algorithms. The Receiver Operating Characteristic (ROC) curve was used to check the accuracy of the model by plotting the False positive rate (x-axis) versus True positive rate (y-axis) for each class with values ranging between 0.0 and 1.0.

4 Results and Discussion

4.1 Delineation of Crop Boundaries

The object-based crop classification was performed for the Morbi District of Gujarat State, India. Time-series data of Sentinel-2 were collected for farm boundary generation and crop classification. The variation between farm parcels were captured almost effectively using the standard deviation method among the parcels. The standard deviation method is one of the simple and effective ways of identifying variation at pixel level. Effectiveness of the result was analyzed based on comparison of the actual farm boundary and the farm generated boundary. The python code is written that executes our methodology. Different libraries were used such as NUMPY, PANDAS, GDAL, SCIKIT-LEARN. NUMPY, PANDAS, and GDAL were helpful for raster, and vector data reading and processing operations while SCIKIT-LEARN was used for crop classification purposes. Automation has been done for our methodology. Cloud-free data were wisely chosen so that the change in the phenology of the crop can be easily seen in false color composite from one date of image to another date of image. The response from the directional filter at 0°, when convolved over the SD image, is shown in Fig. 6a. Dangle is removed from the image generated using logical OR operation. The resultant output raster image is shown in Fig. 6b. A cleaned vector polygon for farm parcels was generated by using post-processing methods and depicted the output in Fig. 6c. Reference farm boundaries were generated by digitizing manually the edge of the farm in False Color Composite (FCC) of sentinel-2 data as shown in Fig. 6d with the help of QGIS software. This was based totally on the visual aspect.

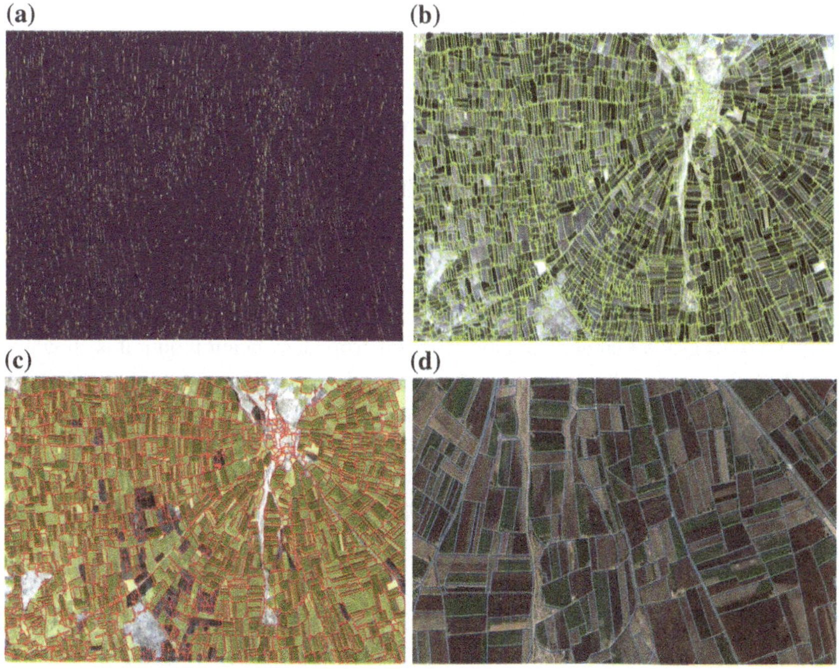

Fig. 6 **a** Image output when directional filter operator at 0° convolved on SD image; **b** Farm boundary generated after dangle removal from the vector file; **c** Farm parcels generated are overlaid on the FCC of satellite images; **d** Actual crop boundaries taken as reference were generated by digitizing the edge of farms in the FCC of satellite data

The result of the boundary generation by the algorithm using the standard deviation method is shown in Fig. 7a. Through visual interpretation, the actual crop boundaries and resulting crop boundaries were overlaid on one another and the results show that our method captures every small farm that was present during the time (Fig. 7b).

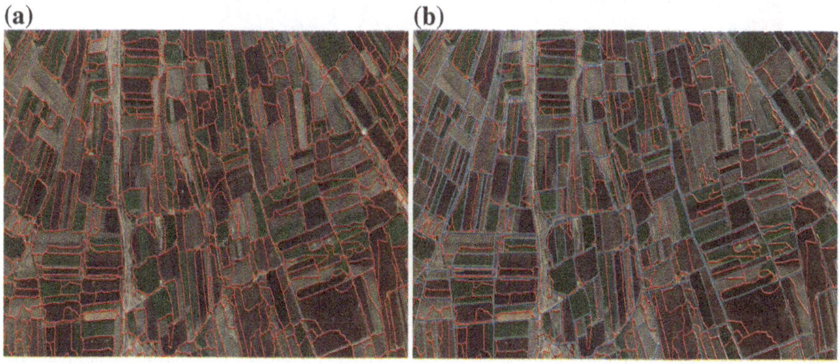

Fig. 7 **a** Resulting (Red) farm boundaries generated using algorithms; **b** Overlay of actual (Blue) versus resulting (Red) farm boundaries

4.2 Classification Accuracy

The classification algorithm implemented here is random forest. This algorithm performs well even with minimum data. In this study, the random forest algorithm replaces the use of powerful models for such huge data and performs well even with imbalanced classes. The classification results were measured crops wise and also the quantitative effectiveness was evaluated by comparing actual area for the crop farms with the resulting crop areas.

The classification algorithm resulted in approx. 85% accuracy with a validation error of 0.21 and F1-score of 0.84. Crop classes are accurately classified and are checked with ground label data. Object-based crop classification output is shown in Fig. 8.

The confusion matrix and True Positive, False Positive, True Negative, and False Negative values were shown in Tables 3 and 4, respectively. A total of 1204 points were predicted using the model. These points contain the crop classes of jowar fodder, maize fodder, coriander, cumin, onion, vegetable, wheat, and gram. The Receiver

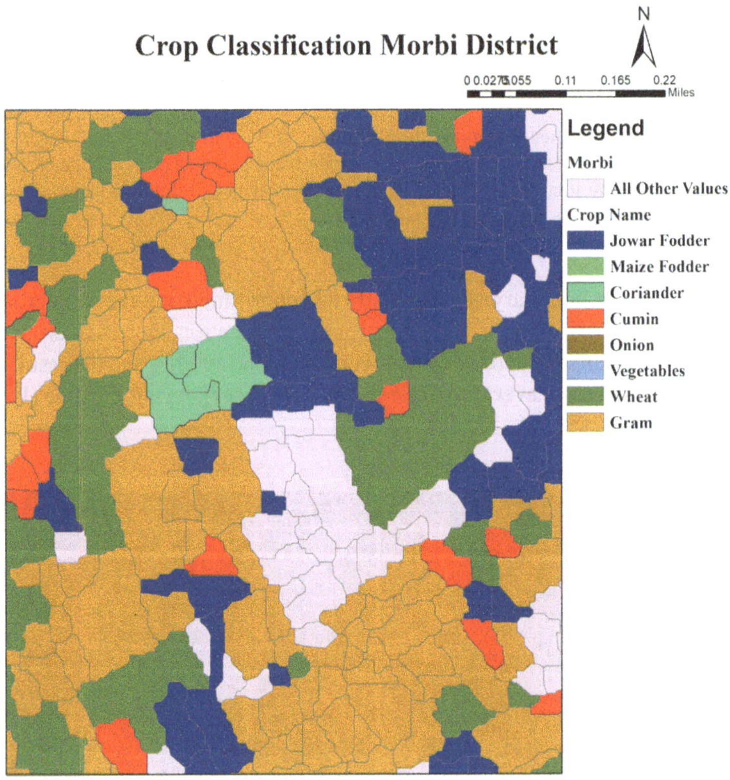

Fig. 8 Object-based crop classification output. Different crop classes were symbolized, and crop labels are seen in the legend section of the map

Table 3 Confusion matrix for crop classification. The X-axis is the predicted class and the Y-axis is the actual class of the crop

	Jowar fodder	Maize fodder	Coriander	Cumin	Onion	Vegetable	Wheat	Gram
Jowar fodder	129	0	1	2	0	0	11	6
Maize fodder	0	16	0	11	0	31	1	11
Coriander	0	0	92	2	0	0	16	2
Cumin	3	7	2	95	0	1	16	1
Onion	0	0	0	4	9	0	1	2
Vegetable	0	29	1	2	0	20	2	7
Wheat	4	0	4	12	0	0	395	14
Gram	3	16	2	4	0	11	12	194

Table 4 Class-wise True Positive, False Positive, True Negative, and False Negative

	Jowar fodder	Maize fodder	Coriander	Cumin	Onion	Vegetable	Wheat	Gram
False Positive	10	52	10	37	0	43	59	43
False Negative	20	54	20	30	7	41	34	48
True Positive	129	16	92	95	9	20	395	194
True Negative	1045	1082	1082	1042	1188	1100	716	919

Operating Characteristic curve (ROC) as shown in Fig. 9 gives the results of the classifier.

Based on the results generated in both the process, boundary generation, and classification, it was observed that object-based crop classification yields more accurate results and preserves farm-level details. The confusion matrix shows that above 80% of samples from each class are correctly identified. The True Positive and False Negative tables show the samples that do not belong to a particular class and are correctly or wrongly classified into a class.

5 Conclusions

In this study, the Sentinel-2 satellite optical time series data was used for object-based crop classification. SD images were generated from these images that hold the

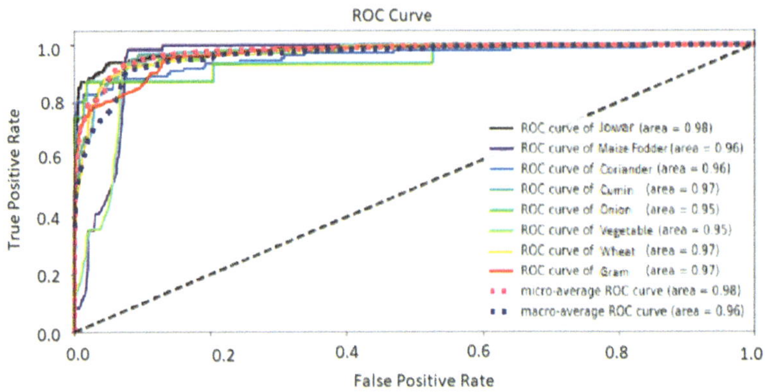

Fig. 9 Graph showing ROC curve for crop classes

features of farm edges. Edges were extracted by using different filters. Raster vector operations were performed to get farm boundaries in shapefile or vector format. Once the base of the object-based crop classification, i.e., farm boundaries were generated, each farm was labeled with different crops using the Random Forest RF algorithm. Ground truths were collected to create a training mode. Using the model, each unknown farm was labeled with different crops. It was observed that the object-based crop classification was successfully implemented. Future work includes improvement in the accuracy of farm boundaries in a more precise way and increased accuracy of the farm generated on a larger scale, i.e., high-resolution data. This method totally depends upon clear optical data and cannot perform in cloudy conditions. Farm boundary generation using Synthetic-Aperture Radar (SAR) data can be another potential option. Crop Knowledge Matrix can be used in crop classification for further classification accuracy improvement.

Acknowledgements The authors would like to thank the Google Earth Engine (GEE) team for providing cloud-computing resources for data visualization and European Space Agency (ESA) for providing open-access data. The authors thank Amnex Infotechnologies Pvt. Ltd. for providing all the needed support and environment to work on this project. Additionally, the authors would like to thank anonymous reviewers for their helpful comments and suggestions.

References

Alemu MM (2016) Automated farm field delineation and crop row detection from satellite images. Master's thesis, University of Twente, The Netherlands.

Belgiu M, Csillik O (2018) Sentinel-2 cropland mapping using pixel-based and object-based time-weighted dynamic time warping analysis. Remote Sens Environ 204:509–523

Crommelinck S, Bennett R, Gerke M, Nex F, Yang MY, Vosselman G (2016) Review of automatic feature extraction from high-resolution optical sensor data for uav-Object Based Crop Classification for Precision Farming 17 based cadastral mapping. Remote Sen 8(8):689

Davidse J (2015) Semi-automatic detection of field boundaries from high-resolution satellite imagery. Wageningen University

Garcia-Pedrero A, Gonzalo-Martin C, Lillo-Saavedra M (2017) A machine learning approach for agricultural parcel delineation through agglomerative segmentation. Int J Remote Sen 38(7):1809–1819

Hakkim VA, Joseph EA, Gokul AA, Mufeedha K et al (2016) Precision farming: the future of indian agriculture. J Appl Biol Biotech 4(06):068–072

Marshall M, Crommelinck S, Kohli D, Perger C, Yang MY, Ghosh A, Fritz S, Bie Kd, Nelson A (2019) Crowd-driven and automated mapping of field boundaries in highly fragmented agricultural landscapes of Ethiopia with very high spatial resolution imagery. Remote Sen 11(18):2082

Maxwell SK (2010) Generating land cover boundaries from remotely sensed data using object-based image analysis: overview and epidemiological application. Spat Spatiotemporal Epidemiol 1(4):231–237

North HC, Pairman D, Belliss SE (2018) Boundary delineation of agricultural fields in multitemporal satellite imagery. IEEE J Sel Top Appl Earth Obs Remote Sen 12(1):237–251

Ok AO, Akar O, Gungor O (2012) Evaluation of random forest method for agricultural crop classification. Eur J Remote Sen 45(1):421–432

Peña JM, Gutiérrez PA, Hervás-Martínez C, Six J, Plant RE, López-Granados F (2014) Object-based image classification of summer crops with machine learning methods. Remote Sen 6(6):5019–5041

Team QD et al (2016) QGIS geographic information system. Open-source geospatial foundation project

Verbaeten S, Van Assche A (2003) Ensemble methods for noise elimination in classification problems. In: International workshop on multiple classifier systems, Springer, pp 317–325

Wang Z, Boesch R (2007) Color-and texture-based image segmentation for improved forest delineation. IEEE Trans Geosci Remote Sen 45(10):3055–3062

Wassie Y, Koeva M, Bennett R, Lemmen C (2018) A procedure for semi-automated cadastral boundary feature extraction from high-resolution satellite imagery. J Spatial Sci 63(1):75–92

Watkins B (2019) Agricultural field boundary delineation using earth observation methods and multi-temporal sentinel-2 imagery. PhD thesis, Stellenbosch: Stellenbosch University.

Yan L, Roy D (2015) Improved time series land cover classification by missing-observation-adaptive nonlinear dimensionality reduction. Remote Sen Environ 158:478–491

Machine Learning Approaches and Sentinel-2 Data in Crop Type Mapping

Pranay Panjala, Murali Krishna Gumma, and Pardhasaradhi Teluguntla

Abstract Crop monitoring becomes essential in attaining food security for implementation of various agricultural serving programs. So, fast and reliable crop monitoring is must. Using traditional methods, crop monitoring maps need high amount of satellite data downloading and processing time. Google Earth Engine (GEE) cloud platform enables us to save time in downloading and processing of time series satellite data, the every satellite imagery is converted into Normalized Difference Vegetation Index (NDVI) image and stacked monthly wise maximum images. The stacked image was used for conducting supervised classification. The main objective of this study is to evaluate the performance of different supervised machine learning (ML) classifiers in GEE platform and Spectral Matching Technique (SMT) using Sentinel-2 10 m satellite imagery in specific crop type classification. The crop classification for the year 2018–19 (*rabi* season) was carried for Jhansi District using supervised classifiers like Random Forest (RF), Support Vector Machine (SVM) and Classification and Regression Trees (CART) in GEE platform and also with SMT with the help of ground data. It was attained nearly 81.8% accuracy for RF, 68.8% for SVM, 64.9% for CART and 88% for SMT. The results obtained using RF classifier were nearly relative to SMT classification map. The study indicates that classifier's performance depends on the quality of ground data used, RF can reduce the error samples in ground samples and produce satisfactory results. This study compared results obtained from all the above classifiers with agricultural statistics and also compared crop-wise accuracies. In the study, it was observed that RF classification is outperformed when compared with other classifiers considered in the study.

P. Panjala · M. K. Gumma (✉)
Geospatial Sciences and Big Data, Global Research Program on Resilient Farm and Food Systems, International Crops Research Institute for the Semi-Arid Tropics (ICRISAT), 502 324 Patancheru, India
e-mail: M.Gumma@cgiar.org

P. Teluguntla
Western Geographic Science Center, U.S. Geological Survey (USGS), AZ 86001 Flagstaff, USA

Bay Area Environmental Research Institute (BAERI), NASA Research Park, Moffett Field, San Jose, CA 94035, USA

G. P. O. Reddy et al. (eds.), *Data Science in Agriculture and Natural Resource Management*, Studies in Big Data 96, https://doi.org/10.1007/978-981-16-5847-1_8

Keywords Crop type mapping · Sentinel-2 · Dryland agriculture · Agriculture · Google earth engine

1 Introduction

Monitoring agricultural changes, ecosystem dynamics, climate change scenarios and other human-induced changes on earth surface require information on land use and land cover (LULC). In agriculture, monitoring of crop changes, land use changes and other physical parameters helps in estimating agriculture statistics, crop yield and drought risk, which in turn helps solving problems faced by farming community. The above estimation requires satellite imagery for LULC mapping. The coarse resolution satellite data like Moderate Resolution Imaging Spectroradiometer (MODIS) 250 m has been used to generate LULC and changes maps, also used for global cropland extent (Gumma et al. 2016). The entry of high-resolution imagery like Sentinel-1, Sentinel-2 and Landsat-8 allows to generate and monitor LULC changes efficiently. But analyzing at large scale, high-resolution data requires high volume of data to download and it is time consuming and cumbersome process. These problems can be overcome by using GEE cloud platform and it contains updated time series satellite imagery data.

Earth observations and spatial sciences have high potential for sustainable agriculture and environmental development (Im 2020). However, cropland spatial maps combined with weather information and social factors to identify areas where productivity is affected and areas is supported by insurance companies. Furthermore, spatial maps could be used as important input parameter for crop yield assessment. One example is the development of short duration or stress-tolerant varieties that reduce the risk of drought exposure. High spatial-temporal imagery serves rapid, large-scale, cost-efficient and continuous monitoring of agriculture crops lands (Brown et al. 2007; Anderson et al. 2012; Thenkabail et al. 2012; McCarthy et al. 2018). Numerous studies have been conducted on LULC monitoring using various resolutions of satellite imagery with various methods at field to Global scale. Remote sensing is a powerful tool to monitor spatial distribution of agriculture croplands along with LULC classes (Pittman et al. 2010). Several land cover products at coarse and high resolution, such as Global Land Cover 2000, Global Land Cover by National Mapping Organizations (GLCNMO) (Tateishi et al. 2014) and Glob Cover (Bicheron et al. 2008). Mapped irrigated areas across the globe (GIAM) (Thenkabail et al. 2009). Several methods and techniques were used for crop classification, mainly temporal vegetation-based algorithms (Jeganathan et al. 2014; Dong et al. 2015; Pan et al. 2015; Xiong et al. 2017a), classification decision tree and regression tree algorithms (Friedl and Brodley 1997; Ozdogan and Gutman 2008; Deng and Wu 2013; Egorov et al. 2015), discrete Fourier harmonic analysis (Harris, 1978; Platas-Garza and de la O Serna 2010; Boche and Mönich, 2020), SMT (Thenkabail et al. 2007; Gumma et al. 2014; Gumma et al. 2018), RF algorithms (Millard and Richardson, 2015; Teluguntla et al. 2018; Oliphant et al. 2019; Xie et al. 2019). These studies were conducted regional to global scale, which have some limitations for field scale management. Hence, monitoring croplands at field scale using high resolution is important for decision making and improving agriculture production. Many studies were conducted with supervised and unsupervised classification algorithms for mapping agriculture croplands through time series remote sensing imagery (Xiong et al. 2017a; Feyisa et al. 2020; Gumma et al. 2020). Croplands mapping using time series analysis were proven to map precise crop extent maps rather than single date imagery (Long et al. 2013; Gumma et al. 2014; Gómez et al. 2016; Belgiu and Csillik, 2018).

Above studies were conducted on classification of multi-temporal satellite images for cropland monitoring including LULC mapping. Monitoring major crop types using Earth observation (EO) data is very complex and has many challenges (Griffiths et al. 2019), such as crop changes within and between years according to local demand or water availability. In dryland areas, field size is very small and dynamically changes the cropping pattern and many areas have an intercropping pattern, especially in the present study area. Several studies have been demonstrated to map croplands using Sentinel-2 EO data (Veloso et al. 2017; Xiong et al. 2017b; Belgiu and Csillik, 2018; Lambert et al. 2018). Lambert et al. developed an approach for improved synthesis over fine temporal intervals and mapped national crop and land cover map. Xiong et al. (2017b) mapped 30 m crop extent for entire Africa using pixel and object-based approaches. Belgiu and Csillik (2018) was used pixel and object-based time-weighted dynamic time warping (DTW) analysis to map croplands. Veloso et al. (2017) were used Sentinel-1 time series data with Sentinel-2 NDVI data for monitoring winter and summer crops and this study was conducted in southwest France, where field size is very large. Given the above background, the present study tested various methods of classification in study area, where there is complex cropping pattern and average field size is one hectare.

GEE allows users to run different machine learning algorithms using java and python languages. GEE collects a multi-petabyte catalog of various satellite imagery and geospatial datasets (Gorelick et al. 2017) for cloud-level analysis. Using GEE, it is easy to monitor decadal changes without downloading data, using and comparing different datasets. GEE also allows to import shape files, images for any cloud analysis and also to export data for offline analysis.

GEE has been widely employed in multi-temporal analyses for large areas and many studies were carried out like sprawl mapping using Landsat 5 and 7 images (Patel et al. 2015), map rice-growing areas in Northeast Asia using Landsat 8 (Dong et al. 2016), automated mapping of cropland of continental Africa (Xiong et al. 2017a), cropland extent and areas of South Asia (Gumma et al. 2020), crop yield map (Lobell et al. 2015), flood mapping (Coltin et al. 2016), land cover change detection (Sidhu et al. 2018) and global non-croplands like global surface water changes (Pekel et al. 2016), global urban land use mapping (Liu et al. 2018), forest cover changes (Hansen et al. 2013).

In this study, we demonstrated different ML approaches and SMT techniques to map major crop types using Sentinel-2 time series data and intensive ground data. The main goal of the study is to test different methods to estimate crop areas at village level and identify the gaps in heterogeneous small holder farming systems.

We applied supervised classification with intensive training data collected in winter season 2018–19. GEE is able to run various machine learning algorithms like decision trees algorithms like CART and RF and optimization algorithms like SVM, which helps in classification.

2 Materials and Methods

2.1 Study Area

Jhansi district lies in 25°6″ to 25°56″ N latitude and 78°18″ to 79°25″ E longitude (Fig. 1), part of Bundelkhand agro-climatic region at an average elevation of 285 m and situation between Pahuj and Betwa rivers. This area experiences about 45–47 °C maximum temperatures during summer and 1–3 °C minimum temperatures during winter. This region receives annual rainfall about 884.6 mm out of which 90% is received during southwest monsoon. This area is commonly affected by droughts and long dry spells that affect both *kharif* and *rabi* seasons. Major soil texture available in this area are loamy, clay and loamy skeletal (Fig. 1). The major crops of Jhansi during *kharif* are paddy, sorghum, sesame, groundnut and pigeon pea, etc., while wheat, chickpea, mustard and pea are the crops in *rabi* season.

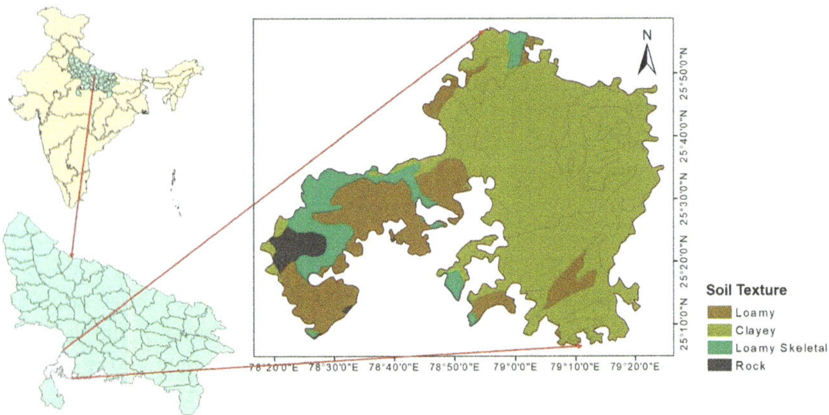

Fig. 1 Location of Jhansi District of Uttar Pradesh state, India with soil textural classes (*Source* NBSS&LUP)

Table 1 Satellite data and their bands with vegetation indices

Satellite imagery/Indices	Bands	Significance
SENTINEL-1	VH	Helps in identification of temporal changes
SENTINEL-2	B2, B4 and B8	
NDVI	(B8 – B4)/(B8 + B4)	Vegetation identification

NDVI—*Normalized Difference Vegetation Index was calculated using band B8 and B4 ranges from* -1 *to 1, which shows the vegetation level and sensitive to changes;* ***VH***—*Vertical–Horizontal polarized median value was extracted from Sentinel-1 imagery*

2.2 Image Acquisition

The present study uses Sentinel-2 and Sentinel-1 satellite imagery for classification (Table 1). The median Sentinel-1 and Sentinel-2 images were collected for November 2018 to March 2019.

2.3 Image Classification

- ***Support Vector Machines (SVM)***

SVM is supervised learning algorithm that analyzes data in classification and regression, which also has an important property of optimization. SVM (Cortes and Vladimir 1995) mainly works on the concept of margin, calculates the minimal distance between the decision boundary and points, followed by creating a margin, which is perpendicular distance between decision boundary and nearby points. The decision boundary is determined by the support vectors (subset of data points). SVM performs both linear and non-linear classification. In linear classification, a line separates the space data points. Non-linear classification is performed using kernel tricks that maps the inputs into high-dimensional feature spaces.

- **Classification and Regression Tree (CART)**

CART is mainly a decision tree supervised classifier that takes the inputs of training data using the theory of information entropy. At every node of the tree, the attribute of training data splits into its subsets and is then enriched into suitable class. The major benefit of using tree models is capturing the non-linearity in the dataset and there is no need for data standardization because they don't calculate distances between the data. The main disadvantage of CART is mostly depending upon training data, small error in training data cause significant change in classification, i.e. high sensitivity.

Table 2 Pros and Cons of SVM, RF and CART classifiers

Classifiers	Pros	Cons
SVM	1. Solo Classes can be easily classified	1. High Processing time
	2. Solving higher dimensions	2. Poor in classifying mixed classes
RF	1. Better correlation of trees, i.e. classes	1. Need predictive power for features
	2. Handling errors of trees	2. Moderate in classifying mixed classes
CART	1. Easy to perform and visualization	1. Likely to be overfitted
	2. Automatic class selection	2. High sensitive to training data

- **Random Forests (RFs)**

RF classifier (Breiman 2001) is also a decision tree supervised classifier, which rectifies the disadvantage of CART classifier, i.e. sensitivity with construction of group of Decision trees. RF assembles the DT at training time and results into classification of the individual trees, which corrects the overfitting of training data.

- **Pros and Cons of Machine Learning Algorithms**

See (Table 2).

- **Spectral Matching Techniques (SMT)**

SMT is one of the well-known techniques in crop classification using NDVI time series data. This technique mainly generates ideal spectral signatures using ground data and correlates with other spectral signatures to label the classes. First, the process starts with downloading the sentinel-2 satellite data (Fig. 2) for November 2018 to March 2019, i.e. *rabi* season and generating maximum NDVI images for every month and stacking in order. Then, run the unsupervised classification on NDVI stacked image with 50 classes, which were grouped based on spectral similarity.

Ideal spectra signatures were prepared using ground data by extraction of spectral values, i.e. NDVI values. 50 classes were compared with ideal spectral signatures, ground data and classes with alike time series were combined into a single class and labelled accordingly. If any disparities in classes, then that class was masked out and the above process is repeated. Accuracy assessment was conducted using independent validation data.

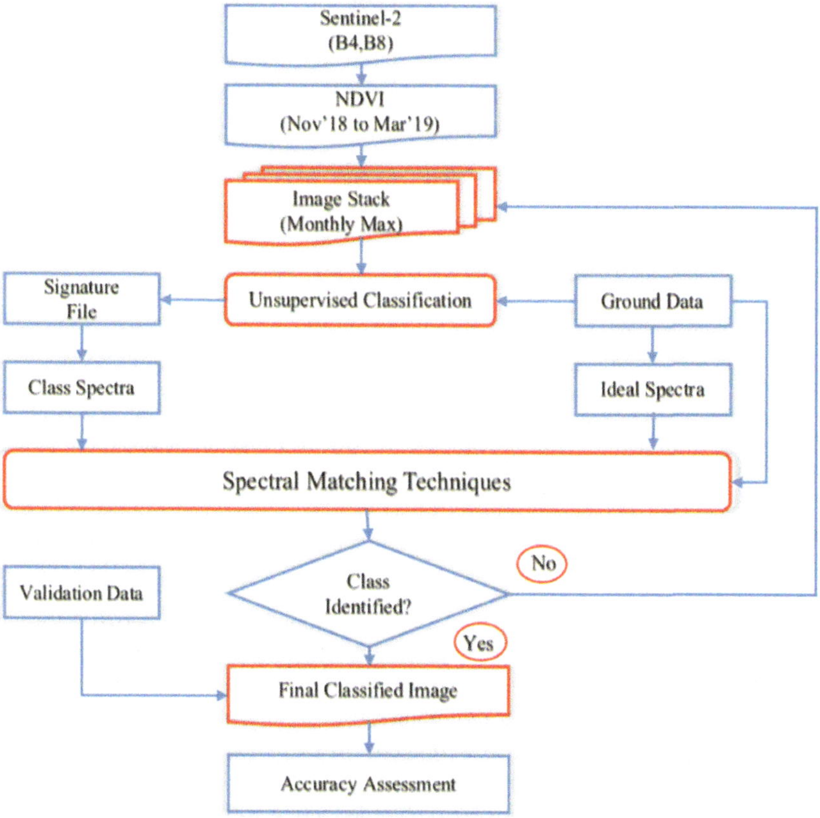

Fig. 2 Flow chart showing methodology of SMT for crop classification and accuracy assessment

2.4 Classification Procedure in GEE Platform

The classification procedure started with preparing image stack from Nov'2018 to Mar'2019 taking median values of NDVI using Sentinel 2 and VH from Sentinel 1 image (Fig. 3). Crop Mask obtained from SMT image was applied on stacked image to extract croplands. Ground data was divided into training and validation data in the ratio of 3:2 randomly and classified the stacked image using ground data with three well-known classification algorithms, i.e. CART, RF and SVM. Accuracy assessment was carried out using independent validation data for every algorithm.

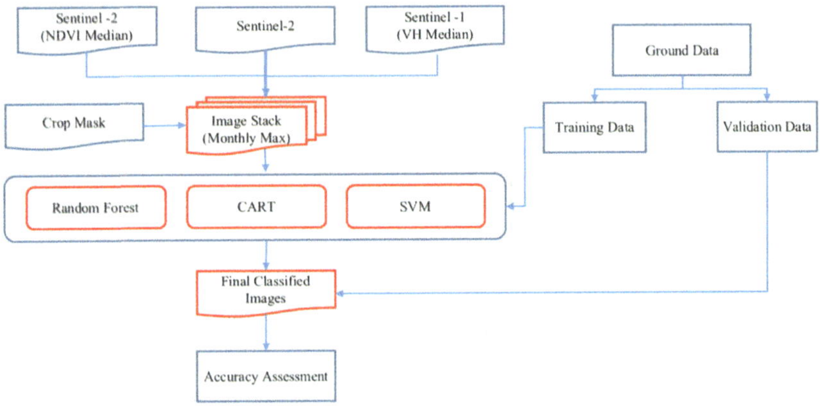

Fig. 3 Flow chart showing methodology used in GEE for crop classification and accuracy assessment

2.5 *Ground Data*

Ground data collection was carried out during the month of December 2018 through Jhansi districts covering major croplands and other LULC. Overall, 195 samples were collected out of which 77 samples were used for accuracy assessment (Fig. 4).

During ground data collection, minimum size of 90 × 90 m homogenous croplands were considered for sampling that were recorded using handheld Garmin GPS unit. Parameters like GPS, Crop parameters like crop intensity, crop pattern and crop growth, and irrigation techniques were recorded along with field photographs. The distribution of crop-wise ground samples is shown in above Table 3.

3 Results and Discussion

3.1 *Spatial Distribution of Crop Classification*

The spatial distribution of crops in Jhansi district using different ML algorithms showed in Fig. 5. The classification maps were obtained from RF algorithm, CART algorithm, SVM algorithm and SMT with classes Wheat, Chickpea, Mustard and Mixed crops.

Using Sentinel-2 (10 m) data, we can easily extract croplands and can easily identify the crop patterns up to field level. It was observed that majority part of crop class was wheat and it spreads across the district followed by chickpea crop. Majority of the mustard is sowed in northern part and some parts of southeast of Jhansi district. In most of the area, it was observed chickpea as mono-crop especially in eastern and central part of Jhansi district. During ground data collection it was observed that

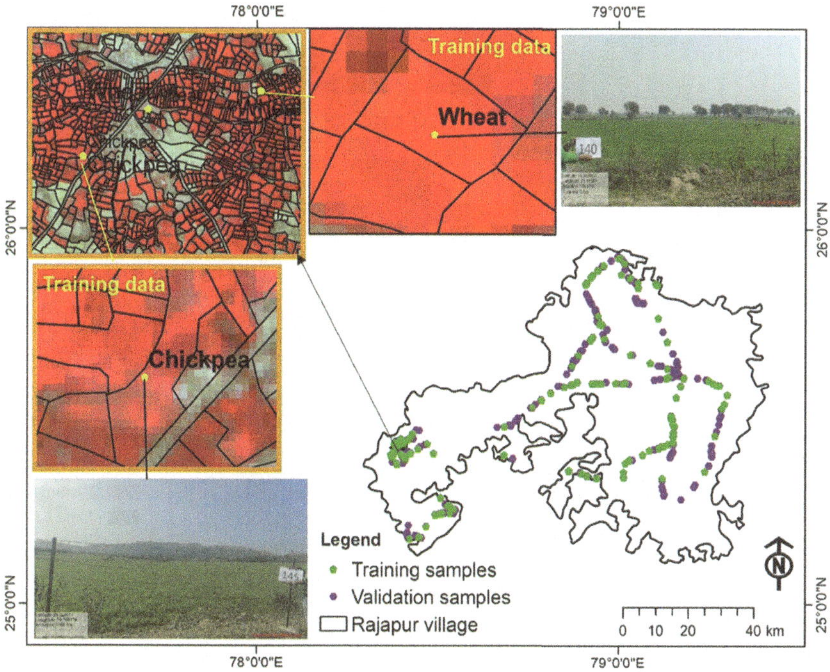

Fig. 4 Ground data collection for Jhansi district; training and validation data used for crop classification

Table 3 Distribution of samples of crops in both training and validation ground Data

Ground data		
Class	Training points	Validation points
Wheat	45	31
Chickpea	39	22
Mustard	20	14
Mixed crops	18	10
Total	122	77

majority of crops in Jhansi district have intermixing crops like chickpea-mustard and wheat-pea.

Figure 6 shows the field level crop indicated red mark in Fig. 5 classification for all classified images and compared with Google Earth High Resolution Image (GEHRI) and nominal Landsat 30 m cropland product. Landsat 30 m cropland contains some other LULC but we successfully classified and eliminated other LULC like built-ups, shrubs and barren, etc., using Sentinel-2 10 m data.

The field level distribution of crops shows that SMT classified mono and mixed crops with good accuracy. But ML algorithms got some gaps in classifying mono

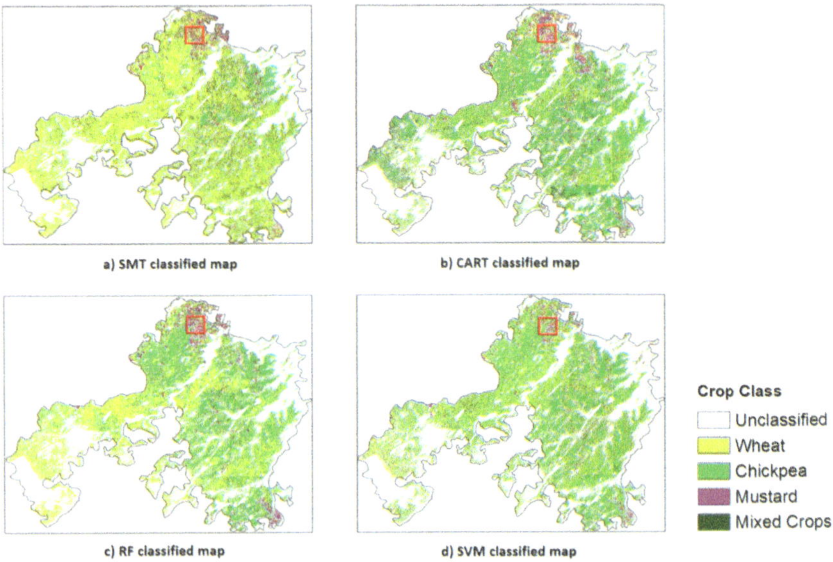

Fig. 5 Spatial Distribution of crops in Jhansi District—SMT classified image (**a**), CART classified image (**b**), RF classified image (**c**) and SVM classified image (**d**) along with selected area for field observation

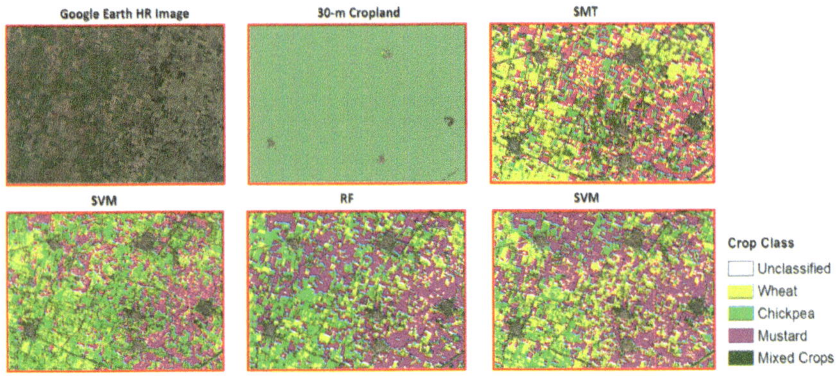

Fig. 6 Field level crop distribution in Jhansi District comparing Google Earth Image, 30 m cropland, SMT classified image, SVM classified image, RF classified image, CART classified image

and mixed crops. RF algorithm classified mustard and chickpea correctly but got some errors in classifying wheat, whereas, CART and SVM algorithms classified all classes with some errors. These errors were mainly due to misperception in deciding the class between mono and mixed crops as both consist of nearly same signatures. These errors can be reduced by using extensive and quality ground data.

3.2 Accuracy Assessment

Accuracy assessment was carried out for classified image obtained using validation data with the help of error matrices. The error matrix of RF classified image obtained overall accuracy of nearly 82% with kappa coefficient 0.735 and obtained good accuracy in both User's and Producers accuracy (Table 4). In RF, some of the chickpea is misclassified as other crops and also mixed crops were misclassified as wheat and chickpea. It indicates that the mixed crops show nearly same signatures of mono-crops in some areas.

SVM algorithm achieved overall accuracy of nearly 69% with kappa coefficient 0.558 and obtained nearly 70% in both user and producers accuracy (Table 5). In SVM, some of the Wheat class was misclassified as mixed crops, whereas, chickpea is misclassified as wheat. Mustard class is misclassified as chickpea class as it was a mixed crop with mustard.

CART algorithm achieved overall accuracy of nearly 65% with kappa coefficient 0.510 and obtained nearly 65% in both user and producers accuracy in every class (Table 6). In CART classification, some of the wheat class was misclassified as chickpea and chickpea crop was misclassified as wheat crop. Some of the mustard crop was misclassified as wheat and chickpea.

Table 4 Accuracy assessment for the crop type map of RF classified image. Map classes are rows and reference classes are columns

Random forest	Wheat	Chickpea	Mustard	Mixed crops	Total	Users
Wheat	30	1	0	0	31	0.968
Chickpea	2	17	1	2	22	0.773
Mustard	2	1	11	0	14	0.786
Mixed crops	3	2	0	5	10	0.500
Total	37	21	12	7	77	
Producers	0.811	0.810	0.917	0.714	**Overall**	**0.818**
					Kappa	**0.735**

Table 5 Accuracy assessment for the crop type map of SVM classified image. Map classes are rows which reference classes are the columns

SVM	Wheat	Chickpea	Mustard	Mixed crops	Total	Users
Wheat	23	2	1	5	31	0.742
Chickpea	5	15	1	1	22	0.682
Mustard	3	2	8	1	14	0.571
Mixed crops	2	1	0	7	10	0.700
Total	33	20	10	14	77	
Producers	0.697	0.750	0.800	0.500	**Overall**	**0.688**
					Kappa	**0.558**

Table 6 Accuracy assessment for the crop type map of CART classified image. Map classes are rows which reference classes are the columns

CART	Wheat	Chickpea	Mustard	Mixed crops	Total	Users
Wheat	20	6	2	3	31	0.645
Chickpea	4	15	2	1	22	0.682
Mustard	1	1	10	2	14	0.714
Mixed crops	1	4	0	5	10	0.500
Total	26	26	14	11	77	
Producers	0.769	0.577	0.714	0.455	**Overall**	**0.649**
					Kappa	**0.510**

Table 7 Accuracy assessment for the crop type map of SMT classified image. Map classes are rows which reference classes are the columns

SMT	Wheat	Chickpea	Mustard	Mixed crops	Total	Users
Wheat	30	1	0	0	31	0.968
Chickpea	2	18	1	1	22	0.818
Mustard	1	1	12	0	14	0.857
Mixed crops	0	1	1	8	10	0.800
Total	33	21	14	9	77	
Producers	0.909	0.857	0.857	0.889	**Overall**	**0.883**
					Kappa	**0.833**

SMT classified image achieved overall accuracy of nearly 88% with kappa coefficient 0.833 and obtained greater than 80% in both user and producers accuracy in every class (Table 7). Every class was classified with good accuracy. Some of the chickpea crop was misclassified as wheat class and also mixed crop into chickpea and mustard.

3.3 Class Wise Assessment in Terms of User's and Producer's Accuracy

The Fig. 7 shows the comparison of user's accuracy for every individual classified class. It was observed that in wheat class, SMT and RF achieved more than 90% and in chickpea and mustard class, it was near 80%. Other classifiers achieved nearby 70% in all classes.

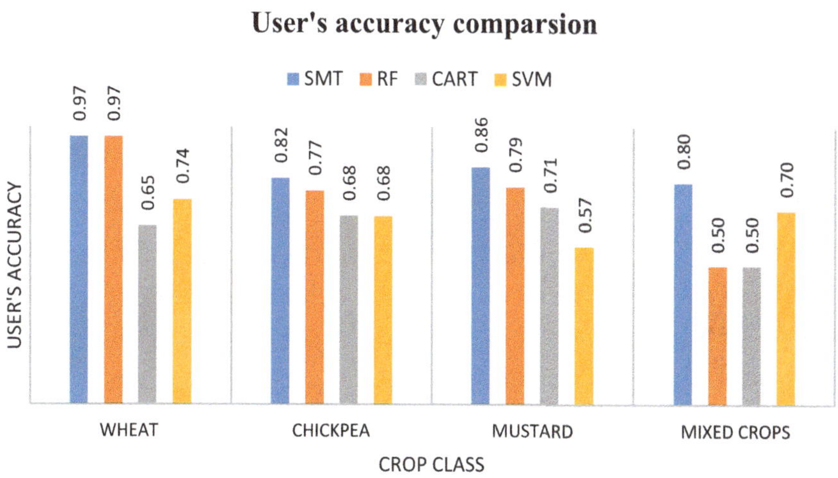

Fig. 7 Comparison of user's accuracy of classified classes, i.e. Wheat, Chickpea, Mustard and Mixed Crops

The Fig. 8 shows the comparison of producer's accuracy for every individual class. We can observe in wheat, chickpea and mustard classes, SMT and RF achieved more than 80%, whereas, other classifiers near to 60% and less.

By looking at the above statistics, it was noticed that most of the classes were classified with good accuracy using SMT and RF algorithm.

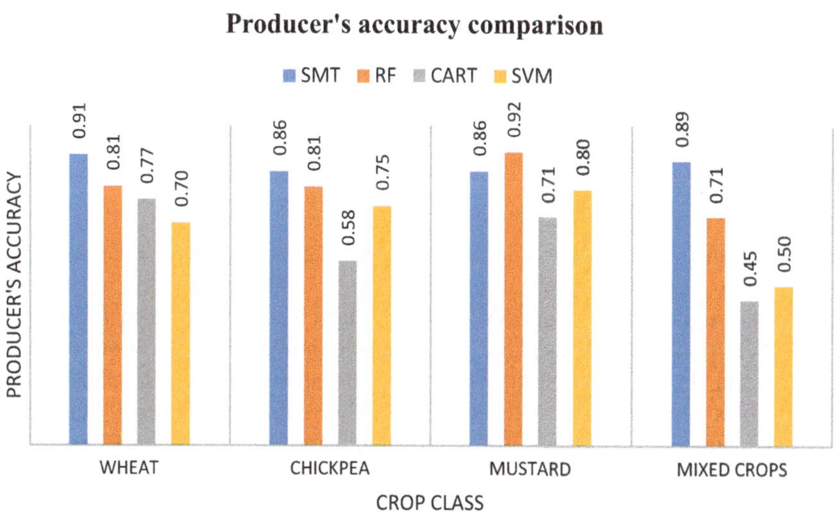

Fig. 8 Comparison of producer's accuracy of classified classes, i.e. Wheat, Chickpea, Mustard and Mixed Crops

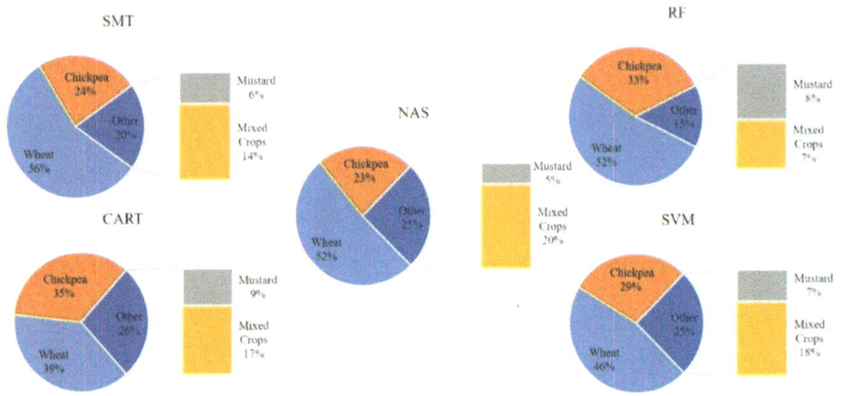

Fig. 9 Comparison of spatial extracted areas using SMT, RF, CART and SVM with National Agriculture Statistics (Meier et al. 2018)

3.4 *Comparing Extracted Areas with NAS Statistics*

We have compared spatial extracted crop areas with National Agriculture Statistics (Meier et al. 2018.). We have noticed that the areas obtained from SMT classified image were highly correlated with NAS areas in terms of all crops (Fig. 9).

In ML algorithms, RF classified image shows good relation with NAS in terms of wheat crop and mustard crop. But in terms of chickpea, it shows higher area compared to NAS because most of the chickpea crop in Jhansi district is intermixing crop. Other algorithms like CART and SVM show less % of areas compared to NAS. SVM classified image shows good correlation with NAS in terms of all crops but very less area in terms of wheat crop.

4 Discussion

The importance of variables in decision tree classifications, i.e. RF and CART can be identified using GEE code whereas it is not possible in identifying variable importance in linear kernel type SVM classifier as it is difficult to extract features.

The selection of variables plays important role in image classification based on theme of map. In this paper, the main theme was to map specific crop classification so, the variables monitoring the vegetation changes were important. Hence, Sentinel-1 VH and Sentinel-2 NDVI bands were considered for crop classification. Considering variable importance in RF Algorithm, VH band plays a vital role, i.e. about 92% importance in classifying image. Bands like B9, B1, B11, B8 and B2 have more than 80% importance in classification and NDVI with 76% (Fig. 10). It is important to select the small number of variables that work well for our theme than selecting large number of variables. Sometimes, selecting large number of variables might reduce

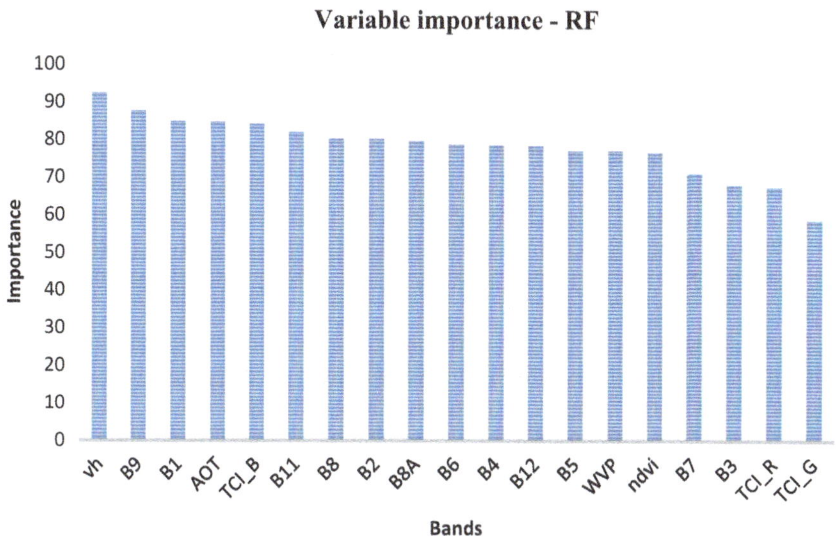

Fig. 10 Bar chart showing variables (Bands) Importance in RF algorithm

accuracy level. The variables with high permutations can train the RF algorithm efficiently results into good image classification.

CART classifier variables and their importance can be observed in Fig. 11. This shows that all the bands were of least importance in classification. B1 with nearly 3.5%% importance and other with less than 1.0%%. This shows that the classification using CART is purely depended on ground data.

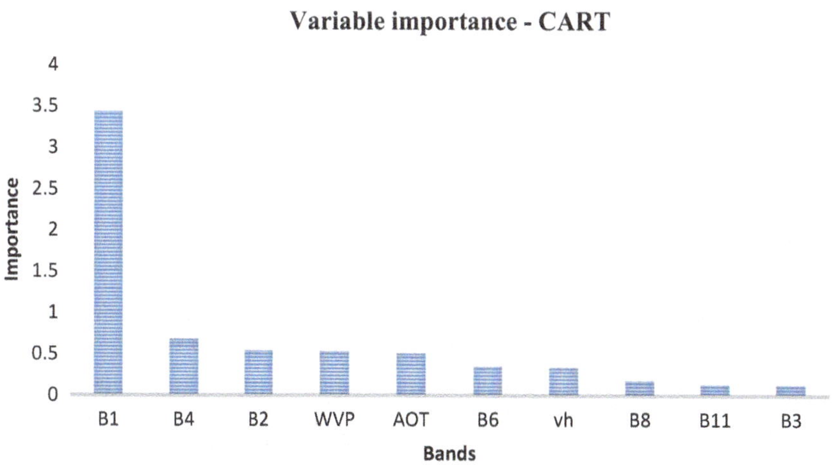

Fig. 11 Bar chart showing variables (Bands) Importance in CART algorithm

Considering accuracy assessment, SMT is better than RF classification. With SMT, initially satellite images were classified into high number of classes using unsupervised classes and later which were simplified into major classes using ground data spectra values but in RF, only classified image based on ground data training points using supervised classification. So, the accuracy of image using RF mostly depends upon equal distribution of training points. SMT needs satellite data to download whereas, using RF worked on GEE cloud and this reduces the time consumption for downloading and processing time. Using RF, it is possible to work for large areas but using SMT, it is a hectic task as it takes a lot of time and reaches computation limit.

In this study, major crop types were mapped using different ML techniques and semi-automated algorithms. SMT technique was used for mapping crop types using Sentinel-2 time series imagery with ground data which was used in ML algorithms. The process starts with unsupervised classification followed by creation of NDVI MVCs. In the process of class identification, the main advantage of this approach is the usage of monthly cloud-free or near-cloud-free images with creation of maximum value composites. In addition, ground data to identified different cropping systems across season. SMT's were predominantly successful in differentiating cropping patterns.

The same training data was used for all three ML algorithms and SMT. Since, ML algorithms work on different statistical approaches, which helps in identifying classes. Here, SVM algorithm calculates the minimum distance between decision boundary and samples, this approach increases the chance of mixing different crop classes. So, the high amount of training data is required to make efficient decision boundary whereas, CART algorithm works on DT where quality of training data plays a role in deciding classes using DT, i.e. error in training sample effect the class. These possibilities make SVM and CART cause less accuracy but in RF algorithm, rectifying the disadvantage of CART in deciding the class. So, RF achieved good accuracy as compared to other algorithms. With SMT, the decision of class was based on spectral signatures comparing with ideal spectral signatures. These help in understanding the signatures and helps in deciding the class. So, high accuracy was achieved using SMT.

5 Conclusions

The present study shows the classification of crop types at small holding farms using different ML and SMT techniques. Each classifier has its advantages in classifying various LULC. RF algorithm classified mono-crops correctly, however, it has its limits in classifying mixed crops. SMT classifies every class and attained good accuracy as it was done by reclassifying class disparities. RF is classified nearly comparable to SMT. Remaining classifiers like CART, SVM were classified satisfactorily and algorithm mistakenly classified other crops. It was observed that ground data plays a key role in classification of crop types. RF classification can run efficiently using some ground points whereas CART requires high amount of ground data. RF classification can neutralize error points using decision trees, but in SVM, error points affect the classification. It is concluded that every classifier with its limits of ground data can efficiently classifies the map. Mapping crop types at high resolution is very important for macro-level planning, yield predictions and crop stresses to support small holder farmers. High-resolution specific crop maps are important inputs for assessing yield predictions at village level and national food security studies. The classification limitation arises when we used satellite data during *kharif* season because of high cloud cover. This limitation can be rectified using Synthetic Aperture Radar data, i.e. Sentinel-1 data. This work was carried out in dryland area in which most of the croplands were small land patches, i.e. small holding farmers and having complex cropping patterns. So, it enables us to generate specific crop maps using four different ML methods and assessed the performance of those methods.

Acknowledgements This research was supported by the RIICE III project funded by the Swiss Agency for Development Cooperation (SDC), the Government of Uttar Pradesh, India, through its 'Doubling Farmers Income' in Bundelkhand, and the CGIAR Research Program on Water, Land and Ecosystems (WLE). The authors are grateful to Divya Kashyap, Arindom Baidya, and other team members for their suggestions and recommendations. We also thank Jameeruddin A and Ismail for their support during ground data collection.

References

Anderson MC, Allen RG, Morse A, Kustas WP (2012) Use of Landsat thermal imagery in monitoring evapotranspiration and managing water resources. Remote Sens Environ 122:50–65

Belgiu M, Csillik O (2018) Sentinel-2 cropland mapping using pixel-based and object-based time-weighted dynamic time warping analysis. Remote Sens Environ 204:509–523

Bicheron P, Huc M, Henry C, Bontemps S, Lacaux J (2008) GlobCover products description manual, Issue 2, Revision 2. European Space Agency (ESA)

Boche H, Mönich UJ (2020) Turing computability of Fourier transforms of bandlimited and discrete signals. IEEE Trans Signal Process 68:532–547

Brown JC. Jepson WE. Kastens JH. Wardlow BD. Lomas JM. Price K. (2007) Multitemporal, moderate-spatial-resolution remote sensing of modern agricultural production and land modification in the Brazilian Amazon. GISci. Remote Sens. 44: 117–148

Breiman L (2001) Random forests. Mach Learn 45(1):5–32

Coltin B, McMichael S., Smith T, Fong T (2016) Automatic boosted flood mapping from satellite data. Int J Remote Sens 37:993–1015

Cortes C, Vladimir V (1995) Support-vector networks. Mach Learn 20(3):273–297

Deng C, Wu C (2013) The use of single-date MODIS imagery for estimating large-scale urban impervious surface fraction with spectral mixture analysis and machine learning techniques. ISPRS J Photogramm Remote Sens 86:100–110

Dong J, Xiao X, Kou W, Qin Y, Zhang G, Li L, Jin C, Zhou Y, Wang J, Biradar C (2015) Tracking the dynamics of paddy rice planting area in 1986–2010 through time series Landsat images and phenology-based algorithms. Remote Sens Environ 160:99–113

Dong J, Xiao X, Menarguez MA, Zhang G, Qin Y, Thau D, Biradar C, Moore III B (2016) Mapping paddy rice planting area in northeastern Asia with Landsat 8 images, phenology-based algorithm and Google Earth Engine. Remote Sens Environ 185:142–154
Egorov A, Hansen M, Roy DP, Kommareddy A, Potapov P (2015) Image interpretation-guided supervised classification using nested segmentation. Remote Sens Environ 165:135–147
Feyisa GL, Palao LK, Nelson A, Gumma MK, Paliwal A, Win KT, Nge KH, Johnson DE (2020) Characterizing and mapping cropping patterns in a complex agro-ecosystem: an iterative participatory mapping procedure using machine learning algorithms and MODIS vegetation indices. Comput Electron Agr 175:105595
Friedl MA, Brodley CE (1997) Decision tree classification of land cover from remotely sensed data. Remote Sens Environ 61:399–409
Gómez C, White JC, Wulder MA (2016) Optical remotely sensed time series data for land cover classification: A review. ISPRS J Photogramm Remote Sens 116:55–72
Gorelick N, Hancher M, Dixon M, Ilyushchenko S, Thau D, Moore R (2017) Google Earth Engine: Planetary-scale geospatial analysis for everyone. Remote Sens Environ 202:18–27
Griffiths P, Nendel C, Hostert P (2019) Intra-annual reflectance composites from Sentinel-2 and Landsat for national-scale crop and land cover mapping. Remote Sens Environ 220:135–151
Gumma MK, Thenkabail PS, Deevi KC, Mohammed IA, Teluguntla P, Oliphant A, Xiong J, Aye T, Whitbread AM (2018) Mapping cropland fallow areas in myanmar to scale up sustainable intensification of pulse crops in the farming system. GISci Remote Sens 55:926–949
Gumma MK, Thenkabail PS, Maunahan A, Islam S, Nelson A (2014) Mapping seasonal rice cropland extent and area in the high cropping intensity environment of Bangladesh using MODIS 500 m data for the year 2010. ISPRS J Photogramm Remote Sens 91:98–113
Gumma MK, Thenkabail PS, Teluguntla PG, Oliphant A, Xiong J, Giri C, Pyla V, Dixit S, Whitbread AM (2020) Agricultural cropland extent and areas of South Asia derived using Landsat satellite 30-m time-series big-data using random forest machine learning algorithms on the Google Earth Engine cloud. GIScі Remote Sens 57:302–322
Gumma MK, Thenkabail PS, Teluguntla P, Rao MN, Mohammed IA, Whitbread AM (2016) Mapping rice-fallow cropland areas for short-season grain legumes intensification in South Asia using MODIS 250 m time-series data. Int J Digital Earth 9(10):981–1003
Hansen MC, Potapov PV, Moore R, Hancher M, Turubanova S, Tyukavina A, Thau D, Stehman S, Goetz SJ, Loveland TR (2013) High-resolution global maps of 21st-century forest cover change. Science 342:850–853
Harris FJ (1978) On the use of windows for harmonic analysis with the discrete Fourier transform. Proc IEEE 66:51–83
Im J (2020) Earth observations and geographic information science for sustainable development goals. GISci Remote Sens 57:591–592
Jeganathan C, Dash J, Atkinson P (2014) Remotely sensed trends in the phenology of northern high latitude terrestrial vegetation, controlling for land cover change and vegetation type. Remote Sens Environ 143:154–170
Lambert M-J, Traoré PCS, Blaes X, Baret P, Defourny P (2018) Estimating smallholder crops production at village level from Sentinel-2 time series in Mali's cotton belt. Remote Sens Environ 216:647–657
Liu X, Hu G, Chen Y, Li X, Xu X, Li S, Pei F, Wang S (2018) High-resolution multi-temporal mapping of global urban land using Landsat images based on the Google Earth Engine Platform. Remote Sens Environ 209:227–239
Lobell DB, Thau D, Seifert C, Engle E, Little B (2015) A scalable satellite-based crop yield mapper. Remote Sens Environ 164:324–333
Long JA, Lawrence RL, Greenwood MC, Marshall L, Miller PR (2013) Object-oriented crop classification using multitemporal ETM+SLC-off imagery and random forest. GISci Remote Sens 50:418–436

McCarthy MJ, Radabaugh KR, Moyer RP, Muller-Karger FE (2018) Enabling efficient, large-scale high-spatial resolution wetland mapping using satellites. Remote Sens Environ 208:189–201

Meier J, Zabel F, Mauser W (2018) A global approach to estimate irrigated areas–a comparison between different data and statistics. Hydrol Earth Syst Sci 22:1119

Millard K, Richardson M (2015) On the importance of training data sample selection in random forest image classification: a case study in peatland ecosystem mapping. Remote Sens 7:8489–8515

Oliphant AJ, Thenkabail PS, Teluguntla P, Xiong J, Gumma MK, Congalton RG, Yadav K (2019) Mapping cropland extent of Southeast and Northeast Asia using multi-year time-series Landsat 30-m data using a random forest classifier on the Google Earth Engine Cloud Int J Appl Earth Obser Geoinf 81:110–124

Ozdogan M, Gutman G (2008) A new methodology to map irrigated areas using multi-temporal MODIS and ancillary data: an application example in the continental US. Remote Sens Environ 112:3520–3537

Pan Z, Huang J, Zhou Q, Wang L, Cheng Y, Zhang H, Blackburn GA, Yan J, Liu J (2015) Mapping crop phenology using NDVI time-series derived from HJ-1 A/B data. Int J Appl Earth Obser Geoinf 34:188–197

Patel NN, Angiuli E, Gamba P, Gaughan A, Lisini G, Stevens FR, Tatem AJ, Trianni G (2015) Multitemporal settlement and population mapping from Landsat using Google Earth Engine. Int J Appl Earth Obser Geoinf 35:199–208

Pekel J-F, Cottam A, Gorelick N, Belward AS (2016) High-resolution mapping of global surface water and its long-term changes. Nature 540:418–422

Pittman K, Hansen MC, Becker-Reshef I, Potapov PV, Justice CO (2010) Estimating global cropland extent with multi-year MODIS data. Remote Sens 2:1844–1863

Platas-Garza MA, de la O Serna JA (2010) Dynamic harmonic analysis through Taylor–Fourier transform. IEEE Trans Instrum Meas 60:804–813

Sidhu N, Pebesma E, Câmara G (2018) Using Google Earth Engine to detect land cover change: Singapore as a use case. Eur J Remote Sens 51:486–500

Tateishi R, Hoan NT, Kobayashi T, Alsaaideh B, Tana G, Phong DX (2014) Production of global land cover data-GLCNMO2008. J Geog Geol 6:99

Teluguntla P, Thenkabail PS, Oliphant A, Xiong J, Gumma MK, Congalton RG, Yadav K, Huete A (2018) A 30-m landsat-derived cropland extent product of Australia and China using random forest machine learning algorithm on Google Earth Engine cloud computing platform. ISPRS J Photogramm Remote Sens 144:325–340

Thenkabail P, GangadharaRao P, Biggs T, Krishna M, Turral H (2007) Spectral matching techniques to determine historical land-use/land-cover (LULC) and irrigated areas using time-series 0.1-degree AVHRR pathfinder datasets. Photogramm Eng Remote Sens 73:1029–1040

Thenkabail PS, Biradar CM, Noojipady P, Dheeravath V, Li Y, Velpuri M, Gumma M, Gangalakunta ORP, Turral H, Cai X (2009). Global irrigated area map (GIAM), derived from remote sensing, for the end of the last millennium. Int J Remote Sens 30:3679–3733

Thenkabail PS, Knox JW, Ozdogan M, Gumma MK, Congalton RG, Wu Z, Milesi C, Finkral A, Marshall M, Mariotto I (2012) Assessing future risks to agricultural productivity, water resources and food security: how can remote sensing help? PE&RS, Photogramm Eng Remote Sens 78:773–782

Veloso A, Mermoz S, Bouvet A, Le Toan T, Planells M, Dejoux J-F, Ceschia E (2017) Understanding the temporal behavior of crops using Sentinel-1 and Sentinel-2-like data for agricultural applications. Remote Sens Environ 199:415–426

Xie Y, Lark TJ, Brown JF, Gibbs HK (2019) Mapping irrigated cropland extent across the conterminous United States at 30 m resolution using a semi-automatic training approach on Google Earth Engine. ISPRS J Photogramm Remote Sens 155:136–149

Xiong J, Thenkabail PS, Gumma MK, Teluguntla P, Poehnelt J, Congalton RG, Yadav K, Thau D (2017a) Automated cropland mapping of continental Africa using Google Earth Engine cloud computing. ISPRS J Photogramm Remote Sens 126:225–244
Xiong J, Thenkabail PS, Tilton JC, Gumma MK, Teluguntla P, Oliphant A, Congalton RG, Yadav K, Gorelick N (2017b). Nominal 30-m cropland extent map of continental Africa by integrating pixel-based and object-based algorithms using Sentinel-2 and Landsat-8 data on Google Earth Engine. Remote Sens 9:1065

Big Data Analytics for Climate-Resilient Food Supply Chains: Opportunities and Way Forward

Navin Kumar C. Twarakavi, Kamal Das, Mohamed Akram Zaytar, Fred Otieno, Jitendra Singh, Bruno Silva, Komminist Weldemariam, and Shantanu Godbole

Abstract A reliable forecast of food production in a given region, under the effects of climate change and increased occurrence of extreme events, is a prerequisite to developing resilience in the future food supply. As the climate is changing, an increasing occurrence of extreme events combined with shift in seasonal weather pattern is rendering traditional agricultural practices a high level of risk. Currently, strategies to plan for an upcoming season are based on data from the recent seasons. Current methods to forecasting food production levels and derisk an upcoming season in any given region, are rudimentary and at best, not scalable. The advent of big data and new data sources such as weather forecasts, remote sensing, scalable machine-learning methods and cloud computing has created new opportunities for understanding the impact of an upcoming growing season. In order to demonstrate the usefulness of the current data sources and methods, this chapter presents a methodology that combines seasonal weather forecasts, geo-spatial information derived from remote-sensing, risks posed by extreme events and crop growth models to estimate production risk

N. K. C. Twarakavi (✉) · K. Das · M. A. Zaytar · F. Otieno · J. Singh · B. Silva · K. Weldemariam · S. Godbole
Future of Climate, IBM Research, Yorktown Heights, United States
e-mail: twarakavi@gmail.com

K. Das
e-mail: kdas3@in.ibm.com

M. A. Zaytar
e-mail: Mohamed.Akram.Zaytar@ibm.com

F. Otieno
e-mail: Fred.otieno@ibm.com

J. Singh
e-mail: jitens@in.ibm.com

B. Silva
e-mail: sbruno@br.ibm.com

K. Weldemariam
e-mail: k.weldemariam@ke.ibm.com

S. Godbole
e-mail: Shantanu.godbole@in.ibm.com

G. P. O. Reddy et al. (eds.), *Data Science in Agriculture and Natural Resource Management*, Studies in Big Data 96, https://doi.org/10.1007/978-981-16-5847-1_9

at a regional scale. The method was validated for multiple growing seasons in some counties in Iowa.

1 Introduction

Much of the global food demand is dependent on a complex web of inter-dependent supply chains. Currently, food supply that is sourced from diverse geographies around the world finds its way to most of the global communities, with efficiency and speed. But, as the impact of climate change is increasingly felt on global systems, questions about risks to these food supply chains as well as the resilience of the same are being raised (Lobell and Gourdji 2012). Food production is a key node in the food supply chains and perhaps has the largest effect on their resilience. Therefore, understanding the vulnerability of food production in the future is a key step for food supply chains to continue to serve the world's population (Ray et al. 2019). Food production is currently vulnerable to climate change through the following pathways: (a) increased occurrence of extreme events that lead to production loss (i.e., yield loss), (b) modal shifts in predicted seasonal weather because of climate change that can potentially, create unfavorable conditions for crop growth (Beddow and Pardey 2015). There is a need to understand the combined effect of these aforementioned pathways on food production to adequately capture the risk posed to the food supply chains.

Increased occurrence of extreme events induced by climate change is having a significant effect of the state of agriculture and food security in the world. Some of these extreme events include drought, heat wave, cold wave, high winds, pluvial flooding (rain-based inundation) and fluvial flooding (flooding due to rivers breaching their banks).

Climate change is also resulting in systemic shifts in seasonal weather patterns. Agriculture around the world is dependent on the fact that the seasonal weather pattern that is most suitable for a crop repeats itself annually albeit with some minor differences. For example, rice growers look forward to a seasonal weather pattern that has moderate temperatures, reasonable rainfall early in the season to meet crop water requirements of the crop and a dry period in the later part of the season to allow for grain to dry. Climate change is shifting these seasonal weathers or changing the pattern of the weather altogether, leading to non-optimal sowing of the crop or unsuitability of the crop respectively. Therefore, understand these effects of climate change has also become very important.

Understanding the risk to food production at macro-scales has long been attempted by industries, various governmental and inter-governmental organizations. The US Department of Agriculture's (USDA) Foreign Agriculture Service (FAS) is probably, among the most well-known institutions that updates on the state of agriculture around the world (USDA 2021). However, the services provided inform on current status through remote sensing or through ground truthing. Similarly, the Food and

Agricultural Organization (FAO) provides statistics on the state of world food production (FAO 2021). But, these statistics are at a country-level and can only be used for hind-cast analysis.

In this chapter, a novel approach to forecast the production risk of crops at the regional level, using geo-spatial data, seasonal weather forecasts, probability of extreme events and crop-growth modeling frameworks is presented. The approach balances the need to provide accurate estimates of production risk with the computational requirements. The proposed approach is demonstrated for forecasting Corn production in Iowa, USA. For the sake of validation, we consider the years 2016–2019, as this enables us to compare with the actual estimates. The chapter is organized as follows:

The second section of the chapter introduces related work to the subject of interest. The third section introduces the proposed methodology as well as the study area of interest. The fourth section discusses some of the results and finding from the application of the methodology to the study area and the fifth section provides some conclusions and scope for future research.

2 Related Work

Regional crop yields have typically been assumed to be used as a proxy for production risk. Estimating regional crop yields has been done through the following approaches: (a) ML models that use weather, satellite imagery and other ancillary data sources to predict crop yields, (b) process-based models that simulate the crop growth processes to predict the crop yields and (c) a hybrid of (a) and (b).

A literature review of the ML-based crop yield prediction models revealed many studies that vary based on scale, geological position and crop. Among the methods used to predict crop yields, deep learning algorithms were shown to be the most effective (Klompenburg et al. 2020). The reason behind the effectiveness of deep learning algorithms lies in the fact that crop growth and resulting yield is highly non-linear function of agroecological and weather. Deep learning algorithms perform best when the underlying relationship between inputs and output is a highly non-linear function. For these reasons, deep learning methods perform better than other ML models. Within deep learning algorithms, Convolutional Neural Networks (CNN) is the most widely used deep learning algorithms (Tedesco-Oliveira et al. 2020; Yang et al. 2019; Nevavuori et al. 2019) and the other widely used deep learning algorithms are Long-Short Term Memory (LSTM) (Ju et al. 2020; Jiang et al. 2020; Schwalbert et al. 2020). While ML models have shown considerable success, their applicability across regions is highly dependent on calibration and availability of input data. Process-based models such as WOFOST (Boogaard et al. 1998), DSSAT (Jones et al. 2003) and Aquacrop (Steduto et al. 2009) have shown applicability across regions, crops and conditions to predict crop yield. By mathematically characterizing the interactions in the soil-plant-atmosphere continuum, process-based models predict crop growth and yield. While process-based models can be applicable across multiple regions, it suffers from scalability issues and poor uncertainty quantification. The proposed approach uses geo-spatial modeling techniques to find clusters of similar risk and agro-climatic properties upon which the process-based models can be applied to achieve accuracy and efficiency at scale. The proposed approach provides the right trade-off between accuracy of prediction and ability to perform forecasts at scale.

3 Approach

The objective of our research is to estimate the production risk for a crop in a given region. We define production risk pt for a crop during a given season *t*, as a yield that can be expected from the region *yt*, as compared to the yield during benchmark season *yb* as shown in Eq. 1. We consider the growing season in a pre-selected year as the benchmark season for our estimation.

$$pt = \frac{y_t}{y_b} \tag{1}$$

Figure 1 shows our methodology describing the steps to calculate the production for a crop, region and year. The workflow consists of the following steps: (a) generation of long-term seasonal forecasts (dataset 1) as well as maps showing probability

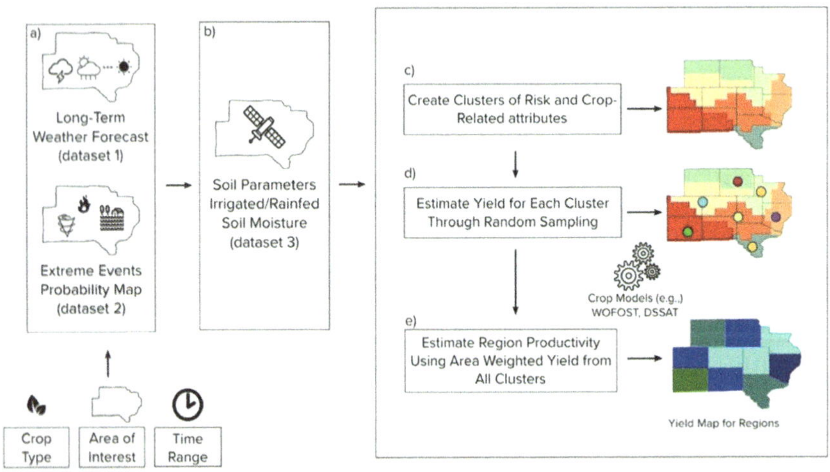

Fig. 1 Production forecasting estimation (ProdEstimator) approach overview. In Step **a**, extract weather forecast and probability maps for extreme events during the growing season; In Step **b**, we extract crop-related parameters such as soil data, irrigation/rainfed statistics; and, Steps **c**–**e** collectively describe estimation of production for the region

for occurrence of extreme events (dataset 2), relevant to the crops; (b) deriving soil parameters, irrigation/rainfed distribution, soil moisture maps from remote sensing imagery (dataset 3); (c) create spatially contiguous clusters that have distinct risk and crop-related attributes, using datasets 2 and 3; (d) estimate the expected yield for each cluster through random sampling using dataset 1 and crop models and (e) estimate production for the region as the area-weighted yield from all the clusters. Each of the steps as well as the datasets is described below.

3.1 Forecasting Seasonal Weather and Extreme Events

Since we are hind-casting to prove the applicability of our approach, we used the ERA5 reanalysis data to develop seasonal weather forecast as well as maps showing the occurrence of extreme events. ERA5 reanalysis data is used because of its global coverage as well as the availability of weather parameters needed for crop modeling. Seasonal weather forecast data consisted of daily records comprising the following: maximum and minimum temperature, solar radiation, dew point temperature and total precipitation. Extreme event maps were generated for the region that indicates the probability of occurrence of the following: (a) inundation, (b) drought, (c) heat wave and (d) cold wave. Inundation is described as standing water on a farm that inhibits a crop from performing its biological activities. This depth is crop-dependent and also dependent on the stage of crop growth. For corn, we assume this to be 10 cm of water. Drought is described as a prolonged duration of no rainfall. For corn, drought is defined as 2 weeks of no rainfall. Heat wave is described as prolonged duration of daily mean temperatures above 25 °C, that last at least a week. Cold-wave is described as a prolonged duration of daily mean temperatures below 15 °C, that lasts at least a week. There are several definitions of the extreme events described above and we choose the above definitions for simplicity.

3.2 Deriving Agro-Climatic Maps

Using remote sensing data imagery, we derived maps of soil parameters, irrigation/rainfed status and climatic zones maps. SoilGrids data was derived from Hengl et al. (2017), irrigation and rainfed area maps were derived from Monfreda et al. (2008), agro-climatic maps were derived from climatology information from ERA5 Hersbach et al. (2020). All of the maps were processed using the IBM PAIRS platform (Lu et al. 2016), a spatial-data platform in order to enable computational efficiency and ease of access.

IBM PAIRS platform provides the following services: (a) a data curation service, where one can upload their geospatial data. The benefit of that is that the user's data becomes available for programmatic querying with all the already curated PAIRS data. (b) a data service, where user can query different datasets programmatically,

(c) search/discovery service for geospatial, where users can query PAIRS to identify or find all locations in a certain geographic area, which certain pre-defined requirements and (d) analytics platform service, which enables users to customize analytics on the cloud.

3.3 Creating Spatially Contiguous Clusters

Using the extreme event maps and crop-related map data, that were derived as described in the previous section, we created spatially contiguous clusters. These clusters are developed such that they have similar agroeconomic and weather patterns. For ensuring spatial contiguity, we used the clustering method described in Guo (2008). The clustering method is a contiguity-constrained agglomerative clustering method. Spatially contiguity of clusters is important in order to scale the solution. Clusters that are scattered can lead to estimates that are non-representative of agroecological zones. During the agglomeration process, we assumed complete linkage with full-order constraining to ensure spatial contiguity (Full-order Complete Linkage (CLK). This is because Full-order CLK produces much better regionalized clusters than other assumptions. The number of clusters were selected based on the silhouette metric (Rousseeuw 1987). Silhouette metric checks consistency within clusters of data. The silhouette metric is a measure of how similar an object is to its own cluster (cohesion) compared to other clusters (separation).

3.4 Production Estimation

The first step to estimate the production in a region is to estimate an expected yield for each of the clusters that were generated. In particular, in order to provide a trade-off between computational requirement and how well the estimated yield represents the cluster, we selected samples through sampling from each cluster (Smith et al. 2017). Here, we used a stratified random sampling approach to ensure that variations within each cluster are quantified.

Once the samples were selected within a cluster j, we ran crop simulation models for each of the samples. For each sample, soil, weather forecast for the year t, and other agronomic information were extracted and provided as inputs to the crop models. We used the WOFOST to estimate the yield. WOFOST was preferred over other models such as DSSAT and AquaCrop, because it (a) achieves reliable estimates with relatively less number of parameters, (b) the required parameters are readily available through global geo-spatial databases. More details on the model comparison can be found in Boogaard et al. (1998).

Once crop yields were estimated for each sample in a cluster, an expected yield for the cluster, y_{tj}, was estimated as the average of the individual sample's yield. It is important to note the difference between yield and production estimation, as referred in this work. Yield estimation refers to an estimate at a point (or a cluster of points), while production estimate, refers to the entire region being considered. Using the expected yields from each of the clusters j, each with an area a_j ($j = 1...k$), we estimate the production of the crop under consideration for the region and year of interest, t. One could apply the same approach for estimating a production for a benchmark year yb.

4 Experiment and Findings

For our experiment, we estimated the corn production risk for counties in the southeast of Iowa (USA) for the years 2017–2019. The benchmark year, b was chosen as 2016, for the purposes of demonstration. We choose the benchmark year following FAO/USDA recommendations (between 2014 and 2016) and used the available county yield data for estimating the cultivar parameters for the crop models. Seasonal weather data for the benchmark year as well as the year of interest were derived from ERA5 reanalysis data. In the study area, sowing activity follows a normal distribution between April and May, while harvesting follows similarly between September and November. Our research, therefore, gathered weather information only relevant to the growing season. Figure 2 describes the climatological characteristics of the counties chosen for our study.

In the study area, we considered flood as the singular extreme event during production risk estimation. However, the presented approach is general enough to evaluate

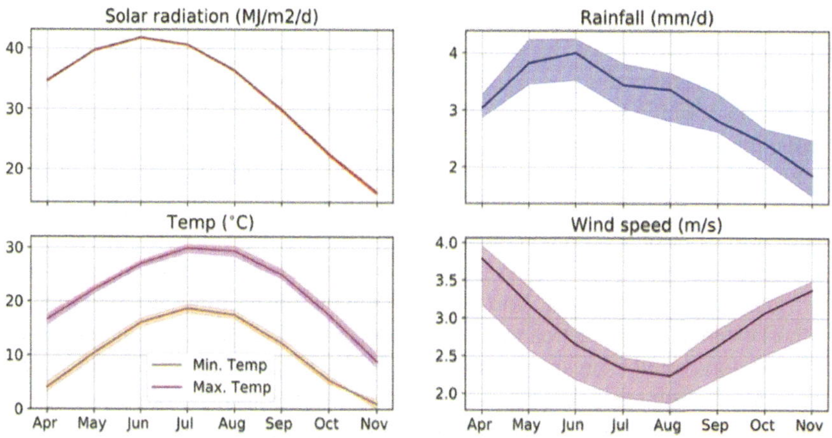

Fig. 2 Median estimates (and ranges) of climatological variables for the chosen counties in southeast Iowa during the Corn growing season. For the entire growing season, the optimum growing conditions for Corn are: minimum daily temperature of 16.6 °C, maximum temperature of 31 °C and an average daily rainfall of at least 3 mm/d. The counties chosen for our study present ideal conditions for growing

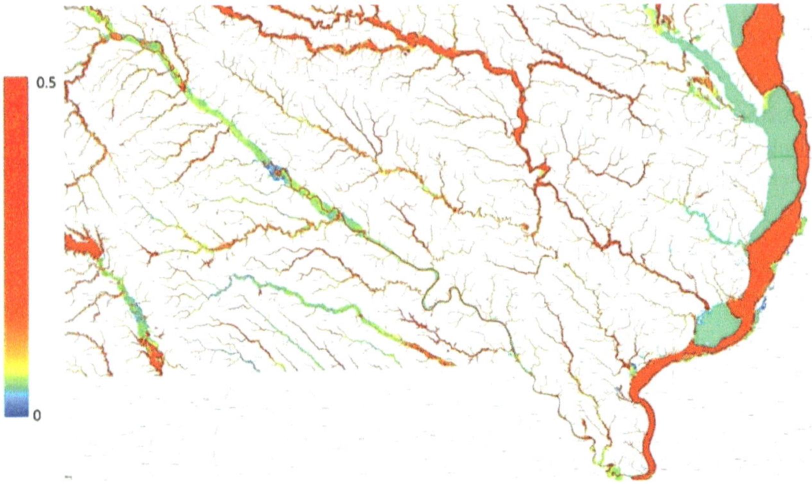

Fig. 3 Probability of flood during corn cultivation for the counties chosen in our study. Note that elevation gradients as well as flow levels in the rivers keep the flood plains closer to the river banks

other extreme events. Flood risk could be dynamic when considering climate change and the increasing non-stationarity nature of its effects. However, for the sake of simplicity, we considered a similar flood risk for both the benchmark and year of interest. We used the flood plain maps from US FEMA, as the proxy for flood risk (Fig. 3).

Based on the procedure described in Sects. 3.2 and 3.3, we created clusters in our study area that possess distinct risk and agro-climatic profile. Figure 4 shows the clusters generated for the selected counties. For our study area, we derived 7 clusters. It is worth noting that the cluster boundaries do not follow county boundaries and this observation has key implications for planning long-term resilience of food production. Most governmental organizations, such as the USDA, plan strategies for the future, based on administrative boundaries. The non-optimality of such a strategy as demonstrated in our result, will become more apparent, as the impact of climate change increases on food production.

According to the procedure described in Sect. 3.4, we estimated the production of Corn in the counties of interest. The number of samples for each cluster that resulted in a robust estimate of production was between 70 and 100 samples. Figure 5 shows the estimates forecasted using our methodology when compared with USDA estimates. The results indicate that our proposed approach can reasonably predict the regional production levels. Observe the USDA survey is available after the growing season and tends to be incomplete on a year-to-year basis. Differently, our approach can enable prediction before the completion of growing season and can be completed for all areas of interest, while allowing for significant reduction in cost.

Using methodology and county-level production forecasted previously, we estimated the production risk at county-level. Figure 6 shows a strong correlation

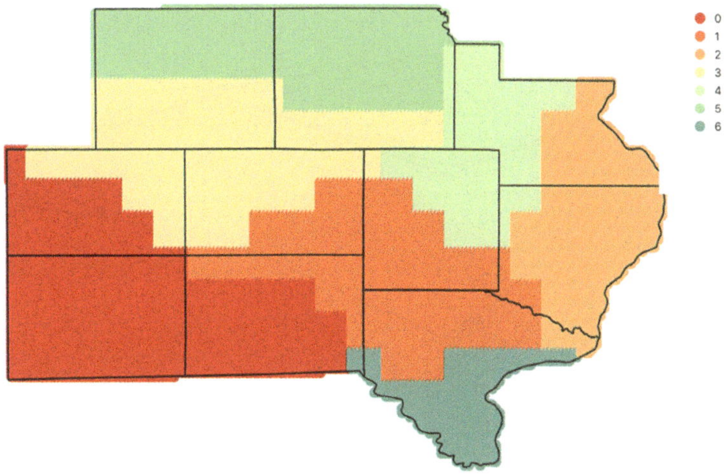

Fig. 4 Clusters based on similarity of risk and agro-climatic profile for Corn cultivation in the study area. The clusters represent similar risk and agro-climatic zones in the study area. Note the orientation of some clusters along the major rivers (Fig. 3) indicating influence of the flood risk

between the predicted and estimated production risks. Although not shown here, an important observation during our study was the balance achieved between computational requirement and accuracy of the production risk estimate. Our approach performed similar production risk estimates, while consuming only 15% of the time it takes to run a comprehensive process-based model across the entire region. One can further reduce computation time by allowing some more uncertainty in the estimates. This is important, because it enables scalability for performing computations at continental scales, which in turn, allows for a novel solution to assist resiliency planning by relevant stakeholders.

5 Conclusions, Limitations and Way Forward

The proposed methodology to estimate production risks at regional level seems to achieve the right trade-off between computational requirement and accuracy. The usefulness of estimating production risk will become more apparent as climate change-induced risks will lead to a need for strategies that will lead to resilience. By balancing accuracy and scalability, the proposed methodology ensures business impact.

While our methodology showcases one method that uses big data analytics mitigating the impact of climate change on future food supply, we believe that more opportunities are available in the future. Along with the opportunities, there are however some challenges. A key challenge would be the accuracy of weather forecasting models. For example, while we used hindcasting to test the validity of our

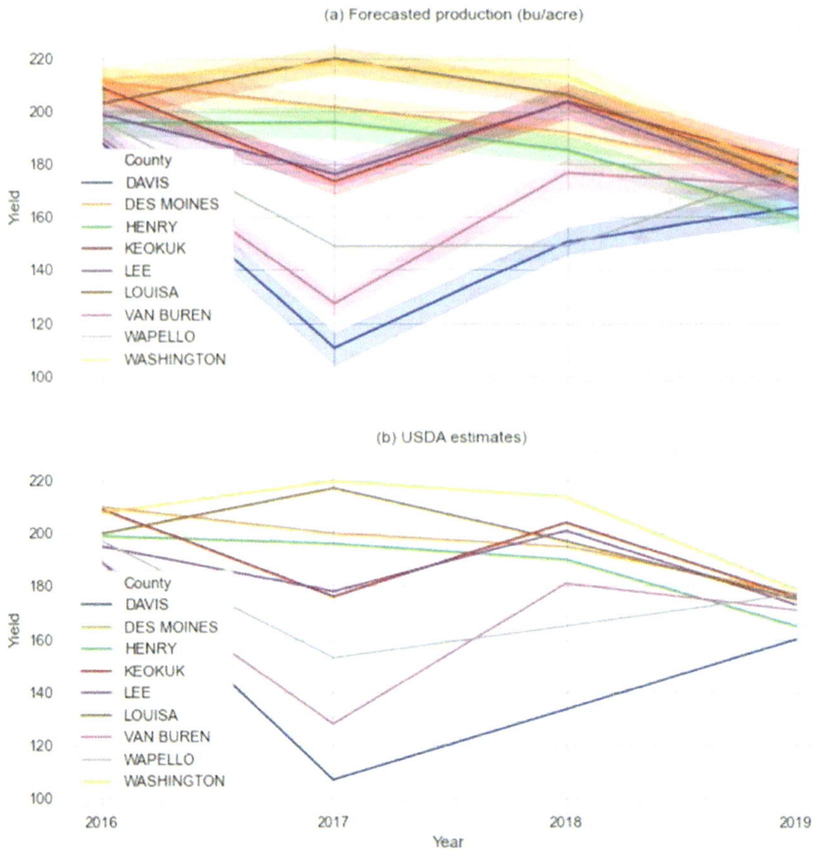

Fig. 5 Comparison between our results and USDA survey estimates (in bu/acre). **a** Production forecast and associated uncertainty for the counties of interest during the growing seasons of 2016–2019 and **b** USDA survey estimates. The standard deviations of our predictions in (**a**) indicate that data and structural uncertainty is well addressed through the approach

approach, we will need to analyze the approach in a forecasting scenario, in order to gain confidence in its applicability. A key hurdle that one can expect to face during forecasting is the increasing uncertainty in weather projections as the lead time increases. Currently, methods have been developed (such as Ham et al. 2019) to extend the commercially available seasonal forecasts to longer periods (up to 1 year).

Another challenge one can foresee is the need to test the forecasting tools in the developing world. For example, our methodology also needs to be tested in the developing world for relevance, as socio-economic factors play a much more influencing role on the production risk. We will incorporate human-resources indices as variables to account for the socio-economic effects.

Thirdly, there is a need to create more complicated models for crop simulation that can consider a variety of exogenous factors such as agronomic skill of the farmers,

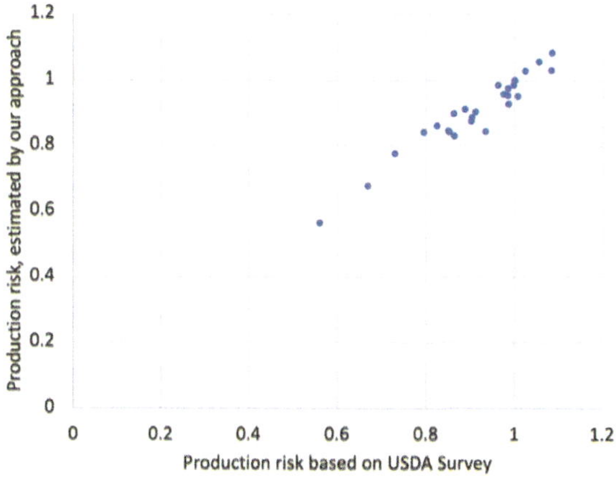

Fig. 6 Predicted versus measured county-level production risk estimates for Corn in the selected counties of Iowa. The R^2 is 0.95 and MAPE is 2.89%, indicating a strong correlation between actual and predicted production risks

among others. For example, in our approach, we use a process-based model. We identified process-based models as a bottleneck in further decreasing the computational requirement of our approach. In this regard, one could use Physics-Informed Neural Networks (PINNs) to replace the current crop simulation models that are process-based models, to enable similar accuracy with reduced computational effort and more input data considerations.

References

Beddow JM, Pardey PG (2015) Moving matters: the effect of location on crop production. J Econ History 75:219–249. https://doi.org/10.1017/S002205071500008X

Boogaard H, Van Diepen C, Rotter R, Cabrera J, Van Laar H (1998) WOFOST 7.1; user's guide for the WOFOST 7.1 crop growth simulation model and WOFOST Control Center 1.5. SC-DLO

FAO: FAOSTAT (2021) http://www.fao.org/faostat/en/

Guo D (2008) Regionalization with dynamically constrained agglomerative clustering and partitioning (REDCAP). Int J Geograph Inf Sci 22:801–823. https://doi.org/10.1080/13658810701674970

Ham Y, Kim J, Luo J (2019) Deep learning for multi-year ENSO forecasts. Nature 573:568–572

Hengl T, Mendes J, Heuvelink GB, Ruiperez Gonzalez M, Kilibarda M, Shangguan W et al (2017) SoilGrids250m: global gridded soil information based on machine learning. PLoS one. 12:e0169748

Hersbach H, Bell B, Berrisford P et al (2020) The ERA5 global reanalysis. Q J R Meteorol Soc 146:1999–2049

Jiang H, Hu H, Zhong R, Xu J, Xu J, Huang J, Wang S, Ying Y, Lin T (2020) A deep learning approach to conflating heterogeneous geospatial data for corn yield estimation: a case study of the US Corn Belt at the county level. Global Change Biol 26:1754–1766. https://doi.org/10.1111/gcb.14885

Jones JW, Hoogenboom G, Porter CH, Boote KJ, Batchelor WD, Hunt L, Wilkens PW, Singh U, Gijsman AJ, Ritchie JT (2003) The DSSAT cropping system model. Eur J Agron 18:235–265

Ju S, Lim H, Heo J (2020) Machine learning approaches for crop yield prediction with MODIS and weather data. Presented at the January (2020)

Lobell DB, Gourdji SM (2012) The influence of climate change on global crop productivity. Plant Physiol 160:1686. https://doi.org/10.1104/pp.112.208298

Lu S, Shao J, Freitag M, Klein LJ et al (2016) IBM PAIRS curated big data service for accelerated geospatial data analytics and discovery. In: IEEE big data. IEEE, pp 2672–2675

Monfreda C, Ramankutty N, Foley JA (2008) Farming the planet: 2. Geographic distribution of crop areas, yields, physiological types, and net primary production in the year 2000. Global Biogeochem Cycles 22. https://doi.org/10.1029/2007GB002947

Nevavuori P, Narra N, Lipping T (2019) Crop yield prediction with deep convolutional neural networks. Comput Electron Agric 163:104859. https://doi.org/10.1016/j.compag.2019.104859

Ray DK, West PC, Clark M, Gerber JS, Prishchepov AV, Chatterjee S (2019) Climate change has likely already affected global food production. PLoS one 14:1–18. https://doi.org/10.1371/journal.pone.0217148

Rousseeuw PJ (1987) Silhouettes: a graphical aid to the interpretation and validation of cluster analysis. J Comput Appl Math 20:53–65. https://doi.org/10.1016/0377-0427(87)90125-7

Schwalbert RA, Amado T, Corassa G, Pott LP, Prasad PVV, Ciampitti IA (2020) Satellite-based soybean yield forecast: integrating machine learning and weather data for improving crop yield prediction in southern Brazil. Agric Forest Meteorol 284:107886. https://doi.org/10.1016/j.agrformet.2019.107886

Smith ANH, Anderson MJ, Pawley MDM (2017) Could ecologists be more random? Straightforward alternatives to haphazard spatial sampling. Ecography 40:1251–1255. https://doi.org/10.1111/ecog.02821

Steduto P, Hsiao TC, Raes D, Fereres E (2009) AquaCrop—the FAO crop model to simulate yield response to water: I. Concepts and underlying principles. Agron J 101:426–437

Tedesco-Oliveira D, da Silva RP, Maldonado W, Zerbato C (2020) Convolutional neural networks in predicting cotton yield from images of commercial fields. Comput Electron Agric 171:105307. https://doi.org/10.1016/j.compag.2020.105307

USDA: PSD Online (2021). https://apps.fas.usda.gov/psdonline/app/index.html#/app/home

Van Klompenburg T, Kassahun A, Catal C (2020) Crop yield prediction using machine learning: a systematic literature review. Comput Electron Agric 177:105709. https://doi.org/10.1016/j.compag.2020.105709

Yang Q, Shi L, Han J, Zha Y, Zhu P (2019) Deep convolutional neural networks for rice grain yield estimation at the ripening stage using UAV-based remotely sensed images. Field Crops Res 235:142–153. https://doi.org/10.1016/j.fcr.2019.02.022

Part III
Data Science Applications in Natural Resource Management

Machine Learning Algorithms for Optical Remote Sensing Data Classification and Analysis

G. P. Obi Reddy and K. C. Arun Kumar

Abstract In recent years, Machine Learning (ML) algorithms have gained much attention and found a profound importance in processing, classification as well as analysis of multispectral, and hyperspectral remotely sensed data. The core objectives of this chapter are firstly to provide a critical review on important advanced ML algorithms in remote sensing data classification, and analysis; secondly, examine the performance of widely used important supervised ML algorithms namely Random Forest (RF), Support Vector Machine (SVM), and Classification and Regression Tree (CART) in satellite image classification, and analysis on Google Earth Engine (GEE) platform to derive distinct Land Use/Land Cover (LULC) classes. ML algorithms are being extensively used in optical remote sensing data analysis it includes the image classification algorithms to precisely allocate objects to a distinct set of known classes, the clustering algorithms to group the objects into classes based on a given set of input variables, the regression algorithms to forecast a response variable from a given a set of covariates, and the dimensionality reduction algorithms to build a small set of new variables that includes most of the information available in the input set of numerous variables. In the study, among the three tested supervised ML algorithms in LULC classification, CART algorithm shows relatively better performance than the RF, and SVM algorithms. The study concludes that advanced ML algorithms have immense potential in optical remote sensing data classification, and analysis to attain the higher classification accuracy.

Keywords Google Earth Engine · Image classification · Machine learning algorithms · Remote sensing data

G. P. O. Reddy (✉) · K. C. A. Kumar
ICAR-National Bureau of Soil Survey and Land Use Planning, Amravati Road, Nagpur 440033, India
e-mail: GPO.Reddy@icar.gov.in

G. P. O. Reddy et al. (eds.), *Data Science in Agriculture and Natural Resource Management*, Studies in Big Data 96, https://doi.org/10.1007/978-981-16-5847-1_10

1 Introduction

Earth observation (EO) data is collected from various platforms like ground-based, airborne, and satellite-based sensors, which offers vital information on various physical, chemical, and biological systems of the planet earth. Remote sensing has the potential to capture, and model various earth's surface processes, and their interactions with the overhead climate (Lillesand et al. 2008). Remote sensing collects the data of earth resources over a large geographic area at different spatio-temporal resolutions, and obtained images can be analysed to derive valuable information by adopting a systematic way of image classification (Miao et al. 2014). Recent trends in remote sensing technologies like robust satellite platforms with advanced sensor capabilities, and open data policies tremendously augmented the access of EO datasets for use in various applications (Toth and Jó´zków 2016; Reddy 2018). The accessibility of satellite data at various spectral, radiometric, spatial, and temporal resolutions has greatly transformed the satellite image classification, tools and techniques to be applied (Fauvel et al. 2013). Since the demand is increased for site-specific information from remotely sensed images in various EO applications (Reddy 2012; Reddy and Sarkar 2012; Andualem et al. 2018; Reddy and Singh 2018), it necessitates the use of advanced algorithms in image classification, and analysis (Reddy et al. 2012; Asokan et al. 2020).

Optical remote sensing systems can be broadly divided into multispectral, and hyperspectral. The multispectral remote sensing systems are designed with an aim to acquire data in the visible to short-wave infrared wavelength range. These systems generally have 4 to less than 20 spectral bands with the spatial resolution varying from coarse to very high. Various operational multispectral remote sensing systems like Landsat-8 OLI, SPOT, Sentinal-2, Cartosat, GeoEye, WorldView-3 are producing wide range of satellite data for use in various EO applications. On the other hand, hyperspectral remote sensing systems designed with hyperspectral sensors having a higher number of spectral bands generally range from more than 20 to 1000. The important multispectral, and hyperspectral satellites, sensors and their characteristics are shown in Table 1. With the advancements in multispectral and hyperspectral satellite sensors technology, these systems are capable of producing a wide range of satellite data. To process, and analyse such versatile remote sensing data needs advanced image processing, and analysis algorithms from machine learning (ML) domain in precisely mapping and monitoring of the earth resources.

In recent times, considerable progress has been made in evolving ML-based algorithms especially in EO applications (Lary 2010). ML algorithms offer vast potential in efficiently classification of remotely sensed images and aids in identification of hidden patterns in the dataset (Alpaydın 2014). Broadly, ML algorithms can be grouped into supervised or unsupervised depending upon the presence of labels in the training data. ML algorithms are being applied extensively in remote sensing especially for image classification, regression, and clustering. In recent years, a wide variety of ML algorithms like Random Forest (RF), Support Vector Machine (SVM), and Classification and Regression Tree (CART), Artificial Neural Networks (ANN),

Table 1 Important multispectral and hyperspectral satellites, sensors and their characteristics

	Satellite	Sensors*	Year of launch	Spatial resolution (m)	Swath (kms)	Revisit (in days)
Multispectral	Terra	MODIS, ASTER, CERES, MOPITT, MISR	1999	250, 500 & 1000	2330	16
	Aqua	MODIS, AIRS, AMSR-E, AMSU-A, CERES, HSB	2002	250, 500 & 1000	2330	16
	Landsat 8	OLI, TIRS	2013	15, 30 & 100	185	16
	Sentinel-2A	MSI	2015	10, 20 & 60	290	5
	Sentinel-2B	MSI	2017	10, 20 & 60	290	5
	RESOURCESAT-2A	AWiFS, LISS-3, LISS-4	2016	5.8, 23.5 & 56	370, 140 & 23.9	5 & 24
	RESOURCESAT-2	AWiFS, LISS-3, LISS-4	2011	5.8, 23.5 & 56	370, 140 & 23.9	5 & 24
	SPOT-6	NAOMI	2012	1.5	10	1
	SPOT-7	NAOMI	2014	1.5	10	1
	CartoSat-2B	PAN	2012	0.8	9.6	4
	GeoEye-1	GIS-MS, GIS-PAN	2008	0.41 & 1.65	15.2	
	WorldView-3	CAVIS, MSS, PAN, SWIR	2014	0.31, 1.24, 3.7 & 30	13.1 & 13.2	1
	CartoSat-3	MX, PAN	2019	0.28	16	4
Hyperspectral	GEO-KOMPSAT-2B	GEMS, GOCI-II	2020	250 & 7000	2500	
	HysIS	HySI	2018	30	30	
	PRISMA	PRISMA	2019	5 & 30	30	29
	EnMAP	HSI	2020	30	30	27
	HISUI	HISUI	2019	20	20	

Source: Authors compilation from different sources
MODIS Moderate Resolution Imaging Spectroradiometer, *ASTER* Advanced Spaceborne Thermal Emission and Reflection Radiometer, *CERES* Clouds and Earth's Radiant Energy System, *MOPITT* Measurement of Carbon Monoxide in the Troposphere, *MISR* Multi-angle Imaging Spectroradiometer, *AIRS* Atmospheric Infrared Sounder, *AMSR-E* Advanced Microwave Scanning Radiometer for EOS, *AMSU-A* Advanced Microwave Sounding, *HSB* Humidity Sounder for Brazil, *OLI* Operational Land Imager, *TIRS* Thermal Infrared Sensor, *MSI* Multispectral Imager, *AWiFS* Advanced Wide Field Sensor, *LISS* Linear Imagine Self Scanning System, *NAOMI* New AstroSat Optical Modular Instrument, *PAN* Panchromatic, *GIS-MS* Visible and NIR, *GIS-PAN* Panchromatic, *CAVIS* Cloud, Aerosol, water Vapor, Ice, Snow, *MSS* Multi Spectral Sensor, *SWIR* Shortwave Infrared Sensor, *MX* Multispectral VNIR, *GEMS* Geostationary Environment Monitoring Spectrometer, *GOCI* Geostationary Ocean Color Imager, *HySI* Hyper Spectral Imager, *PRISMA* Prism Spectrometer, *HSI* Hyperspectral Instrument, *HISUI* Hyperspectral Imager Suite

and Decision Tree (DT) classifiers are being widely used in satellite image classification (Petropoulos et al. 2012). The most popular SVM algorithm was widely used to classify multispectral, and hyperspectral satellite data of medium to high spatial resolution (Son et al. 2017). The efficiency, and computational capabilities of remote sensing image classification depends on various factors, which includes classification algorithms (Duro et al. 2012), sensor types (Duro et al. 2012), training samples (Mathur and Foody 2008) input features (Huang and Zhang 2013), pre-and post-processing techniques (Phiri et al. 2018), ancillary data (Corcoran et al. 2013) and target classes (Bangira et al. 2019). Hence, these factors influence significantly, while classifying the satellite images to enhance the overall accuracy of classification. Lu and Weng (2007) reported that the efficiency of satellite image classification primarily depends on input data, its resolution, quality of ancillary data used, objectives of classified map, and image classifier selected to perform the task. Shao and Lunetta (2012) reported that SVM algorithm has superior simplification capability and showed robust performance than ANN classifiers, and CART on small training datasets. In comparison to traditional parametric classifiers, ML algorithms have the ability to process and analyse high dimensionality data with higher accuracy in mapping of different classes having complex characteristics (Ghimire et al. 2012).

The chapter aimed firstly to provide a review on important ML algorithms in optical remotely sensed image processing, and analysis, secondly, to examine the performance of RF, SVM, and CART supervised ML algorithms in LULC classification on Google Earth Engine (GEE) platform. According to their analytical capabilities, and objectives, ML algorithms were grouped as classification, clustering, regression, and dimensionality reduction algorithms. Under the image classification category, supervised classification algorithms are extensively applied for remote sensing image classification, and analysis. These algorithms segment the spectral values of an image into regions, which reflects the distinct classes of interest on ground in a specific application. In clustering, unsupervised classification algorithms are extensively used in classification of remotely sensed data as these systems follow clustering procedures that are extremely fast with minimum operational parameters. With adoption of suitable procedures for ground-truth accuracy assessment, clustering algorithms have the ability to provide remarkably fast means of quality results on a continuous basis. Whereas, regression algorithms can be applied to estimate or predict a response variable based on a given set of covariates. Dimensionality reduction algorithms are used to seek a low-dimensional representation of either all or most of the numerical information of original input variables.

2 ML Algorithms in Remote Sensing Image Classification and Analysis

Image classification is the essential, and widely used approach in remotely sensed data processing, and analysis to extract the maximum information, and generate thematic maps for various applications. Accessibility of a variety of satellite image

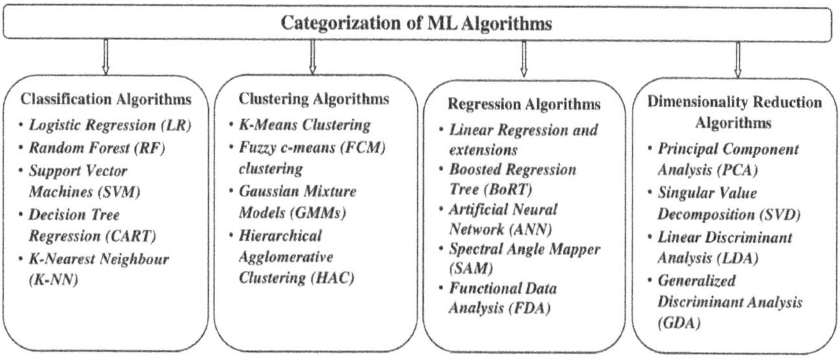

Fig. 1 ML algorithms based on their analytical capabilities in image classification

data at various resolutions, advanced computer processing, and data analytical capabilities have significantly improved the opportunities to extract, and analyse image information by using appropriate algorithms. The important ML algorithms categorized under classification, clustering, regression, and dimensionality reduction are shown below (Fig. 1).

2.1 Classification Algorithms in Remote Sensing Data Analysis

Supervised classification algorithms are being widely used in classification, analysis of remotely sensed images. These algorithms work on the principle of segmenting the spectral values of input image into regions/groups, which represents the ground cover classes of interest in a given application. Classification algorithms are generally applicable when the objective is to precisely assign objects into a distinct set of known classes with the help of input variables or covariates. Dataset comprising input, and output variables were needed to 'train' the model, and subsequently, applied on test datasets that comprised only the input variables. Like in crop type mapping, classification algorithms categorize the pixels in remotely sensed image into different crop categories depending upon a given training dataset that comprises variables derived from input images, and identified sample sites on the ground. Then the developed model can be run on the rest of the satellite image(s), where sample data is not available to classify different crop types.

2.1.1 Logistic Regression (LR)

As a supervised learning classification algorithm, LR is being widely applied to define the relationship of many independent variables on a dichotomous dependent variable (David and Mitchel 2010). The LR model can be developed by considering a set of input variables, where the response is known. The LR algorithm has the ability to predict the probability of an event happening by fitting the data into a logistic curve (Hosmer and Lemeshow 2000). LR has proved its usefulness in classification of satellite remote sensed data, particularly when the independent variables do not follow the normal distribution. LR algorithm can achieve the best fitting model to describe the association between the dependent, and independent variable (Pradhan 2010). In LR algorithm, incorporation of more independent variables in the analysis is the better to have a complete model and obtain accurate results (Ayalew et al. 2004). This algorithm can also apply to forecast the absence, or presence of an event based on various predictor variables (Chau and Chan (2005). By using LR algorithm, the spatial prediction of an event can be modelled (Shirzadi et al. 2012) by considering the dependent, and independent variables. In the LR algorithm, a dependent variable represents either presence, or absence of an event, if the dependent variables are binary, then the logistic function is valid for event susceptibility analysis (Atkinson and Massari 1998). LR was widely applied in remote sensing data analysis for change detection in LULC (Fraser et al. 2005).

2.1.2 Random Forest (RF)

As a supervised ML algorithm, in recent years, RF was extensively applied in remotely sensed data analysis (Lary et al. 2016) for classification as well as regression tasks (Breiman 2001). As an ensemble classifier, RF uses a set of classification trees in order to make a prediction (Belgiu and Drăgut 2016). RF uses bagging (or bootstrap aggregation) and also uses random selection of features for splitting. Bagging uses the original training dataset and makes duplicate subsets of the training dataset for training by choosing randomly with replacement. In RF, by utilizing many decision trees, predictive models are established. RF algorithms have ability in handling a massive number of correlated input variables including skewed dataset, and avoid model overfitting. These abilities make RF algorithms more robust in mapping of features on diverse landscapes (Attarchi and Gloaguen 2014). RF algorithm needs the description of the ensemble size, the maximum size of the individual trees, and the number of variables arbitrarily chosen at each node to generate a prediction model (Yang et al. 2016). In classification of crop/no crop through analysis of Landsat imagery, compared a number of methods, and reported that RF algorithms provide better accuracy, and robustness in model output (Schmidt et al. 2016). Mellor et al. (2015) established a relation between the class area, and allocation of training samples and the study reported that the larger areas of land cover class need a greater number of training samples in order to obtain the better classification accuracy. Jin et al. (2014) reported that the RF algorithm provides robust performance, if training sample data is well represented. The object and pixel-based RF classifications were used to analyse fused satellite images, and achieved overall accuracy of 80% in both the cases (Fu et al. 2017).

2.1.3 Support Vector Machines (SVM)

In satellite image classification, and data analysis, as a supervised classification algorithm SVM (Vapnik 1995) was found to be efficient as it produces highly accurate classification results with lesser training sample dataset (Mantero et al. 2005). Based on the training data, SVM depicts the input data into a high-dimensional feature space, where it develops a hyperplane (Kavzoglu et al. 2014). In SVM algorithm, in order to separate the original input space, the optimal linear hyperplane is used, and the kernel function helps to transform the data into two classes consisting of training, and validation dataset. Hence, SVM algorithm abilities depend on suitable kernel functions. SVM is being widely used because of its robust performance in classification, (Pal and Mather 2005), and regression (Burges 1998) to obtain least regression error, particularly in applications like time-series predictions (Müller et al. 1997). By using Sentinel-2 image data, the robustness of RF, SVM, and kNN algorithms was compared in LULC classification, and reported that SVM algorithm produced higher overall accuracy with minimum sensitivity to the given training datasets, followed by RF, and kNN algorithms (Noi and Kappas 2018). Breiman (2001) reported that SVM and RF algorithms depicted insensitivity to overtraining, which explains their capabilities in handling unstable data. Thus, in non-parametric algorithms, SVM, and RF algorithms are gaining prominence in satellite data classification. SVM algorithms are found to be vital in remote sensing images analysis as they have the potential in handling small training datasets in generating better classification accuracy as compared to the traditional methods (Maryantica and Lin 2017). Shafri and Zeen (2011) used hyperspectral data in mapping of urban areas and reported that SVM is the most effective method in classification of hyperspectral data. Despite these advantages, the challenges linked with SVM algorithm performance are the selection of a suitable kernel, and optimum kernel parameters.

2.1.4 Classification and Regression Tree (CART)

As an advanced technique of DT classifiers, CART can be built from a set of training data. CART has abilities in processing of continuous as well as minor attributes for both classification, and regression. In CART algorithm by using the binary partitioning, a regression tree is constructed, wherein every node is iteratively divided into two partitions. Firstly, all the input variables of algorithm were assembled into a similar partition. The critical phases of the CART algorithm include defining the set of rules to divide each node of the tree, choosing when the tree is complete, correlating each terminal node to a class to perform the regression (Breiman et al. 1984). The CART algorithm can handle both numerical and categorical data. To assess the robustness of CART algorithm, the mean squared error (MSE) can be computed and a split decision takes place depending on Greedy minimization. In the algorithm, the split, which actually minimizes the MSE can be designated for an additional split of that specific node into two off-springs, and this splitting rule can be applied to each of the new offsprings. In the algorithm, each tree grows to the largest possible extent and it helps to achieve better regression accuracy of the algorithm. The disadvantage of this process is that it can generate many complex trees, such trees unable to generalize the data effectively, and leads to 'overfitting' the algorithm. It has notable sensitivity to the given training datasets, any minor changes made in the training dataset can result in very different sets of subsets (Bishop 2006; Breiman 2001). The CART algorithm is being extensively used in remote sensing data analysis because of its simple interpretation abilities, high precision, and its robustness in characterization of complex interactions among variables. Foody and Mathur (2004) reported that in LULC classification by using remote sensing images, ML algorithms such as RF, SVM, and CART performed better in providing highly accurate results.

2.1.5 K-Nearest Neighbour (kNN)

As a nonparametric algorithm, kNN algorithm is being widely used in many data mining tasks (He and Wang 2007). kNN uses all contributing cases of the given dataset, and it classifies new cases effectively depending on their resemblance indices. The algorithm classified the cases by voting for neighbour classes, and the ideal case is the one that has the maximum resemblance indices (Wu et al. 2008). The best choice of the selected number of neighbours (k) in kNN algorithm depends on the metrics used for the tasks of classification as well as regression. In the algorithm, the k-value is sensitive to the selected dataset and it can also differ between datasets. In the algorithm k can be defined as 'equal to the square root of the number of samples' (Duda et al. 2012), hence, fine-tuning of k-value is a bit difficult for complex applications (Naidoo et al. 2012). The information of the neighbouring pixels helps kNN's algorithm to understand the underlying pattern in the training dataset (Calvo-Zaragoza et al. 2015). Algorithms like Tree ensemble, Naïve Bayesian, SVM, kNN, and ANN were tested to understand the role of algorithm selection, reference sample size, and class distribution as well as scene heterogeneity to achieve higher classification accuracy through analysis of Landsat data and reported that SVM as well as kNN algorithms were performed better in Landsat data classification (Heydari and Mountrakis 2018).

2.2 Clustering Algorithms in Remote Sensing Data Analysis

Clustering algorithms basically perform unsupervised classification tasks, which scrutinize a large number of unknown pixels, and divide them into a number of classes depending upon the actual groupings that exist in the input image values, where it doesn't require a definite training data (Hastie et al. 2008). In these algorithms, the computer fixes a spectrally separable class, and automatically defines their information value (Lilles and Kiefer 1994). In clustering analysis, algorithms assign data points to clusters, where items of similar clusters are as alike as possible, at the same time items of dissimilar clusters are as different as possible. Clustering algorithms can be applied to group objects into classes by using a set of input variables. In satellite image analysis, the pixel allocation into clusters depends on a set of input variables derived from input image, then obtained clusters are examined, labelled, and compared their characteristics.

2.2.1 K-Means Clustering

In ML, k-means clustering algorithm is a best example of non-hierarchical clustering, and is widely applied in clustering analysis especially in remotely sensed image processing (Zhong and Zhang 2010). The k-means algorithm accepts the data derived from k different clusters, and subsequently, allocates each unlabelled point to the closest cluster until no changes occur. This algorithm classifies the dataset based on the number of k-clusters, and for every cluster, k centroids are being defined. Usman (2013) applied a k-means clustering algorithm to process the high-resolution remote sensing data into different classes and then labelled them as distinct classes. Napoleon and Ramaraj (2014) reported that k-means clustering was found to be the best algorithm to classify and segment the images of high-resolution QuickBird satellites.

2.2.2 Fuzzy c-Means (FCM) Clustering

FCM as an unsupervised clustering algorithm was widely used to segment images into clusters with similar spectral properties. FCM has the ability to allow pixels belonging to several clusters with varying degrees of membership (Hartigan and Wong 1979). It uses a membership level in defining the association between points, and clusters (Bezdek 1984). FCM subdivides the given dataset into c-clusters, or classes (Foody 1996). It starts by randomly assigning pixels into different classes, and iteratively placing the pixels to other classes, while minimizing the generalized least-squared error. FCM uses the distance between pixels, and cluster centers in the spectral domain of image to calculate the membership function. In FCM clustering, every data point belongs to a cluster with some degree that is defined by a membership grade, where the sum of the memberships for each pixel must be equal to unity. FCM algorithm and its variants were successfully applied in applications like image classification, pattern recognition, data mining, and image segmentation. FCM algorithm performs well where difficulty exists in making assumptions on the statistical distributions of sample data, particularly if training sets of pure pixels are small in size in satellite image.

2.2.3 Gaussian Mixture Models (GMMs)

As a probabilistic model, GMM is used for clustering data with real-valued components, and fitting complex statistical distribution by using a set of simple density distribution weighted combinations. GMMs are widely used to characterize normally scattered sub-populations within the given total population. The advantage of these mixture models is that they do not need information on which subset of a data point belongs to and it allows the model to learn the sub-populations automatically. This algorithm does not count the spatial correlations, and it classifies satellite image data purely on spectral information of the image. Though GMMs play an important role in satellite image segmentation, it shows the difficulty in modelling asymmetric and multimodal dispersions of pixel intensities. GMMs are widely used with a Gaussian distribution as its component in satellite image segmentation tasks (Ji et al. 2017; Ban et al. 2018).

2.2.4 Hierarchical Agglomerative Clustering (HAC)

The HAC approach is being widely used to construct embedded classification schemes. In the HAC algorithm, every observation has its own cluster and then pairs of clusters merged sequentially to generate the hierarchy till left out with one cluster. To execute the process, the HAC algorithm requires a metric (distance) as well as a linkage (Hastie et al. 2005). In the HAC algorithm, the metric measures the degree of dissimilarity between the pairs of observations, whereas, the linkage measures the dissimilarity between the clusters. HAC has the ability to perform directly with the dissimilarity matrix, which is a depiction of dissimilarity between the pairs of objects. Medina et al. (2013) analysed hyperspectral data by using HAC algorithm and reported that it can effectively explain the spectral variability of vegetation, and such robust outputs help to effectively evaluate plant species diversity.

2.3 Regression Algorithms in Remote Sensing Data Analysis

Regression algorithms are applicable if the objective is to estimate or predict a response variable by using a set of given covariates, where the response is continuous in nature instead of categorical. Similar to the other classifiers, the regression algorithm is trained based on a given set of input variables, where the response is known. These algorithms range from simple linear regression to neural networks. The neural networks algorithms like artificial, convolutional, and deep neural networks, which are commonly used in satellite image classification at pixel level to obtain higher mapping accuracy (Maggiori 2017).

2.3.1 Linear Regression and Extensions

Under linear regression (LR), the Simple Linear Regression (SLR) model assumes there is a linear relationship between the input variables (x), and the single output variable (y). On the other hand, the Multiple Linear Regression (MLR) assumes the relationship between several independent variables, and a dependent variable. Huang et al. (2013) applied SLR on satellite spectral data to predict rice yields by considering the yield, and the Vegetation Index (VI) and reported a significant positive relationship between Normalized Difference Vegetation Index (NDVI), and the rice yields. Ryu et al. (2011) used MLR, and Partial Least Squares Regression (PLSR) algorithms with airborne, and satellite spectral data to predict the protein, and nitrogen contents of rice and concluded that both MLR, and PLSR algorithms showed higher accuracy of two-year models than that of single-year models. Zhou et al. (2017) applied linear regression models on multispectral time-series imagery acquired by Unmanned Aerial Aehicles (UAVs) combining with VIs to predict the rice yield, and reported highest positive correlation between VIs and rice yield. Matus-Hernández et al. (2018) applied MLR to predict the Chlorophyll-a concentration based on time-series Landsat imageries and found that MLR performed better in predicting log-transformed Chlorophyll-a. The major theoretical constraints of all regression algorithms are that only relationships can be achieved effectively, however, the underlying contributing mechanism can never be achieved with optimism. MLR cannot explicitly interpret the multi-collinearity between controlled variables (Yiging et al. 2013).

2.3.2 Artificial Neural Network (ANN)

Neural networks are basically a sequence of algorithms, which mimic the actions of a human brain to identify the relationships between big datasets. ANN can adapt to changing input; consequently, the network creates the best possible result without redesigning the output criteria. In ANN, the elements are structured in layers, where the algorithm has minimum input, and output layers with a neuron for every input variable, and output class. ANNs are extensively used in various remote sensing data classification especially in forecasting, and prediction of natural events and risk assessment. ANNs algorithms have efficient capabilities in modelling compound processes like rainfall prediction with accurate approximation (Abbot and Marohasy 2014). Kavzoglu and Mather (2003) explained the use of ANN with respect to satellite images and proved that ANNs produce higher classification accuracies by employing fewer training samples. The algorithms like ANN, SVM, RF, and k-NN were applied in predicting SOC with the help of DEM, and Landsat 8 data and reported that ANN showed the best performance in predicting SOC as compared to other algorithms (Taghizadeh-Mehrjardi et al. 2016).

2.3.3 Boosted Regression Tree (BoRT)

BoRT is a regression technique developed in combination with a regression tree and boosting algorithm (Friedman 2001), it has superior elasticity, and predictive ability in modelling the data. BoRT was widely used to quantify the relationship between dependent, and independent variables (Robinzonov 2013), determination of independent variables, classification, and forecasting studies (Aertsen et al. 2011). In BoRT, boosting algorithm is used to enhance the model accuracy by building suitable new trees for residual error (Rahmati et al. 2018), whereas, the regression trees are effectively used to label the classification system from the decision tree groups in the algorithm (Naghibi et al. 2018). In the BoRT algorithm, image sub-division can be designated as a tree-like structure, it consists of nodes, which represent the selected factors, branches act as if-else connectors between the nodes, and leaves represent the terminal nodes contains the subdivisions of responses (Robinzonov 2013). Generally, three parameters like total of boosting tree iterations, interface tree depth, and reduction are essential to optimize the BoRT model (Kuhn and Johnson 2013). Besides the computational performance and precision in estimation, BoRT can define complex nonlinearities, and relations between variables, and has the ability to accommodate missing data.

2.3.4 Spectral Angle Mapper (SAM)

Image classification is a vital component in remote sensing especially in extracting information, and pattern recognition from the imagery. The transition zone of any two classes poses a challenge to image classification, and hard classifiers, it poses difficulty in classifying these mixed pixels as the information content in a transition zone belongs to more than one class (Shivakumar and Rajashekararadhya 2017). SAM algorithm has the ability to fix the spectral resemblance between the given two spectra through computing the angle between them, by considering them as vectors in a space with dimensionality equal to the number of bands (Kruse et al. 1993). SAM classifier has many advantages, which includes, it is easy to use, rapid, and powerful classification algorithm to map the spectral comparison of image spectra to reference spectra and it also has the ability to suppresses the impact of shading effects while highlighting the target reflectance characteristics (Kruse et al. 1993).

2.3.5 Functional Data Analysis (FDA)

As a nonparametric algorithm, FDA algorithm is being widely used to describe, and classify curves, where every sample point in the time-series data is used to be a function perceived along an underlying continuum. FDA was mainly used in image classification, and feature reduction from remote sensing datasets. FDA algorithm was applied for a logistic regression in analysis of environmental data (Escabias et al. 2005). Cardot et al. (2003) tested a functional logistic regression algorithm to predict land use changes by using the time-series remotely sensed images. Liu et al. (2013) applied rotating functional factor analysis in order to improve the assessment of temporal changes by using time series vegetation indices derived from remote sensing images. FDA algorithms like functional Principle Component Analysis (PCA), bases smoothing, and linear representations have potential to analyse the variance of reflectance data in satellite remote sensing data.

2.4 Dimensionality Reduction Algorithms in Remote Sensing Data Analysis

In the recent past, noticeable increase is being observed in the availability of numerous of hyperspectral sensors, which efficiently collect information in many spectral bands through on-board satellites/airborne platforms. The high dimensional nature of hyperspectral image data of numerous spectral bands with huge size of information poses many challenges in image classification, and analysis (Lira 2010). In the field of hyperspectral image classification, the processing of redundant information, and dimensionality reduction are necessary (Thenkabail et al. 2012). These algorithms have the ability to effectively identify, and remove redundant information to improve the classifier performance, and accuracy (Feng et al. 2017). Dimensionality reduction algorithms are applicable if many variables exist in input remote sensing dataset to extract, and develop a small set of new variables that comprise all or most of the information contained in input variables. Extracted such new variables could be used as inputs in other image analysis activities or they could be considered as end results of that specific image analysis task.

2.4.1 Principal Component Analysis (PCA)

PCA works on a principle to reduce the dimensionality of a dataset, while preserving as much 'variability' as possible in the input dataset. PCA algorithm is widely used in analysis of hyperspectral images especially for dimensionality reduction (Theodoridis and Koutroumbas 2003). Potgieter et al. (2007) applied PCA on Moderate Resolution Imaging Spectroradiometer (MODIS) time-series data of Enhanced Vegetation Index (EVI) products to assess the crop area for various crops in North East Australia. PCA was widely used as a pre-processing technique in remote sensing for different purposes, which includes satellite image enhancement (Cheng et al. 2017), feature recognition (Bajwa et al. 2009), and change detection with multi-temporal images (Munyati 2008), evaluation of interannual vegetation anomalies (Lasaponara 2006), and image compression (Báscones et al. 2018).

2.4.1.1. PCA Applied on Landsat 8 Data in Dimensionality Reduction

The PCA is being widely used as a dimension reduction algorithm in satellite data processing and analysis to maximize the amount of information from the original bands into the least number of principal components. It is the optimal linear scheme for reducing a set of high dimensional vectors into a set of lower dimensional vectors. In the study, PCA has been applied on multi-temporal satellite data of Landsat-8 level 1 images of eight bands (Blue, Green, Red, Near Infrared, SWIR 1, SWIR2, and Cirrus) at 30 m spatial resolution captured on 14 April 2021. By using PCA, a set of correlated variables (original bands) have been transformed in new uncorrelated

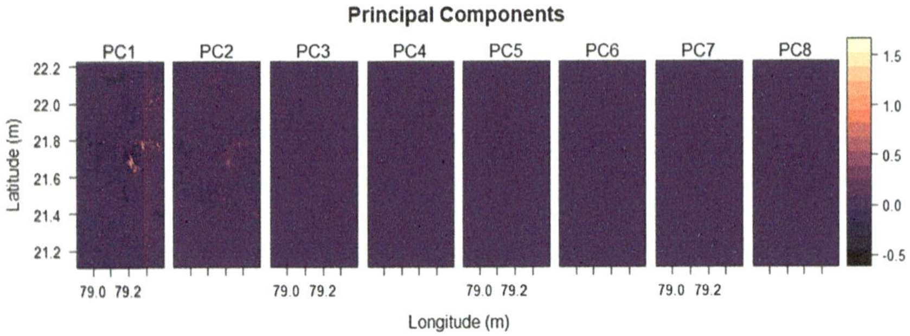

Fig. 2 PCA applied on Landsat 8 images in dimensionality reduction

variables (principal components), which contain the maximum original information. The PCA analysis performed on Landsat 8 images shows that the first three principal components contain about 98.25% of the information in the original eight bands (Fig. 2). The PC1 has the highest percentage of variance (78.21%) with a standard deviation of 0.096, followed by PC2 and PC3 with percentage of variance of 15.70 and 4.34% with a standard deviation of 0.043 and 0.022, respectively (Table. 2). PC1 of Landsat-8 was greatly associated with the bands represent green, red, and near-infrared (NIR) and shortwave infrared (SWIR) bands.

2.4.2 Singular Value Decomposition (SVD)

SVD algorithm is being used to decompose a matrix into several component matrices, revealing the useful, and interesting characteristics of the original matrix. SVD is a robust and reliable orthogonal decomposition technique widely used for image fusion as it has advantages like conceptuality, and stability. SVD can also adjust the variations that exist in the local statistics of an image (Yang and Lu 1995). SVD can be used to identify, and order the dimensions, where data points have the most variations. SVD algorithm is also widely used to fix the rank of the matrix, and measure the sensitivity of a linear system to numerical error. SVD analysis can be performed to obtain pairs of spatial patterns between two variables that are optimally interrelated (Bretherton et al. 1992) and such correlated spatial patterns can be used to develop statistical prediction models.

2.4.3 Linear Discriminant Analysis (LDA)

The LDA algorithm is commonly used in dimensionality reduction, and classification tasks (Duda et al. 2001). It was also applied to classify the patterns between two classes and LDA has the ability to extend in classification of multiple patterns. In the case of two classes, the LDA algorithm creates one hyperplane and projects the data to maximize the separation of the two categories. As a supervised feature extraction

Table 2 Importance of components obtained from Landsat 8 data

Components	PC1	PC2	PC3	PC4	PC5	PC6	PC7	PC8
Standard deviation	0.096061	0.043047	0.022622	0.011370	0.007825	0.003182	0.002003	0.001095
Proportion of variance	0.782117	0.157060	0.043376	0.010958	0.005190	0.000858	0.000340	0.000102
Percentage of variance	78.211691	15.706024	4.337584	1.095756	0.518965	0.085804	0.034017	0.010159
Cumulative proportion	0.782117	0.939177	0.982553	0.993511	0.998700	0.999558	0.999898	1.000000

algorithm, LDA has the ability to optimize the between-class scatter matrix, and diminish the within-class scatter matrix (Fukunaga 1990). Though LDA performance is good in many tasks, it poses challenges in processing hyperspectral image data like ill-posed problem, and calculation of the inverse matrix of the within-class scatter, as a result it cannot handle many spectral bands of hyperspectral image (Lu et al. 2005). Of late, many variations of the LDA algorithm have been developed, which have the ability to augment its robustness. LDA used for identification of tropical rainforest tree species by using hyperspectral images (Clark et al. 2005), Regularized LDA (RLDA) has the ability to address the ill-posed problem (Bandos et al. 2009), Semi-supervised Discriminant Analysis (SDA) helps to extend the training set (Cai et al. 2007) and classical LDA has the potential in recognition of conifer tree species with hyperspectral images (Peng et al. 1997).

2.4.4 Generalized Discriminant Analysis (GDA)

GDA is the nonlinear extraction technique (Baudat and Anouar 2000) and it overcomes the limitations of LDA. GDA has the ability to condense the high dimensional data vectors, and identification can be held in a low dimensional space with higher competence. GDA can map input vectors into a high dimensional feature space and has potential to work with nonlinear discriminant analysis by applying kernel function operators (Baudat and Anouar 2000). By applying a diverse kernel function, a wide range of nonlinearities can be covered. GDA seeks a projection space that transforms features into a lower dimensional space, and maximizes the ratio of the between-class variance with respect to the within-class variance. The GDA space preserves the most valuable information and reduces the training time of the used classifier that shows the high classification efficiency of the algorithm (Zhang and Yeung 2011).

3 Evaluation of Supervised ML Algorithms in LULC Classification on GEE Platform

LULC models provide a wide range of interactions between surface change, and human activities (Yesuph et al. 2019). Accurate LULC maps help to show how land cover change affects the ecosystem services (Zhang et al. 2020). In the chapter, the performance of RF, SVM, and CART algorithms was evaluated by using cloud-free high resolution Sentinal-2 multispectral imageries at 10 m spatial resolution of heterogeneous landscape of central India on cloud-based GEE platform. These Sentinel-2 imageries were selected for the study by using the date range of (filterDate) single image and lastly sorting by cloud cover ('CLOUD_COVER') and extracted in a false colour composite (FCC). By using geometry tools of GEE, training set samples were generated with the help of earth engine code editor. In evaluation, about 126 training samples were generated to represent the reflectance spectra for each LULC class, i.e., arable land, wastelands, open forest, settlements, and water bodies. The selected RF, SVM, and CART algorithms were executed by using the '*ee.classifier* package' on GEE platform to model the association between training features, and response land use classes. The derived outputs from the algorithms were used to develop the LULC maps for the selected area by using Geographic Information System (GIS) (Fig. 3).

After execution of three selected algorithms, the overall, producer, and user accuracies were computed (Congalton 1991) from the confusion matrix (or error matrix) to evaluate the classification robustness of three algorithms. The analysis shows that the CART algorithm attained the best performance in LULC classification with 98.11% of overall accuracy, next to that RF, and SVM algorithms with relatively lower overall accuracy of 95.58 and 88.80%, respectively. Results indicate that all three algorithms could classify the settlements, and waterbodies well with more than 80% of user's accuracies (Table 3). The CART algorithm provided a producer's accuracy of above 95% for all five LULC classes. However, it gave the highest producer's accuracy for the LULC classes of arable land, settlements, and open forest among the three algorithms. The study shows that CART is a reliable algorithm for LULC change modelling using remotely sensed images (Shafizadeh-Moghadam et al. 2017).

4 ML Algorithms in Remote Sensing Data Analysis-Overfitting and Bias

Overfitting of ML algorithms is often encountered, it needs to be taken care while developing the models in predicting/classifying images. The overfitting of the model involves development of a complex model to elucidate peculiarities, and outliers in the data. In other words, if we use the same kind of data on which data was trained in assessing the model, the user gets very high prediction/classification accuracy. Any little modifications in the input data affects the prediction/ classification accuracy. This overfitting of models can be corrected by utilizing a larger dataset, and properly segmenting the input dataset. When a model is developed to classify satellite images, a big dataset helps for proper training, and testing algorithms. In algorithm building, users generally divide the dataset with 75:25 ratio, where 75% of the total data was used to build the model and after it was trained, the remaining 25% was used to assess the robustness of the model. However, it is not always necessary to follow a 75:25 ratio, users can decide any other dataset division suitable appropriately for that particular dataset analysis. The training set should have an impartial depiction of the total dataset, in other words, it should not be too small a representation with respect to the testing set of the whole dataset. In case, the training dataset considered is too small, then predictions of the model may not be reliable, since the model has not been trained completely on an unbiased representative training set.

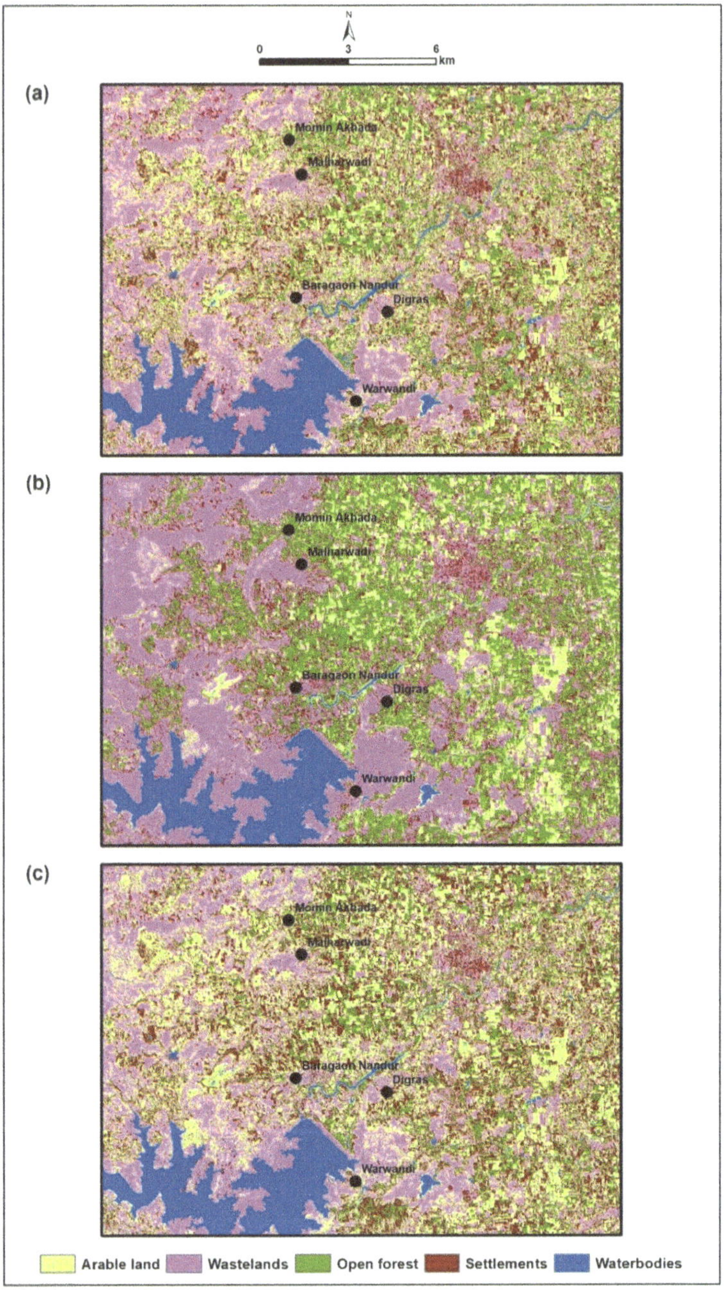

Fig.3 Supervised ML algorithms **a** random forest, **b** support vector machine and **c** CART used in classification of Sentinal-2 high-resolution data on GEE platform

Table 3 Confusion matrix of ML algorithms used in LULC classification

LULC class	Arable land	Wastelands	Open forest	Settlements	Waterbodies
(a) Random Forest Algorithm					
Arable land	**2**	0	0	2	1
Wastelands	0	**28**	0	0	0
Open forest	0	0	**12**	0	0
Settlements	2	0	0	**26**	0
Waterbodies	0	1	0	0	**28**
Producer's accuracy	93.03%	97.4%	98.52%	93.8%	97.9%
User's accuracy	90.13%	98.79%	98.85%	93.36%	97.86%
Overall accuracy			**95.58%**		
(b) Support Vector Machine Algorithm					
Arable land	**25**	3	0	1	0
Wastelands	0	**26**	2	0	0
Open forest	0	0	**12**	0	0
Settlements	1	3	0	**24**	0
Waterbodies	0	1	3	0	**25**
Producer's accuracy	96.5%	79.6%	71.5%	96.6%	98.52%
User's accuracy	89.65%	93.69%	98.52%	86.91%	86.78%
Overall accuracy			**88.80%**		
(c) CART Algorithm					
Arable land	**28**	0	0	0	1
Wastelands	0	**28**	0	0	0
Open forest	0	0	**12**	0	0
Settlements	0	1	0	**27**	0
Waterbodies	0	0	0	0	**29**
Producer's accuracy	98.59%	97.7%	98.32%	98.12%	97.4%
User's accuracy	96.01%	98.92%	98.67%	96.69%	98.12%
Overall accuracy			**98.11%**		

5 ML Algorithms in Remote Sensing Data Analysis—Accuracy Assessment

Performing the accuracy assessment is an essential part in evaluating the performance of various algorithms. In accuracy assessment of the algorithms, Overall accuracy is a widely used metric in interpretation and assess the effectiveness of algorithms in accuracy estimation (Plourde and Congalton 2003). It can be defined as the percentage of sample data, which was properly classified by the classifier. To assess the class-level performance of a classifier, tools like confusion matrix, user's accuracy, and producer's accuracy are widely used (Stehman 2009). GEE platform provides various methods such as error matrix, producer's accuracy, user's accuracy, and accuracy to accomplish the accuracy assessment for various classifiers. Another important dimension is to understand the relative performance of the same category like supervised algorithms of RF, SVM, and CART through comparison of their performance with each other. Congalton and Green (2010) and Foody (2009) were evaluated different supervised ML algorithms in relative comparison by using Z-Score. In general, the accuracies of ML algorithms are mainly influenced by the training samples selected, sampling size, and heterogeneity of samples. Foody (2009) explained the influence of small change in sampling size and its impact on overall accuracy of algorithms performance and outputs. Through understanding the effect of influencing factors greatly assist in minimizing their negative impact on model output, and enhance the model efficiency. When using the ML algorithms, the spectral, and spatial resolutions of input data also play a critical role in order to achieve the high mapping accuracy.

6 Conclusions

The recent trends in remote sensing technologies in spatial data acquisition through versatile platforms, sensors, and information infrastructures have transformed the field of image classification, and analysis. The core aim of any image classification is to classify the pixels into useful information to use in various real-world applications. Supervised ML algorithms require pre-specifications of maximum settings to obtain the desired outcome with high model performance. Clustering algorithms basically perform unsupervised classification tasks, which inspect a huge number of unknown pixels, then divide into a number of clusters based on the heterogeneity present in the image values. On the other hand, the regression algorithm depends on a set of input variables, where the response is known. The deep learning models with robust feature extraction, and independent learning capabilities significantly enhance the features detection accuracy in satellite image classification. The response in the regression method is continuous instead of categorical. Clustering algorithms are well-known algorithms for classifying the satellite images. The value of a particular clustering algorithm depends on how relatively the reference points represent the data as well as how fast the program runs. The performance evaluation of RF, SVM, and CART algorithms by using high-resolution Sentinal-2 imageries of heterogeneous landscape of central India on cloud-based GEE platform shows that the CART model attained the best performance in LULC classification with an overall accuracy of 98.11%, then the RF (95.58%), and SVM (88.80%) algorithms. The chapter concludes that ML algorithms have immense potential, and abilities in classification and analysis of optical remote sensing data more effectively to attain the higher classification accuracy.

References

Abbot J, Marohasy J (2014) Input selection and optimisation for monthly rainfall forecasting in Queensland, Australia, using artificial neural networks. Atmos Res 138:166–178

Aertsen W, Kint V, Van Orshoven J, Muys B (2011) Evaluation of modelling techniques for forest site productivity prediction in contrasting ecoregions using stochastic multicriteria acceptability analysis (SMAA). Environ Model Softw 26:929–937

Alpaydın E (2014) Introduction to machine learning. MIT Press, Cambridge, MA

Andualem TG, Belay G, Guadie A (2018) Land use change detection using remote sensing technology. J Earth Sci Clim Chang 9:1–6

Asokan A, Anitha J, Ciobanu M, Gabor A, Naaji A, Hemanth DJ (2020) Image processing techniques for analysis of satellite images for historical maps classification—An overview. Appl Sci 10:4207

Atkinson PM, Massari R (1998) Generalised linear modelling of susceptibility to land sliding in the central Apennines. Italy. Comput Geosci 24(4):373–385

Attarchi S, Gloaguen R (2014) Classifying complex mountainous forests with L-Band SAR and Landsat data integration: a comparison among different machine learning methods in the hyrcanian forest. Remote Sens 6:3624–3647

Ayalew L, Yamagishi H, Ugawa N (2004) Landslide susceptibility mapping using GIS-based weighted linear combination, the case in Tsugawa area of Agano River, Niigata Prefecture, Japan. Landslides 1:73–81

Bajwa IS, Naweed M, Asif MN, Hyder SI (2009) Feature based image classification by using principal component analysis. ICGST Int J Graph vis Image Process GVIP 9:11–17

Ban Z, Liu J, Li C (2018) Superpixel segmentation using Gaussian mixture model. IEEE Trans Image Process 27:4105–4117

Bandos TV, Bruzzone L, Camps-Valls G (2009) Classification of hyperspectral images with regularized linear discriminant analysis. IEEE Trans Geosci Remote Sens 47:862–873

Bangira T, Alfieri SM, Menenti M, van Niekerk A (2019) Comparing thresholding with machine learning classifiers for mapping complex water. Remote Sens 11(11):1351

Báscones D, González C, Mozos D (2018) Hyperspectral image compression using vector quantization, PCA and JPEG2000. Remote Sens 10:907

Baudat G, Anouar F (2000) Generalized discriminant analysis using a kernel approach. Neural Comput 12(10):2385–2404

Belgiu M, Drăgut L (2016) Random forest in remote sensing: a review of applications and future directions. ISPRS J Photogramm Remote Sens 114:24–31

Bezdek JC, Ehrlich R, Full W (1984) FCM: The fuzzy c-means clustering algorithm. Comput Geosci 10:191–203

Bishop CM (2006) Pattern Recognition and Machine Learning. Springer. pp. 1–758

Breiman L (2001) Random forests machine learning; Springer: Berlin/Heidelberg. Germany 45:5–32

Breiman L, Friedman JH, Olshen RA, Stone CJ, (1984) Classification and regression trees. The Wadsworth Statistics/Probability Series. Chapman and Hall, New York.

Bretherton CS, Smith C, Wallace JM (1992) An intercomparison of methods for finding coupled patterns in climate data. J Clim 5:541–560

Burges CJ (1998) A tutorial on support vector machines for pattern recognition. Data Min Knowl Discov 2(2):121–167

Cai D, He X, Han J (2007) Semi-supervised discriminant analysis. Proceedings of the IEEE International conference on computer vision, Rio De Janeiro, Brazil 14–21:1–7

Calvo-Zaragoza J, Valero-Mas JJ, Rico-Juan JR (2015) Improving kNN multi-label classification in prototype selection scenarios using class proposals. Pattern Recognit 48(5):1608–1622

Cardot H, Faivre R, Goulard M (2003) Functional approaches for predicting land use with the temporal evolution of coarse resolution remote sensing data. J Appl Stat 30(10):1185–1199

Chau KT, Chan JE (2005) Regional bias of landslide data in generating susceptibility maps; case of Hong Kong Island. Landslides 2:280–290

Cheng Q, Jing L, Panahi A (2017) Principal component analysis with optimum order sample correlation coefficient for image enhancement. Int J Remote Sens 27(16):3387–3401

Clark ML, Roberts DA, Clark DB (2005) Hyperspectral discrimination of tropical rain forest tree species at leaf to crown scales. Remote Sens Environ 96(3/4):375–398

Congalton RG (1991) A review of assessing the accuracy of classifications of remotely sensed data. Remote Sen Environ 37(1):35–46

Congalton RG, Green K (2010) Assessing the accuracy of remotely sensed data: principles and practices, vol 25. CRS Press, Boca Raton

Corcoran JM, Knight JF, Gallant AL (2013) Influence of multi-source and multi-temporal remotely sensed and ancillary data on the accuracy of random forest classification of wetlands in northern Minnesota. Remote Sens 5(7):3212–3238

David K, Mitchel K (2010) Logistic regression: a self-learning text, 3rd edn. Springer-Verlag Inc., New York

Duda RO, Hart PE, Stork DG (2001) Pattern classification, second ed, Wiley Interscience

Duda RO, Hart PE, Stork DG (2012) Pattern classification. Wiley, Hoboken, NJ, USA

Duro DC, Franklin SE, Dubé MG (2012) A comparison of pixel-based and object-based image analysis with selected machine learning algorithms for the classification of agricultural landscapes using SPOT-5 HRG imagery. Remote Sens Environ 118:259–272

Escabias M, Aguilera AM, Valderrama MJ (2005) Modeling environmental data by functional principal component logistic regression. Environmetrics 16:95–107

Fauvel M, Tarabalka Y, Benediktsson JA, Chanussot J, Tilton JC (2013) Advances in spectral-spatial classification of hyperspectral images. Proc IEEE 101(3):652–675

Feng F, Li W, Du Q, Zhang B (2017) Dimensionality reduction of hyperspectral image with graph-based discriminant analysis considering spectral similarity. Remote Sens Environ 9(4):323–336

Foody GM (1996) Approaches for the production and evaluation of fuzzy land cover classification from remotely-sensed data. Int J Remote Sens 17:1317–1340

Foody GM (2009) Sample size determination for image classification accuracy assessment and comparison. Int J Remote Sen 30:5273–5291

Foody GM, Mathur A (2004) A relative evaluation of multiclass image classification by support vector machines. IEEE Trans Geosci Remote Sens 42:1335–1343

Fraser RH, Abuelgasim A, Latifovic R (2005) A method for detecting large-scale forest cover change using coarse spatial resolution imagery. Remote Sens Environ 95:414–427

Friedman JH (2001) Greedy function approximation: a gradient boosting machine. Inst Math Stat 29:1189–1232

Fu B, Wang Y, Campbell A, Li Y, Zhang B, Yin S, Xing Z, Jin X (2017) Comparison of object-based and pixel-based random forest algorithm for wetland vegetation mapping using high spatial resolution GF-1 and SAR data. Ecol Indic 73:105–117

Fukunaga K (1990) Introduction to statistical pattern recognition. Academic Press Professional, Inc., San Diego, California

Ghimire B, Rogan J, Rodríguez-Galiano V, Panday P, Neeti N (2012) An evaluation of bagging, boosting, and random forests for land-cover classification in cape cod, Massachusetts, USA. Gisci Remote Sen 49(5):623–643

Hartigan JA, Wong MA (1979) A K-means clustering algorithm. Appl Stat 8(1):100–108

Hastie T, Tibshirani R, Friedman J (2008) The elements of statistical learning, 2nd edn. Springer, Berlin, Germany

Hastie TJ, Tibshirani RJ, Friedman JH (2005) The elements of statistical learning: data mining, inference, and prediction. Math Intell 27:83–85

He QP, Wang J (2007) Fault detection using the k-nearest neighbor rule for semiconductor manufacturing processes. IEEE Trans Semicond Manuf 20:345–354

Heydari SS, Mountrakis G (2018) Effect of classifier selection, reference sample size, reference class distribution and scene heterogeneity in per-pixel classification accuracy using 26 Landsat sites. Remote Sens Environ 204:648–658

Hosmer DW, Lemeshow S (2000) Applied logistic regression. Wiley series in probability and mathematical statistics. Wiley, New York, p 307

Huang J, Wang X, Li X, Tian H, Pan Z (2013) Remotely sensed rice yield prediction using multi-temporal NDVI data derived from NOAA's-AVHRR. PLoS ONE 8:e70816

Huang X, Zhang L (2013) An SVM ensemble approach combining spectral, structural, and semantic features for the classification of high resolution remotely sensed imagery. IEEE Trans Geosci Remote Sens 51(1):257–272

Ji Z, Huang Y, Yong X, Zheng Y (2017) A robust modified Gaussian mixture model with rough set for image segmentation. Neurocomputing 266:550–565

Jin H, Stehman SV, Mountrakis G (2014) Assessing the impact of training sample extraction on accuracy of an urban classification: a case study in Denver, Colorado. Int J Remote Sens 35:2067–2081

Kavzoglu T, Mather PM (2003) The use of backpropagating artificial neural networks in land cover classification. Int J Remote Sens 15(23):4907–4938

Kavzoglu T, Sahin EK, Colkesen I (2014) Landslide susceptibility mapping using GIS-based multi-criteria decision analysis, support vector machines, and logistic regression. Landslides 11(3):425–439

Kruse FA, Boardman JW, Lefkoff AB, Heidebrecht KB, Shapiro AT, Barloon PJ, Goetz AFH (1993) The spectral image processing system (SIPS)—interactive visualization and analysis of imaging spectrometer data. Remote Sens Environ 44:145–163

Kuhn M, Johnson K (2013) A short tour of the predictive modeling process. In Applied Predictive Modeling. Springer, Berlin/Heidelberg, Germany, pp. 19–26

Lary DJ, Alavi AH, Gandomi AH, Walker AL (2016) Machine learning in geosciences and remote sensing. Geosci Front 7:3–10

Lary DJ (2010) Artificial intelligence in geoscience and remote sensing. In: Imperatore P, Riccio D (eds) Geoscience and remote sensing, new achievements. IN-TECH, Vukovar, Croatia, pp. 1e24

Lasaponara R (2006) On the use of principal component analysis (PCA) for evaluating interannual vegetation anomalies from SPOT/VEGETATION NDVI temporal series. Ecol Modell 194:429–434

Lillesand TM, Kiefer RW, Chipman J (2008) Remote sensing and image interpretation, 6th edn. Wiley

Lillesand T, Kiefer R (1994) Remote sensing and image interpretation, 3rd ed, Wiley, Inc. p. 750

Lira J (2010) Tratamiento digital de imágenes multiespectrales, 2nd ed. Universidad nacional autónoma de México, México D.F

Liu L, Tang H, Caccetta P, Lehmann EA, Hu Y, Wu X (2013) Mapping afforestation and deforestation from 1974 to 2012 using Landsat time-series stacks in Yulin District, a key region of the three-north shelter region, China. Environ Monit Assess 185:9949–9965

Lu D, Weng Q (2007) A survey of image classification methods and techniques for improving classification performance. Int J Remote Sens 28(5):823–870

Lu J, Plataniotis K, Venetsanopoulos A (2005) Regularization studies of linear discriminant analysis in small sample size scenarios with application to face recognition. Pattern Recognit Lett 26:181–191

Maggiori E, Tarabalka Y, Charpiat G, Alliez P (2017) Convolutional neural networks for large-scale remote-sensing image classification. IEEE Trans Geosci Remote Sens 55:645–657

Mantero P, Moser G, Member S, Serpico SB, Member S (2005) Partially supervised classification of remote sensing images through SVM-based probability density estimation. IEEE Transact Geosci Remote Sens 43(3):559–570

Maryantica N, Lin C (2017) Exploring changes of land use and mangrove distribution in the economic area of Sidoarjo District, east java using multi-temporal landsat images. Inform Process Agri 4:321–332

Mathur A, Foody GM (2008) Multiclass and binary SVM classification: implications for training and classification users. IEEE Geosci Remote Sens Lett 5(2):241–245

Matus-Hernández MÁ, Hernández-Saavedra NY, Martínez-Rincón RO (2018) Predictive performance of regression models to estimate chlorophyll-a concentration based on Landsat imagery. PLoS ONE 13(10):e0205682
Medina O, Manian V, Chinea JD (2013) Biodiversity assessment using hierarchical agglomerative clustering and spectral unmixing over hyperspectral images. Sensors 13:13949–13959
Mellor A, Boukir S, Haywood A, Jones S (2015) Exploring issues of training data imbalance and mislabelling on random forest performance for large area land cover classification using the ensemble margin. ISPRS J Photogramm Remote Sens 105:155–168
Miao L, Zang S, Zhang B, Li S, Wu C (2014) A review of remote sensing image classification techniques: The role of spatio-contextual information. Eur J Remote Sens 47(1):389–411
Müller K-R, Smola AJ, Rätsch G, Schölkopf B, Kohlmorgen J, Vapnik V (1997) Predicting time series with support vector machines. In: International conference on artificial neural networks. Springer. pp. 999–1004
Munyati C (2008) Use of principal component analysis (PCA) of remote sensing images in wetland change detection on the Kafue flats. Zambia Geocarto Int 6(3):15–26
Naghibi SA, Pourghasemi HR, Abbaspour K (2018) A comparison between ten advanced and soft computing models for groundwater qanat potential assessment in Iran using R and GIS. Theor Appl Climatol 131(3–4):967–984
Naidoo L, Cho MA, Mathieu R, Asner G (2012) Classification of savanna tree species in the greater Kruger national park region, by integrating hyperspectral and LiDAR data in a random forest data mining environment. ISPRS J Photogramm Remote Sens 69:167–179
Napoleon D, Ramaraj E (2014) An efficient segmentation of remote sensing images for the classification of satellite data using K-Means clustering algorithm. Int J Innov Res Sci Tech 1(6):314–319
Noi PT, Kappas M (2018) Comparison of random forest, k-nearest neighbor, and support vector machine classifiers for land cover classification using sentinel-2 imagery. SensS SensS 18:18
Pal M, Mather PM (2005) Support vector machines for classification in remote sensing. Int J Remote Sens 26(5):1007–1011
Peng G, Ruiliang P, Bin Y (1997) Conifer species recognition: an exploratory analysis of in situ hyperspectral data. Remote Sens Environ 62(2):189–200
Petropoulos GP, Arvanitis K, Sigrimis N (2012) Hyperion hyperspectral imagery analysis combined with machine learning classifiers for land use/cover mapping. Expert Syst Appl 39(3):3800–3809
Phiri D, Morgenroth J, Xu C, Hermosilla T (2018) Effects of preprocessing methods on Landsat OLI-8 land cover classification using OBIA and random forests classifier. Int J Appl Earth Obs Geoinf 73:170–178
Plourde L, Congalton RG (2003) Sampling method and sample placement: how do they affect the accuracy of remotely sensed maps. Photogramm Eng Remote Sens 69(3):289–297
Potgieter AB, Apan A, Dunn P, Hammer G (2007) Estimating crop area using seasonal time series of enhanced vegetation index from MODIS satellite imagery. Aus J Agri Res 58(4):316–325
Pradhan B (2010) Flood susceptible mapping and risk area delineation using logistic regression, GIS and remote sensing. J Spat Hydrol 9:1–18
Rahmati O, Naghibi SA, Shahabi H, Bui DT, Pradhan B, Azareh A, Rafiei-Sardooi E, Samani AN, Melesse AM (2018) Groundwater spring potential modelling: Comprising the capability and robustness of three deferent modeling approaches. J Hydrol 565:248–261
Reddy GPO (2012) Principles of digital image processing, In: remote sensing and GIS in digital terrain analysis and soil–landscape modeling, Reddy GPO, Sarkar D (eds) NBSS&LUP Publ No 152, pp 86–97
Reddy GPO (2018) Satellite remote sensing sensors: principles and applications. In:Reddy GPO, Singh SK (eds) geospatial technologies in land resources mapping, monitoring and management. geotechnologies and the environment, vol 21. Springer, Cham, pp 21–43
Reddy GPO, Kumar N, Sarkar D (eds.) (2012) Training manual on GIS and digital image processing, NBSS&LUP Publ No 153, pp. 166

Reddy GPO, Sarkar D (Eds.) (2012) Remote sensing and GIS in digital terrain analysis and soil-landscape modeling, NBSS&LUP Publ No 152, p 300

Reddy GPO, Singh SK (2018) Geospatial technologies in land resources mapping, monitoring, and management, geotechnologies and the environment 21, Springer, p 638

Robinzonov N (2013) Advances in boosting of temporal and spatial models. Ludwig-Maximilians-Universität München

Ryu C, Suguri M, Umeda M (2011) Multivariate analysis of nitrogen content for rice at the heading stage using reflectance of airborne hyperspectral remote sensing. Field Crop Res 122:214–224

Schmidt M, Pringle M, Devadas R, Denham R, Tindall D (2016) A framework for large-area mapping of past and present cropping activity using seasonal Landsat images and time series metrics. Remote Sens 8:312

Shafizadeh-Moghadam H, Asghari A, Tayyebi A, Taleai M (2017) Coupling machine learning, tree-based and statistical models with cellular automata to simulate urban growth. Comput Environ Urban Syst 64:297–308

Shafri HZM, Zeen RM (2011) Mapping Malaysian urban environment from airborne hyperspectral sensor system in the VIS-NIR (0.4–1.1 μm) spectrum. Res J Environ Sci 5(6):587

Shao Y, Lunetta RS (2012) Comparison of support vector machine, neural network, and CART algorithms for the land-cover classification using limited training data points. ISPRS J Photogramm Remote Sen 70:78–87

Shirzadi A, Saro L, Joo OH, Chapi K (2012) A GIS-based logistic regression model in rock-fall susceptibility mapping along a mountainous road: Salavat Abad case study, Kurdistan. Iran Nat Hazards 64:1639–1656

Shivakumar BR, Rajashekararadhya SV (2017) Performance evaluation of spectral angle mapper and spectral correlation mapper classifiers over multiple remote sensor data, in 2017 Second IEEE International conference on electrical, computer and communication technologies. 677–684

Son NT, Chen CF, Chen CR, Minh VQ (2017) Assessment of sentinel-1A data for rice crop classification using random forests and support vector machines. Geocarto Int 1–15

Stehman SV (2009) Sampling designs for accuracy assessment of land cover. Int J Remote Sens 30:5243–5272

Taghizadeh-Mehrjardi R, Nabiollahi K, Kerry R (2016) Digital mapping of soil organic carbon at multiple depths using different data mining techniques in Baneh region. Iran. Geoderma 266:98–110

Thenkabail PS, Lyon JG, Huete A (2012) Hyperspectral remote sensing of vegetation, 1st edn. CRC Press Taylor & Francis Group, New York

Theodoridis S, Koutroumbas K (2003) Pattern recognition, 2nd edn. Elsevier Academic Press, Greece

Toth C, Jó´zków G, (2016) Remote sensing platforms and sensors: a survey. ISPRS J Photogramm Remote Sens 115:22–36

Usman B (2013) Satellite imagery land cover classification using K-means clustering algorithm computer vision for environmental information extraction. Elixir Comput Sci Eng 63:18671–18675

Vapnik VN (1995) The nature of statistical learning theory. Springer, New York

Wu X, Kumar V, Quinlan JR, Ghosh J, Yang Q, Motoda H, McLachlan GJ, Ng A, Liu B, Philip SY (2008) Top 10 algorithms in data mining. Knowl Inf Syst 14:1–37

Yang JF, Lu CL (1995) Combined techniques of singular value decomposition and vector quantization for image coding. IEEE Trans Image Process 4(8):1141–1146

Yang R-M, Zhang G-L, Liu F, Lu Y-Y, Yang F, Yang F, Yang M, Zhao Y-G, Li D-C (2016) Comparison of boosted regression tree and random forest models for mapping topsoil organic carbon concentration in an alpine ecosystem. Ecol Indic 60:870–878

Yesuph AY, Dagnew AB (2019) Land use/cover spatiotemporal dynamics, driving forces and implications at the Beshillo catchment of the blue Nile basin, north eastern highlands of Ethiopia. Environ Syst Res 8:21

Yiging G, Jianhui W, Danong Z, Mingjuan Z, Liru Z (2013) To identify the important of soil properties affecting Dinoseb adsorption with statistical analysis. The Sci World J 3(13):1–7
Zhang M, Huang H, Li Z, Hackman KO, Liu C, Andriamiarisoa RL, Raherivelo NAN, T, Li Y, Gong P, (2020) Automatic high-resolution land cover production in Madagascar using Sentinel-2 time series, tile-based image classification and google earth engine. Remote Sen 12(21):3663
Zhang Y, Yeung D (2011) Semi-supervised generalized discriminant analysis. IEEE Trans Neural Netw 22(8):1–11
Zhong YF, Zhang LP (2010) Initialization methods for remote sensing image clustering using K-Means algorithm. Syst Eng Electron 32(9):2009–2014
Zhou X, Zheng H, Xu X, He J, Ge X, Yao X, Cheng T, Zhu Y, Cao W, Tian Y (2017) Predicting grain yield in rice using multi-temporal vegetation indices from UAV-based multispectral and digital imagery. ISPRS J Photogramm Remote Sens 130:246–255

Geo-Big Data in Digital Augmentation and Accelerating Sustainable Agroecosystems

Gopal Krishna and Chandrashekhar Biradar

Abstract Digital technologies are penetrating into the agriculture sector rapidly but lack dynamics, granularity and demand-driven advisory. At the same time, agri-food systems transformation marching towards economically viable and ecologically sustainable options for healthy people and a healthy planet. Such transformation requires a systematic quantification and characterization of farming system dynamics and farm typology at much higher spatial and temporal granularity with real-time analytics and advisory. The rise of digital augmentation to quantify land use pattern, farming systems and site-specific recommendation on a real-time basis has become possible by recent advances in big-data analytics, cloud computing, along with smartphone-enabled decisions making at the farm level to address the gaps at multiple domain levels in demand-driven agroecological interventions. In the chapter, we discussed ongoing digital augmentation efforts for accelerating agroecological transition through sustainable intensification of rice fallows in India. Systematic characterization of these rice fallows can be sustainably intensified to grow short-duration legume crops, vegetables and oilseeds between two crops. Near-real-time mapping of active rice-growing areas, their dynamics in terms of length of fallows, corresponding soil moisture to map suitable areas for specific crops and varieties for cropping system diversification and sustainable intensification to increase land/water productivity, nutrition, crop yield and overall system-level resilience. The capabilities and limitations of mapping rice-based farming systems using Sentinel series of optical and microwave satellite imagery during the *kharif* season in monsoon climate have been discussed. The chapter shows a new methodological approach for multi-scale mapping at the macro to micro levels to assist informed decision making.

Keywords Digital augmentation · Sustainable intensification · Geospatial big data · Legumes · Rice fallow

---

G. Krishna
International Center for Agricultural Research in the Dry Areas (ICARDA), New Delhi, India

C. Biradar (✉)
International Center for Agricultural Research in the Dry Areas (ICARDA), Cairo, Egypt
e-mail: c.biradar@cgiar.org

G. P. O. Reddy et al. (eds.), *Data Science in Agriculture and Natural Resource Management*, Studies in Big Data 96, https://doi.org/10.1007/978-981-16-5847-1_11

1 Introduction

The sustainable agroecosystems are the vital foundation for ensuring food, nutrition and environmental security (Biradar and Xiao 2009; Biradar et al. 2019). To meet the demand of increasing population, we need to grow more quantity and quality food from the same land and resources. It is estimated that nearly 50% more production increase will be required to feed the projected 9.15 billion population of the world by 2050 (Alexandratos and Bruinsma 2012). To meet this increasing demand with limited land areas or same arable land areas, cropping intensification is required (Tilman et al. 2011, 2002; Tripathi et al. 2019). Remote Sensing's wide area coverage, synoptic view and high temporal data availability makes it an ideal and pow2020erful tool that can provide fallow crop land estimates and map its dynamics over large areas through an independent approach (Chandna and Mondal 2020; Thenkabail et al. 2009). Remote sensing has been widely used for rice fallows mapping (Chandna and Mondal 2020; Krupnik et al. 2017) and sustainable intensification (Krupnik et al. 2017; Godfray et al. 2010; Pretty and Bharucha 2014) for policy-making and decision-making. Cropping system frequency/intensity mapping has also been used to contribute to crop production and food security (Liu et al. 2020; Wu et al. 2018). Using moderate resolution data, several research studies have been conducted to map the cropping intensity (Quarmby et al. 1992). Higher spatial resolution-based mapping would be proved much more efficient to map the cropping intensity. In this order, integration of Sentinel-2 and Landsat data have been found to be effective in temporal data availability gap filling (Li and Roy 2017). The spatial mapping of the rice-based cropping system has been executed by exploiting various types of satellite remote sensing that includes coarse resolution, i.e. Advanced Very High Resolution Radiometer (AVHRR) 1.1 km data (Huang et al. 2013; Fang et al. 1998), moderate resolution, i.e. Moderate Resolution Imaging Spectrometer (MODIS) 250–500 data (Biradar and Xiao 2009; Zhang et al. 2015; Xiao et al. 2006), high resolution, i.e. Landsat series 30 m and Sentinel-2 10 m (Qiu et al. 2017; Ramadhani et al. 2020; Zhang et al. 2017; Jiang et al. 2020) and very high resolution Planet 3 m satellite data (Shi et al. 2018). Many studies have recommended that time-series data always yields better results for spatial mapping instead of single day imageries (Gomez et al. 2016; Xiong et al. 2017). The launch of Sentinel-2 (S-2) series satellites, real-time crop monitoring has become comprehensively possible due to the high temporal frequency of imageries. S-2 imageries have NIR, red and green bands of 10 m spatial resolution that makes it an indispensable satellite data for crop dynamics mapping. Extensive research studies have been conducted using optical S-2 satellite imageries for spatial mapping of rice (Cai et al. 2019; Behera et al. 2020; Jiang et al. 2020; Singh et al. 2021). S-2 data can also be used for village/sub-district-level rice mapping due to its good spatial resolution. Subsequently, it can be utilized to map the yield gaps, forecast crop production, and understand inter and intra annual dynamics for number of applications in sustainable management of the agri-food systems.

Since there is almost no scope for increase in agricultural land, there is a need to produce more food from the existing land and water resources (Das et al. 2015; Dias et al. 2016). To ensure both food and nutritional security, sustainable intensification in the crop fallow areas through cereal-based systems is the need of hour (Biradar et al. 2020). In India, the eastern part has 73.77 Mha total (Bandyopadhyaya et al. 2016) geographical area with 33.6 Mha cultivable land, i.e. 45% of the total geographical area with rice has a major *kharif* crop. The state of West Bengal is considered as the major rice-producing state in the eastern region. There are few studies on sustainable intensification of rice-fallow agro-ecosystem (Gumma et al. 2016, 2018; Gatto et al. 2020; Chandna and Mondal 2020; Kumar et al. 2020) but none of them discusses about the monthly dynamics of the rice fallows and residual soil moisture along with the vital rice fallows length. Monthly dynamics mapping provides detailed information about rice fallow and residue soil moisture. To fill this research gap, the present study depicts monthly dynamics of rice fallows and soil moisture during post-*kharif* seasons as well as rice fallow length mapping for in-depth insight for decision-making. Besides this, this study also depicts precise spatial mapping of the rice crop, rice fallow as well as synchronized soil moisture mapping and mapping of annual trends through hotspots mapping by analyzing past 5 years data for the West Bengal state of India. Overall, the broad objective of this chapter is sustainable intensification of rice fallow areas by legume cultivation suitability mapping through utilization of digital augmentation techniques. Under digital augmentation, many techniques can be experimented but this chapter particularly discusses the advanced techniques for digital augmentation through spatial big data analysis for mapping of rice crop, rice fallows, synchronized soil moisture and legume suitability.

2 Site Description

The state of West Bengal, India lies between 21° 39′ 19″ to 27° 14′ 41″ N latitudes and 88° 2′ 37″ to 89° 3′ 33″ E longitudes has been selected for the study being the highest rice-growing state in India (Fig. 1). Climate of West Bengal is typically tropical, hot, humid and monsoon type (except in the northern mountainous region of Jalpaiguri and Darjiling areas). There are mainly hot, monsoon, retreat monsoon and cold seasons as classified by the Indian Meteorological Department (IMD). Soil survey conducted by department of agriculture classifies soils of West Bengal into 6 broad categories, i.e. laterite, red, alluvial, tarai, saline and red-gravel soils. The alluvial soil tracts approximately cover agriculturally important areas. They occur mostly in the districts of Murshidabad, Birbhum, Barddhaman, Puruliya, Hugli and Medinipur (NATMO Maps 2010). Main cropping season in this state is monsoon/*kharif* season. Most of the parts of West Bengal state grow rice during *kharif* season followed by maize, jute and wheat. In the east-southern part, jute crop looks to be predominant. Research work has been carried out by using multiple digital augmentation approaches to accelerate sustainable agroecosystems. High-resolution rice crop mapping was conducted for this state. Same site was evaluated for rice fallow and legume suitability mapping study.

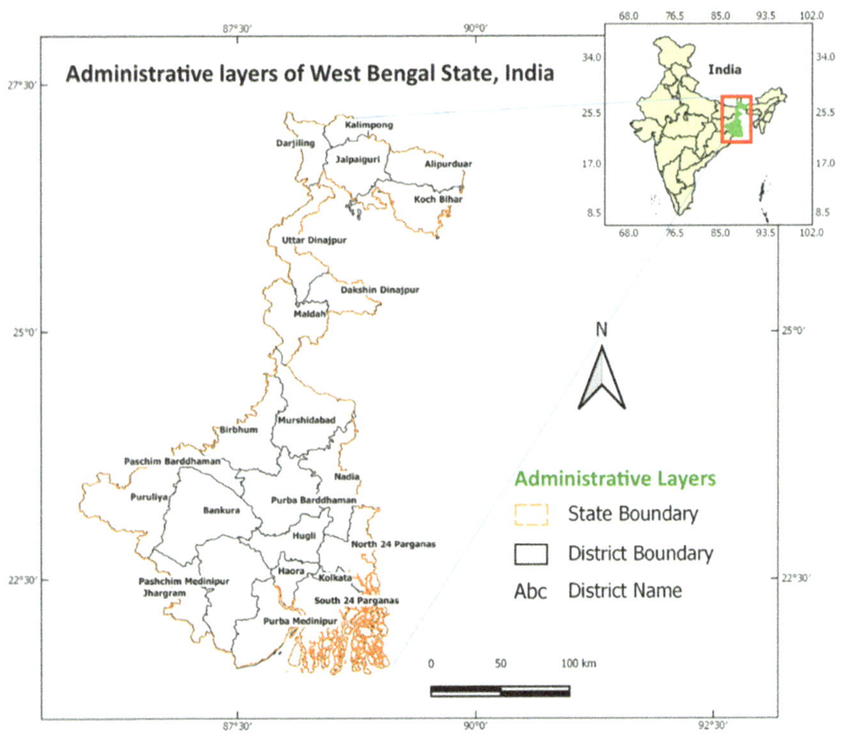

Fig. 1 Spatial location map of area of study state 'West Bengal' for rice crop mapping and rice fallows mapping

3 Materials and Methods

3.1 Data Used

3.1.1 Satellite Data

This chapter depicts rice crop mapping and rice fallows mapping for India's highest rice-growing state of India, i.e. West Bengal. Sentinel-2 (S-2) multispectral satellite data for the study area were collected from Google Earth Engine (GEE) during the cropping season (July/August to November/December) during the period from 2017 and 2019 for rice crop and rice fallows mapping. S-2 imagery has 13 spectral bands covering the visible (vis, band 2–4), Red Edge (RE, band 5–7), Near Infrared (NIR, band 8), and Shortwave Infrared (SWIR, band 11–12) regions of the electromagnetic spectrum with 10 m, 20 m, 10 m, and 20 m spatial resolutions, respectively. The rest of the bands (band 1, 9, and 10) have 60 m spatial resolution. The study area was covered by 187 and 263 scenes for the years 2017 and 2019, respectively. For rice classification, all bands of S-2 composite except 1, 8A, 9 and 10 were utilised. Bands

1, 9 and 10 are for coastal aerosol, water vapour and cirrus, respectively and 8A is for red edge position. Therefore, these were excluded from classification input. Besides these bands, Normalized Difference Vegetation Index (NDVI) and Normalized Difference Water Index (NDWI) were also computed and included as input bands to enhance the feature identification during classification. To overcome cloud cover, Sentinel-1 (S-1) Synthetic Aperture Radar (SAR) Ground Range Detected (GRD) imageries were also utilized for rice mapping. All pre-processing operations, i.e. including orbital correction, thermal noise removal, radiometric calibration, and terrain correction (orthorectification) were carried out using GEE's Sentinel-1 toolkit. Detailed info about Sentinel-1 algorithms at GEE platform can be found at https://developers.google.com/earth-engine/guides/sentinel1. S-1 GRD imageries of the study area were acquired in 'VH' transmitter Receiver Polarization, Interferometric Wide (IW) swath mode for the entire cropping season (August to November). Entire crop growing season was divided into sowing, vegetative and near-maturity periods as bands. Mean of these bands for each ascending and descending orbit pass were then mosaiced to create a three bands SAR imagery. Year 2017 was used for the district-level acreage validation with reference to acreage data reported by Directorate of Economics and Statistics, Ministry of Agriculture & Farmers' Welfare, Govt. of India (DES MoAFW 2020).

3.1.2 Ground Truth Data

To generate reference and validation data, a detailed ground truth (GT) survey was conducted in the West Bengal state. A total of 998 points were collected and farmers were interviewed to assess the cropping pattern adopted by them throughout the year. The survey was conducted through Google Earth Imagery and Open Data Kit (ODK) collect (ODK 2020) Android app. Few random pictures of the fields have been depicted in Fig. 2. The relevant surveyed data was split into 75:25 ratio. The 75% data points were used to generate training samples and the remaining 25% data was used to validate the results.

Besides this GT data, district-level rice acreage figures were also obtained from Govt. of India (DES MoAFW 2020) to validate remote sensing generated district-level rice acreage estimates.

3.2 Rice Crop Mapping Using Geospatial Optical and SAR Data

Using series data of cropping seasons from Sentinel series visual and SAR satellite imagery, spectral signatures were determined for rice crop and other LULC classes. The generation of spectral signatures for rice helped in generating more samples for the validation of classification. Hence, it was useful for accuracy improvement.

Fig. 2 Random field photographs taken during West Bengal ground truth data collection

Sentinel-2 satellite imagery from June to September for *kharif* 2017–18 and 2019–20 was collected from GEE image collection. A weekly image collection of NDVI images for selected months was created from this stack and the same was used for spectral signature determination.

Two machine learning algorithms were exploited to harness the information from optical and SAR satellite imageries. K-Means machine learning algorithm was used to identify the LULC features based on prior expert knowledge and this assisted in feature identification. Another ensemble-based machine learning algorithm random forest (RF) was applied to get the more detailed and near ground-reality rice crop based on reference data. Intensive ground samples were used for class determination. The training classes were established based on 522 ground samples of rice and 100 samples of other crops. The samples were selected from large homogeneous areas of crop types so that rice can be extracted efficiently without confusion. These samples

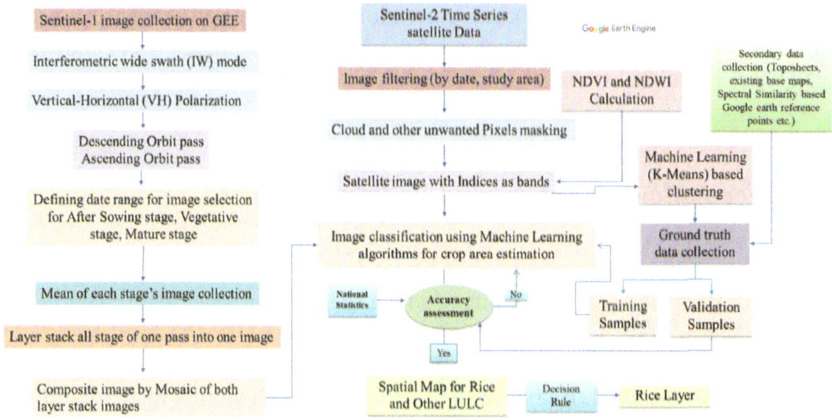

Fig. 3 The flow diagram for spatial mapping of rice using optical and SAR geospatial data

were converted to a feature collection for further use to train the algorithm. A detailed approach has been depicted in Fig. 3.

To train this algorithm, *image.sample Regions()* function was used to get the predictors into the table and create a training dataset. The inputs required are image bands, ground truth as a feature collection and the property label, which the classifier should use for predictors.

Crop mapping for a given area can be proved to be beneficial for decision-making and policy planning. The mapping of crop gives an idea of which areas have not been utilized for cultivation, the reasons for not growing crops on cultivable land, and how policies can be made into their interest so that crop cultivation can be promoted. Similarly, in the current scenario, crop mapping is also being used for crop insurance claim settlements. Using high resolution satellite imageries, crop growing areas can be effectively identified and this can help in decision-making and policy planning at micro level, i.e. farm level as well as for crop insurance claim settlement.

3.3 *Mapping of Rice Fallows Using Big Geospatial Data*

Rice fallows can be mapped through various ways by utilizing remote sensing imageries. By unsupervised classification, supervised classification and computing relevant vegetative indices. One efficient method of delineating the rice fallow is to map through thresholding of Soil Adjusted Vegetative Index (SAVI) (Huete 1988). Mathematically, SAVI can be expressed as follows:

$$\left(\text{SAVI} = \frac{\text{NIR} - \text{RED}}{\text{NIR} + \text{RED} + L}\right) * (1 + L) \tag{1}$$

where L represents the soil brightness correction factor. Value of L ranges between 0 and 1. For very high vegetation regions, L is given as 0 that makes SAVI equivalent to NDVI (Li et al. 2018). Ideally, the $L = 0.5$ is used for most of the conditions.

3.4 Rice Fallows Mapping and Its Importance

In the Eastern Indian scenario, the cultivable land left uncultivated after the main cropping season due to the non-availability of adequate irrigation water for growing cereal crops is called rice fallow land. In recent time, when the world is witnessing increasing demand for food due to an increase in global population, the food growers/farmers need to grow more food. But since the cultivable land cannot be increased in that ratio, we need to grow more food on unutilized cultivable land. Here, fallow rice intensification can play a very crucial role in growing more food as well as a nutritional food. By growing more food, it can also contribute to the increment of the farmers' income.

3.5 Mapping Soil Moisture Dynamics

For accelerating sustainable agroecological intensification, the utilization of available soil moisture is of vital use. Monitoring monthly dynamics of soil moisture can provide an instrumental insight. Weekly, biweekly, and monthly median image collections can be utilized to generate relative soil moisture imageries that can reveal where short-duration crops can be planted without much need of irrigation water. Relative soil moisture (Normalized Difference Moisture Index (NDMI)) can be generated using the following equation:

$$\text{NDMI} = \frac{B8A - B12}{B8A + B12} \tag{2}$$

3.6 Legume Suitability Mapping Based on Available Soil Moisture

Areas with relatively longer rice fallow length and available residue soil moisture after *kharif* season can be utilized to grow short-duration pulses/ legumes that are water-efficient, easy to grow and soil nurturing. The places where rice fallow length is 60 days or more days and have availability to residue soil moisture can be identified as suitable areas for legume intensification. For this readily available suitability measurement, rice fallow length layer with soil moisture layer can be analysed and

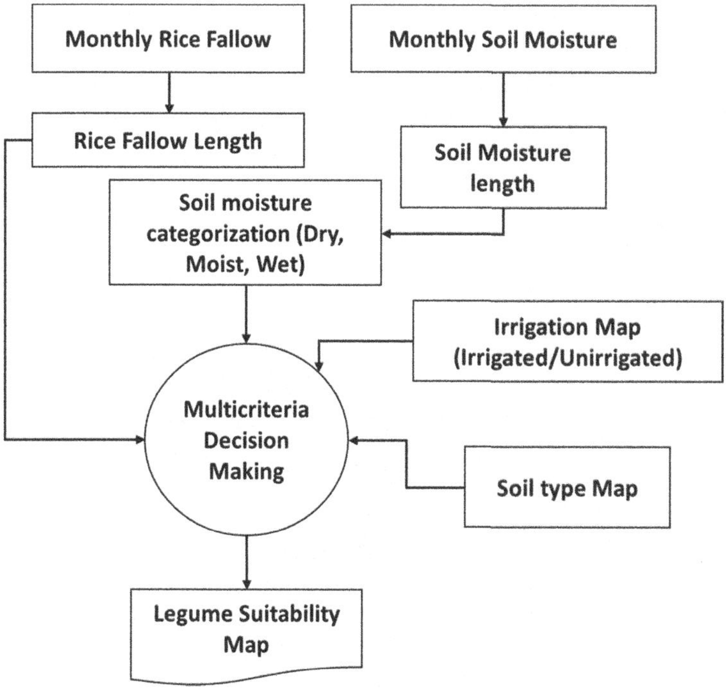

Fig. 4 A broad overview of legume suitability mapping in the rice fallow areas

areas with higher suitability can be identified. Areas with higher rice fallow length and available residue soil moisture may prove to be the best suitable areas for legume cultivation. But on ground, there might be several other important factors responsible for legume cultivation suitability assessment like irrigation facility and soil type. Major two contributing factors are the rice fallow length and residue soil moisture, followed by soil type and irrigation facility. Based on all these parameters, a Multi-criteria Decision Making (MCDM) analysis was carried out (Fig. 4). In the study, all MCDM analysis was carried out in the GEE cloud computing platform by composing a javascript code.

4 Results and Discussion

4.1 Spatial Distribution of Rice Crop in the West Bengal State

Geoinformation coupled with a supervised classification technique was used to map the spatial distribution of rice crops in the West Bengal state. Rice is the major Kharif (June to October) crop in this state. Rice mapping shows that the entire West Bengal state produces rice during this growing season (Fig. 5). In the East-Southern part, Jute and other crops look to be predominant. District-wise analysis shows that the highest rice-growing district is the Medinipur West followed by Purba Bardhhaman district. In contrast, Kalimpong the newly created hilly district found to be the lowest rice-growing district in the state.

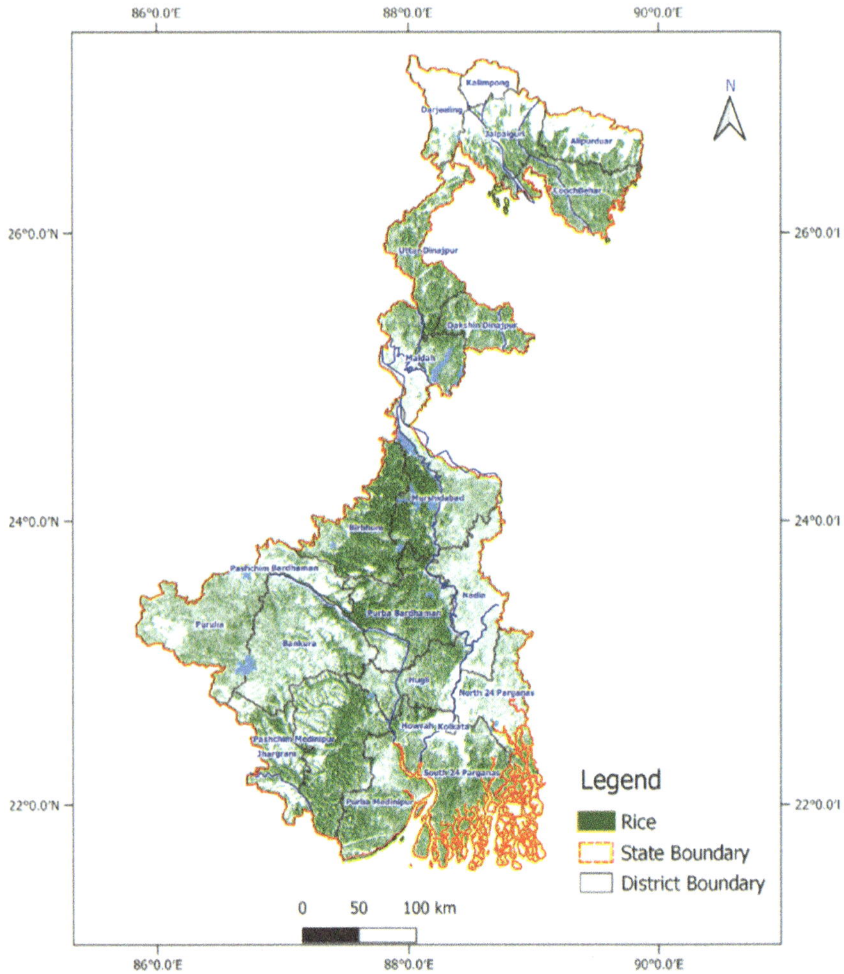

Fig. 5 Spatial distribution of rice crop in the West Bengal state

4.1.1 High-Resolution Crop Mapping

The Sentinel-2 satellite imageries come into 10 m spatial resolution and approximately 5-days temporal resolution. This data can be efficiently utilized for crop mapping at the macro (state and district) to micro-level, i.e. sub-district, village-level, or multiple cadastre-level dispute resolutions. For instance, if we have mapped rice crop accurately, then overlay the cadastre over the classified data, it makes sure that the said crop insurance claimant or group of claimants from that village have grown the rice crop or not. An instance of such an example has been depicted in Fig. 6. Sometimes, this has been observed by insurers that claimants grow some low insurance price crops and later claim for higher insurance price crops. This type of insurance claim dispute can be resolved by using Sentinel-2 satellite imagery-based mono-cropped crop mapping.

4.2 Rice Fallows Mapping

Rice fallow areas for West Bengal state were extracted by evaluating multispectral high-resolution Sentinel-2 satellite data of 10 m resolution. These rice fallows were extracted weekly and aggregated to make a monthly rice fallow image for each month from October 2019 to January 2020 (Fig. 7). Figure 7 shows that October 2019 has

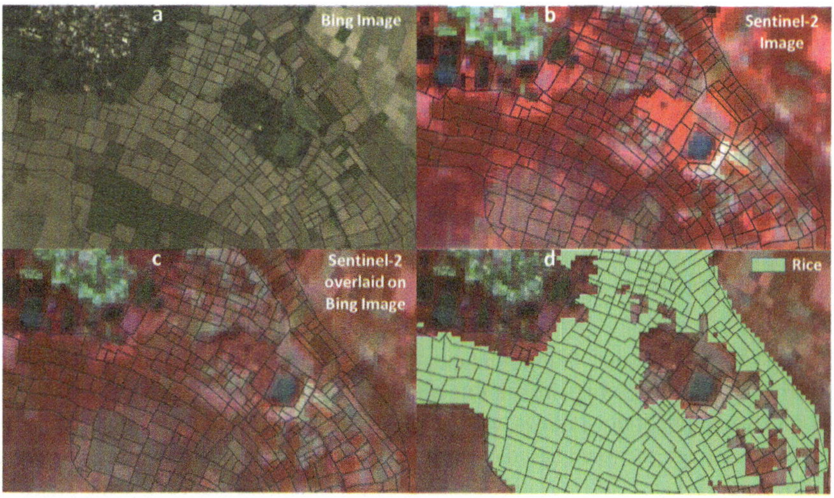

Fig. 6 Mapping of rice in a part of Hooghly district. **a** cadaster shapefile overlaid on Bing satellite imagery, **b** cadaster shapefile overlaid on Sentinel-2 multispectral satellite imagery, **c** cadaster shapefile overlaid on Sentinel-2 sharpened by Bing imagery, **d** classified rice pixels overlaid on cadaster and Sentinel-2 sharpened by Bing imagery. This image shows an example of micro-level mapping that can be used for cadastre-level crop identification for dispute resolution

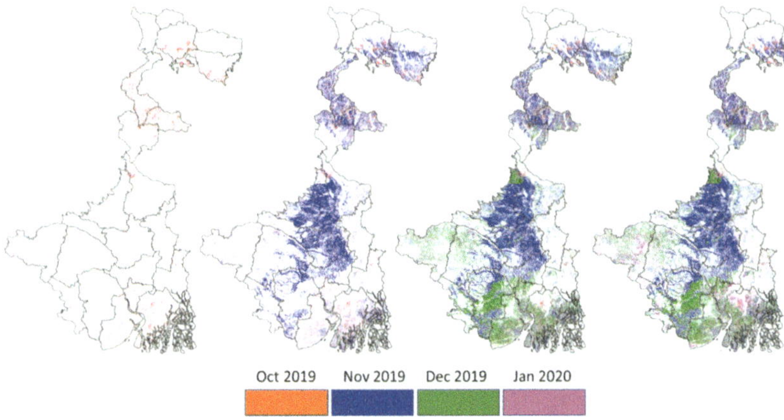

Fig. 7 Rice fallows monthly dynamics mapping through multispectral high-resolution Sentinel-2 satellite imageries

the lowest rice fallow availability, whereas, November followed by December 2019 depicts most of the rice fallow in the West Bengal state. These maps are of high resolution as these are developed by satellite data of 10 m spatial resolution and can be utilized for decision-making. Rice fallow observed during October month shows cultivation of short duration rice varieties, whereas, rice fallows observed during January depict cultivation of long duration rice varieties. November and December months' rice fallows represent normal duration rice varieties. Sometimes delay in monsoon may become a reason for delayed sowing by 15 days to 01 month and becomes the reason for delayed harvesting as well. Hence, it could affect the emergence of rice fallows. This case was observed during *kharif* 2019–20 and this could also be a possible reason for the presence of very less rice fallow during the month of October.

4.3 Rice Fallows Length Mapping and Its Importance

Monthly/bimonthly rice fallow mapping can be performed by creating a median monthly/bimonthly composite satellite image and applying SAVI thresholding on it. To map the length of rice fallows, we need to add monthly rice fallow images into one image and categorize them based on their increasing duration. This will create a rice fallow length image. The rice fallows length (Fig. 8) mapping is needed to identify rice fallows' duration for crop intensification. When the decision maker knows the length of rice fallow, he can decide what action could be taken for intensification. Besides this, rice fallows length mapping plays a significant role in the legume cultivation suitability assessment. The chances of higher suitable land availability are higher if rice fallow length is greater than 60 days. The length of rice fallows shows the

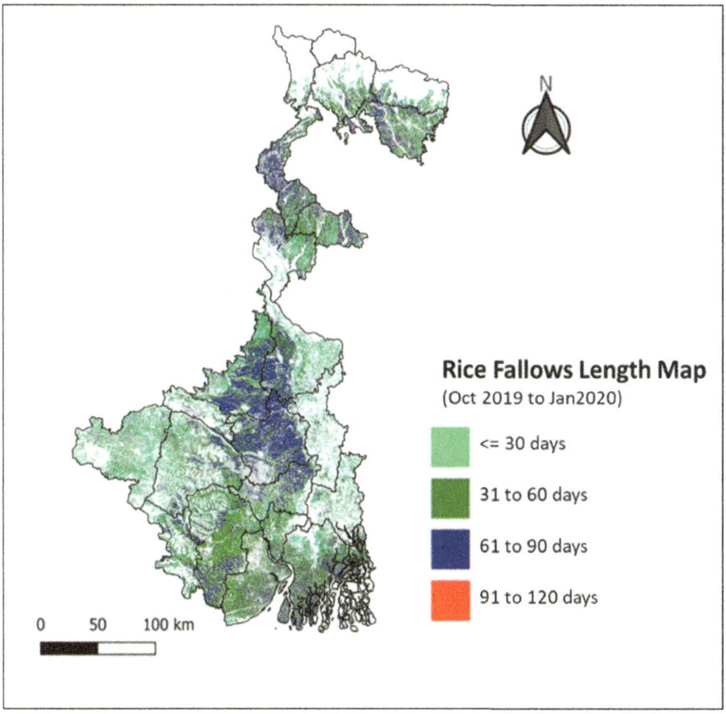

Fig. 8 Rice fallow length map of the West Bengal state showing duration of rice fallow throughout the state

areas where no crop was grown for an approximate number of days. The rice fallow length of <=30 days reflects that immediately after rice harvesting, another crop has been planted, hence, no fallow areas are left. Whereas, 91–120 days rice fallow length shows cultivable land has been left fallow for longer duration. This rice fallow length may prove most favourable for legume intensification if residue soil moisture and/or irrigation facility is available. Short duration legume crops or vegetables can be grown in the rest of the rice fallow length areas, i.e. 31–60 days and 60–90 days. Purba Bardhaman, Birbhum and parts of Murshidbad, Bankura and Uttar Dinajpur districts were observed under higher rice fallows length category. The eastern part of the state where rice is not a predominant crop was observed under the shortest duration of rice fallow.

4.4 Legume Suitability Mapping

Agrointensification on rice fallows has been widely advised because it utilizes cultivable lands left uncultivable. Legume intensification in the rice fallow land has great potential to improve nutrition in food, better fertility in soil and additional income to the growers. Legume crops need less attention/management practices and less irrigation. That's why they fit perfectly in the rice fallow intensification. This has been observed in the rice fallow areas that soil moisture gets depleted immediately after *kharif* crop harvest. Therefore, to utilize the residue soil moisture in suitable areas, sowing of legume crops immediately after rice *kharif* harvest would be highly helpful. It would use the residue soil moisture in the rice fallow areas. Now plant phenomics has supported the development of rice varieties that are salt, flood and drought tolerant. Short duration legume varieties have also been developed. These can be wisely used by utilizing improved agronomy and water management practices in the rice fallow areas to increase the soil productivity, nutrition in food and additional income to farmers. The legume suitability mapping based on rice fallows length, soil moisture, soil type and irrigation facility was conducted for the West Bengal state. The legume suitability map has been depicted in Fig. 9. This map shows that Purba Bardhhaman, Birbhum and Paschim Medinipur districts have higher potential to grow the legume crops in rice fallow land. Whereas, parts of Bankura, Puruliya, Uttar Dinajpur and Koch Bihar have medium potential to grow the legume crops.

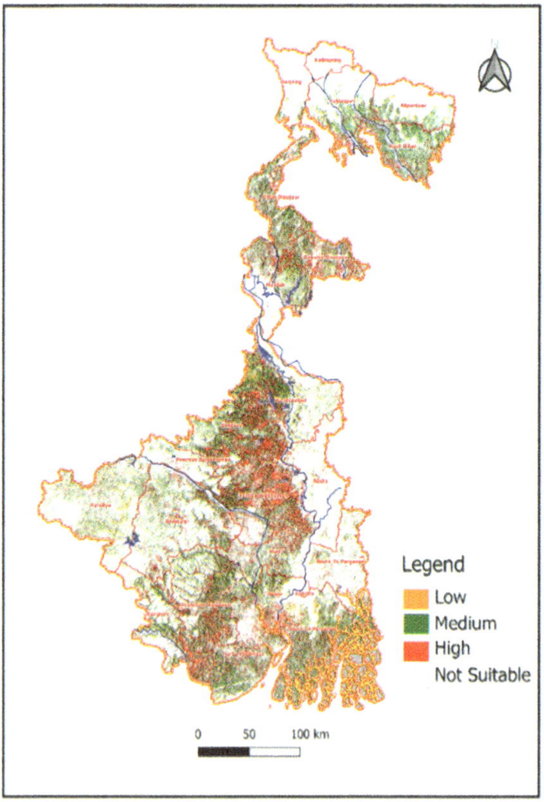

Fig. 9 Legume suitability map of the West Bengal state showing suitable areas for growing legume crops in the rice fallow lands

4.5 Annual Trends and Hotspots of Rice Fallows

The frequency/hot spots mapping for October to January for 2016–17 to 2019–20 was performed using 10 m high spatial resolution data for West Bengal State (Fig. 10). For trends/frequency mapping, Sentinel-2 satellite imagery was analysed on a weekly basis for each year. Weekly basis data was aggregated to a monthly and yearly basis. This analysis generated the hotspot areas for rice fallows and soil moisture. These maps are ready for ground validation. These hot spot areas are the most prominent areas for intensifying legume cultivation on rice fallows because soil moisture condition remains moderate to high during the rice fallow period. The frequency mapping for rice fallows and soil moisture at spatially high resolution during the last four years enables the decision making at micro level (i.e. sub-district/tehsil level). The soil moisture-based legume suitability hotspots mapping provides real-time perception for favourable areas for the legume cultivation intensification. The crux is that the legume suitability hotspots depict the availability of suitable rice fallows duration/length and soil moisture. These hot spot maps have been prepared for field validation. This analysis depicts that Purba Barddhaman, Pashchim Medinipur, Birbhum, Hugli and Bankura districts have high suitable areas for rice fallows intensification through legume cultivation, whereas, Koch Bihar, Maldah, Alipurduar, Uttar Dinajpur, Jalpaiguri and Darjiling districts have lowest suitability.

A substantial increase in nutrition in food and enhancement in the livelihoods of farmers can be achieved by intensifying pulses/legumes on rice fallows. This analysis

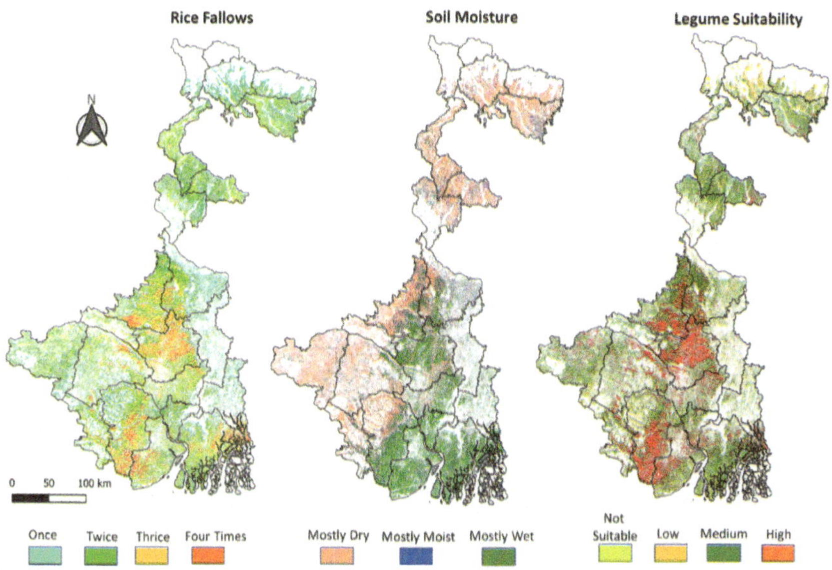

Fig.10 Hotspots mapping (by frequency mapping) for rice fallows, soil moisture and soil moisture-based legume suitability to intensify the Legume cultivation

may provide deep insight for utilizing the rice fallows (based on their frequency) for pulses intensification or growing vegetables. This can ultimately contribute in enhancing the nutritional values in food and the farmers' economic growth.

4.6 *The Accuracy Assessment*

4.6.1 Rice Mapping Accuracy

The accuracy of rice mapping was assessed through two means. The first way of accuracy assessment was classification accuracy assessment through confusion matrix. Another way of accuracy assessment was classified district-level rice acreage comparison with official statistics of district-level rice acreage. In confusion matrix-based estimation, accuracy assessment was computed at each district level. District-wise classification accuracy through confusion matrix yielded the overall accuracy between 75 and 94%, with a kappa value between 0.72 and 0.85.

Remote sensing derived district-level acreage was observed to be in good agreement with govt. provided district-level rice acreage. To assess accuracy level through acreage data rice classification for year 2017 was also mapped as official statistics was available up to 2017 at the time of conducting the analysis. The coefficient of determination between geospatially derived district-level rice acreage with National Rice Statistics (NRS) data of district-level rice acreage was as high as 0.97 (Fig. 11). This shows that the classified results are a very good reflection of the ground reality.

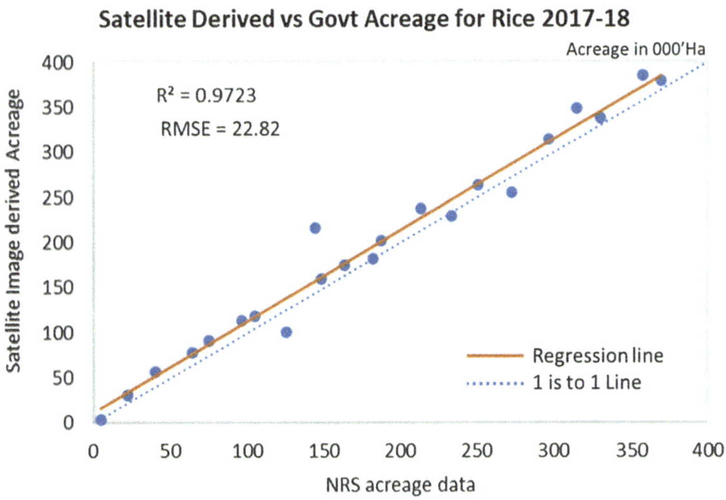

Fig. 11 Comparative analysis between district-level rice acreage statistics derived from remote sensing analysis and National Rice Statistics (NRS) acreage data for the West Bengal state

4.6.2 Rice Fallows Mapping Accuracy

The accuracy assessment shows that the accuracy percentage was between 77 and 99%, whereas, the kappa value for the confusion matrix was between 0.86 and 0.98 (Table 1). This shows that rice fallows have been delineated with very good accuracy. The samples taken for validation were distributed throughout the West Bengal state. October month has the lowest area under rice fallow land that's why a smaller number of samples were selected for October month. November, December and January months depict an accuracy level of more than 90%. This shows good agreement of remote sensing derived rice fallow area with ground reality (see Table 2).

Various digital augmentation efforts driven by geospatial big data analysis could provide highly valuable inputs for sustainable agroecological intensification. For instance, in crop insurance claim settlements, farm to farm manual dispute resolution and claim settlement is a highly tedious and labour-intensive task that may or may not have higher accuracy. But if high-resolution satellite imagery extracted crop mapping would be overlaid over the digital land records, it would provide very accurate outputs without going manually from farm to farm. This digital augmentation for insurance claim resolution would be more precise, less time-consuming as well as economic. Few insurers are already adopting such techniques for their insurance claim settlements. Similarly, digital augmentation through rice fallow mapping is capable of providing vital input for decision mapping. The mapping of rice fallow using big geospatial data is time-efficient, accurate and cost-effective. Using high-resolution geospatial mapping, policy makers can investigate the details at macro (state or district) to micro level (sub-district or village level), increasing their level of understanding for decision making. Rice fallow land identification using satellite imageries could be efficiently used to increase the nutrition in food as well as some addition to the grower's income. The rice fallow land inaccessible or hard to reach can also be mapped using digital augmentation through satellite imagery.

Similarly, crop mapping through satellite imageries could be very efficiently used to identify the cultivated land that is left uncultivated due to some reasons like lack of irrigation water, etc. These areas could be identified after crop mapping and their problem can be addressed through accurate and relevant decision making. Digital augmentation powered by geoinformatics can be used for macro to micro-level crop yield mapping. Another example of sustainable intensification using digital augmentations is suitability mapping for growing short-duration and water-efficient pulses

Table 1 Accuracy matrix of month-wise rice fallows of 2019–20 mapping using Sentinel-2 data

Month	Rice fallow accuracy matrix			Kappa value for confusion matrix of all layers
	Qualified samples	Total samples	Accuracy (%)	
Oct-19	100	130	77	0.86
Nov-19	2962	3222	92	0.98
Dec-19	2611	2684	97	0.91
Jan-20	2694	2727	99	0.89

Table 2 Area figures generated from hotspot mapping using data analysis for October to January for 2016–17 to 2019–20. District wise area (in 000'ha) under rice fallow frequency, soil moisture and legume cultivation suitability (%)

District name	Rice fallow frequency (000'ha)				Soil moisture (000'ha)			Legume cultivation suitability (%)		
	Once	Twice	Thrice	Four times	Mostly dry	Mostly moist	Mostly wet	Low suitability	Medium suitability	High suitability
Uttar Dinajpur	22	136	99	1	143	97	19	44	41	5
Hugli	9	75	142	10	26	156	55	24	40	32
North 24 Parganas	39	54	45	25	1	47	116	27	30	17
Bankura	34	138	223	17	182	193	37	23	38	31
Dakshin Dinajpur	19	109	62	1	25	130	36	32	45	15
Jalpaiguri	25	41	14	1	43	32	7	44	16	3
Alipurduar	13	23	9	0	13	23	10	44	18	2
Koch Bihar	26	137	73	1	94	120	23	47	36	5
Nadia	51	58	15	1	3	55	69	38	15	4
Birbhum	9	94	260	12	141	193	41	18	39	41
Haora	11	41	29	7	5	44	40	38	33	15
Kolkata	–	–	–	–	–	–	–	–	‘–	–
Paschim Barddhaman	3	26	50	3	45	34	3	25	42	30
Purba Barddhaman	11	66	325	28	127	262	42	11	28	58
Maldah	27	127	55	1	48	129	34	46	34	7
Darjiling	8	8	2	0	10	6	3	39	11	0
Kalimpong	0	0	0	0	0	0	0	0	0	0

(continued)

Table 2 (continued)

District name	Rice fallow frequency (000'ha)				Soil moisture (000'ha)			Legume cultivation suitability (%)		
	Once	Twice	Thrice	Four times	Mostly dry	Mostly moist	Mostly wet	Low suitability	Medium suitability	High suitability
South 24 Parganas	24	116	172	45	3	107	248	33	46	12
Murshidabad	49	125	142	7	52	183	89	30	31	23
Jhargram	23	49	65	11	85	56	8	28	33	23
Pashchim Medinipur	44	126	214	91	88	267	120	21	29	41
Puruliya	45	164	137	6	220	117	15	32	37	19
Purba Medinipur	32	87	118	62	10	119	170	27	35	26
Total	524	1800	2251	330	1364	2370	1185	–	–	–

Note Area in 000'Ha for rice fallow and soil moisture

crops in the rice fallow areas. This will nurture the soil for future use, provide more nutrition in food and support the farmers economically. This chapter shows some instances of digital augmentation that can be efficiently utilized for sustainable agrointensification. By applying the digital augmentation techniques, more food can be grown to feed the world's ever-increasing population that would address the Sustainable Development Goals (SDGs) goal of food security. Besides this, these techniques are cost-effective, fast and reliable; hence, could be effectively used to resolve the issues on time.

5 Conclusions

Digital augmentation combines advances in earth observation, in-situ and spatial modelling techniques in the geo-big data domain. The study shows effectively employed for accelerating sustainable intensification, with the options for planning and targeting on the ground intervention. The use case at the state of West Bengal successfully demonstrated the digital augmentation techniques for mapping crop types, crop fallow dynamics and identifying the suitable areas for growing legumes in the short window of the rice fallows. The results show that rice fallows intensification could become quite accommodating in growing more food with high nutritional value and additional income to the farmers. Mapping the spatial extent of rice crop gives an insight in the area of interest and immensely helps in the decision-making process. Digital augmentation techniques coupled with crop simulation or predictive modeling could help to identify potential loss to the individual farmers' fields by analysing yield gap at pixel level. Hence, it can be concluded that digital augmentation techniques powered by geoinformatics for sustainable agrointensification can provide vital inputs for decision-making, planning and implementation of appropriate policies at ground level.

References

Alexandratos N, Bruinsma J (2012) http://large.stanford.edu/courses/2014/ph240/yuan2/docs/ap106e.pdf. ESA working paper. Accessed 4 Jun 2021

Bandyopadhyaya PK, Singh KC, Mondal K, Nath R, Ghosh PK, Kumar N, Basu PS, Singh SS (2016) Effect of stubble length of rice in mitigating soil moisture stress and on yield of lentil (Lens culinaris Medik.) in rice-lentil relay crop. Agri Water Manage 173:91–102

Behera MD, Biradar C, Das P, Chowdary VM (2020) Developing quantifiable approaches for delineating suitable options for irrigating fallow areas during dry season-a case study from Eastern India. Environ Monitor Assess 191:805

Biradar C, Ghosh S, Löw F et al. (2019) Geo big data and digital augmentation for accelerating agroecological intensification in drylands. ISPRS Int Arch Photogramm Remote Sens Spatial Inf Sci XLII-3/W6:545–548

Biradar C, Krishna G, Sarker A, Swain N, Nangia V, Devkota M, Govind A, Swelam A, Yigezu Y, Sahoo R, Wery J (2020). Digital augmentation for sustainable intensification of dryland farming systems. In: 16th ESA congress, 1–4 September 2020, Sevilla, Spain

Biradar CM, Xiao X (2009) Quantifying the area and spatial distribution of double- and triple-cropping croplands in India with multi-temporal MODIS imagery in 2005. Int J Remote Sen 32(2):367–386

Cai Y, Lin H, Zhang M (2019) Mapping paddy rice by the object-based random forest method using time series Sentinel-1/Sentinel-2 data. Adv Space Res 64:2233–2244

Chandna P, Mondal S (2020) Analyzing multi-year rice-fallow dynamics in Odisha using multi-temporal Landsat-8 OLI and Sentinel-1 Data. Gisci Remote Sen 57:431–449

Das B, Singh A, Panda S, Yasuda H (2015) Optimal land and water resources allocation policies for sustainable irrigated agriculture. Land Use Policy 42:527–537

DES MoAFW. (2020) Directorate of economics and statistics, ministry of agriculture & farmers' welfare, govt. of India. https://eands.dacnet.nic.in/. Accessed 04 June 2020

Dias L, Pimenta F, Santos A et al (2016) Patterns of land use, extensification, and intensification of Brazilian agriculture. Glob Chang Biol 22:2887–2903

Fang H, Wu B, Liu H, Xuan H (1998) Using NOAA AVHRR and Landsat TM to estimate rice area year-by-year. Int J Remote Sens 3:521–525

Gatto M, Petsakos A, Hareau G (2020) Sustainable intensification of rice-based systems with potato in Eastern Indo-Gangetic plains. Am J Potato Res 97:162–174

Godfray H, Beddington J, Crute I et al (2010) Food security: the challenge of feeding 9 billion people. Science 327:812–818

Gomez C, White JC, Wulder MA (2016) Optical remotely sensed time series data for land cover classification: a review. ISPRS J Photogramm Remote Sens 116:55–72

Gumma M, Thenkabail P, Deevi K et al (2018) Mapping cropland fallow areas in Myanmar to scale up sustainable intensification of pulse crops in the farming system. Gisci Remote Sen 55:926–949

Gumma M, Thenkabail P, Teluguntla P et al (2016) Mapping rice-fallow cropland areas for short-season grain legumes intensification in South Asia using MODIS 250 m time-series data. Int J Digit Earth 9:981–1003

Huang J, Wang X, Li X et al. (2013) Remotely sensed rice yield prediction using multi-temporal NDVI data derived from NOAA's-AVHRR. PLoS One 8:e70816

Huete AR (1988) A soil-adjusted vegetation index (SAVI). Remote Sens Environ 25:295–309

Jiang Y, Lu Z, Li S et al (2020) Large-scale and high-resolution crop mapping in china using Sentinel-2 satellite imagery. Agriculture 10:433

Krupnik T, Schulthess U, Ahmed Z, McDonald A (2017) Sustainable crop intensification through surface water irrigation in Bangladesh? A geospatial assessment of landscape-scale production potential. Land Use Policy 60:206–222

Kumar R, Mishra JS, Rao KK, Mondal S, Hazra KK, Choudhary JS, Hans H, Bhatt BP (2020) Crop rotation and tillage management options for sustainable intensification of rice-fallow agro-ecosystem in Eastern India. Sci Rep 10(1):1–15

Li J, Roy D (2017) A global analysis of Sentinel-2A, Sentinel-2B and Landsat-8 data revisit intervals and implications for terrestrial monitoring. Remote Sens 9:902

Li P, Xiao C, Feng Z (2018) Mapping rice planted area using a NEW Normalized EVI AND SAVI (NVI) derived FROM Landsat-8 OLI. IEEE Geosci Remote Sen Lett 15(12):1822–1826

Liu L, Xiao X, Qin Y, Wang J, Xu X, Hu Y, Qiao Z (2020) Mapping cropping intensity in China using time series Landsat and SENTINEL-2 images and Google Earth Engine. Remote Sen Environ 239:111624

NATMO (National Atlas and Thematic Mapping Organisation) Maps (2010) Department of Science and Technology (DST), Government of India

ODK (2020) Collect data anywhere. https://getodk.org/. Accessed 4 June 2021

Pretty J, Bharucha ZP (2014) Sustainable intensification in agricultural systems. Ann Bot 114:1571–1596

Qiu B, Lu D, Tang Z et al (2017) Automatic and adaptive paddy rice mapping using Landsat images: case study in Songnen Plain in Northeast China. Sci Total Environ 598:581–592

Quarmby NA, Townshend JRG, Settle JJ, White KH, Milnes M, Hindle TL, Silleos N (1992) Linear mixture modeling applied to Avhrr data for crop area estimation. Int J Remote Sens 13:415–425

Ramadhani F, Pullanagari R, Kereszturi G, Procter J (2020) Mapping of rice growth phases and bare land using Landsat-8 OLI with machine learning. Int J Remote Sens 41:8428–8452

Shi Y, Huang WJ, Ye HC, Ruan C, Xing NC, Geng Y, et al. (2018) Partial least square discriminant analysis based on normalized two-stage vegetation indices for mapping damage from rice diseases using PlanetScope datasets. Sensors 18(6):1901

Singh RK, Behera MD, Rizvi R, Biradar C (2021) Application of time-weighted dynamic time wrapping for crop type mapping using Sentinel-2 data automated crop type mapping using time-weighted dynamic time warping. Curr Res Environ Sustain 3(2021):100032

Thenkabail PS, Biradar CM, Noojipady P, Dheeravath V, Li Y, Velpuri M, Gumma M et al (2009) Global irrigated area map (GIAM), derived from remote sensing, for the end of the last millennium. Int J Remote Sen 30:3679–3733

Tilman D, Balzer C, Hill J, Befort BL (2011) Global food demand and the sustainable intensification of agriculture. Proc Natl Acad Sci 108(50):20260–20264

Tilman D, Cassman KG, Matson PA, Naylor R, Polasky S (2002) Agricultural sustainability and intensive production practices. Nature 418(6898):671

Tripathi AD, Mishra R, Maurya KK, Singh RB, Wilson DW (2019) Estimates for world population and global food availability for global health. The role of functional food security in global health. Academic Press, Elesiver, pp 3–24

Wu WB, Yu QY, You LZ, Chen K, Tang HJ, Liu JG (2018) Global cropping intensity gaps: increasing food production without cropland expansion. Land Use Policy 76:515–525

Xiao X et al (2006) Mapping paddy rice agriculture in South and Southeast Asia using multi-temporal MODIS images. Remote Sens Environ 100:95–113

Xiong J, Thenkabail PS, Gumma MK, Teluguntla P, Poehnelt J, Congalton RG, Yadav K, Thau D (2017) Automated cropland mapping of continental Africa using google earth engine cloud computing. ISPRS J Photogramm Remote Sens 126:225–244

Zhang G, Xiao X, Biradar C, Dong J, Qin Y, Menarguez M, Zhou Y, Zhang Y, Jin C, Wang J, Moore B III (2017) Spatiotemporal patterns of paddy rice croplands in China and India from 2000 to 2015. Sci Total Environ 579:82–92

Zhang G, Xiao X, Dong J, Kou W, Jin C, Qin Y, Zhou Y, Wang J, Menarguez MA, Biradar C (2015) Mapping paddy rice planting areas through time series analysis of MODIS land surface temperature and vegetation index data. ISPRS J Photogramm Remote Sens 106:157–171

Transforming Soil Paradigms with Machine Learning

Kumari Sweta, S. Dharumarajan, Amar Suputhra, M. Lalitha, R. Vasundhara, B. Kalaiselvi, B. N. Shashi Kumar, A. Sathish, and Rajendra Hegde

Abstract Numerous technological advancements have assisted to secure the vital key for scientific predictions in various fields, and soil science is no exception. Soil has always been chosen as an indispensable component by scientists, environmentalists, and policy makers for shaping a sustainable present and a secure future. Evidently, a huge amount of soil spatial data is required in the process to attain the desired outcomes. However, the challenges posed by the paucity of time and resources pose the hurdle in the collection and analysis of soil information. State-of-art technologies like Machine Learning (ML) and Big Data come as saviors to address those challenges. ML is helping to quantify, predict, identify, and classify the soil resources. The advanced algorithms, and models helped to gain better insights into soil mapping along with widening the perspective for its better management. Digital Soil Mapping (DSM), ML integrated with spectroscopic soil studies is gaining momentum among the scientific communities. These innovative approaches have the capabilities to solve global issues like desertification, ecological stability, carbon pool management and climate mitigation in a holistic and integrated way by keeping the soil as one of the key parameters. In this chapter, an attempt has been made to present a comprehensive overview of ML algorithms, which have been adopted by many researchers across the globe in prediction of various soil properties and presented a case study on digital soil mapping by using ML algorithms.

Keywords Machine learning · Digital soil mapping · Soil properties · Spectroscopic soil studies · Prediction · Random forest model · Support vector machine

---

K. Sweta · A. Sathish
Department of Soil Science and Agricultural Chemistry, University of Agricultural Sciences, Bangalore, 560065, India

S. Dharumarajan (✉) · A. Suputhra · M. Lalitha · R. Vasundhara · B. Kalaiselvi · B. N. S. Kumar · R. Hegde
ICAR-National Bureau of Soil Survey and Land Use Planning, Regional Centre, Bangalore 560024, India

G. P. O. Reddy et al. (eds.), *Data Science in Agriculture and Natural Resource Management*, Studies in Big Data 96, https://doi.org/10.1007/978-981-16-5847-1_12

1 Introduction

The high-paced progressing world is experiencing a great diversity of constraints and challenges to address the complex issues ranging from capacity enhancement for feeding the growing population to maintaining the ecological balance. Consequently, the management of limited natural resources has become the epicenter of global discussions and debates. In this regard, soil—the highly precious resource that takes millions of years to develop—is a priority among experts from various fields for its conservation and sustainability. Soil conservation programs need spatio-temporal information on soil properties for measuring soil quality changes and preparing action plans. Soil studies have carried out for quite a long time to gather such soil information and have undergone several modifications in the due course. The initial studies were solely dependent on conventional methods which ranged from transact walks in area to manual mapping of topography and terrain. The diversity of constraints of conventional methods has been elaborated further in this chapter. Altogether, the conductance of soil studies with following the conventional methodologies alone is a herculean task as they are time-consuming, laborious, and expensive (Zeraatpisheh et al. 2020; Lamichhane et al. 2021). The other way out is integration of conventional techniques with the state-of-art technologies of the modern world. One such sufficiently advanced technology is ML, which is potent to address numerous constraints of conventional studies.

In the field of soil studies, supervised ML algorithms are more commonly used than unsupervised ones. This argument can be logically backed up by the fact that soil is a dynamic entity and therefore, to train the model accurately, some amount of labeled data in the form of soil legacy data must be fed to the system. In addition to the inherent dynamic nature of soil, comparatively low popularity of unsupervised learning among soil scientists can also be attributed to their computational complexities. There is a high probability to obtain inaccurate outcomes in unsupervised learning without human interventions. However, with advancements in ML assisted pedological understandings, they may become more popular. Supervised algorithms are mainly being focused here. They are further divided into classification and regression algorithms. The classification algorithms are used for prediction of discrete variables such as soil textural class and taxonomic units, and regression algorithms are utilized in predicting continuous variables like pH, EC, CEC, etc.

Some of the widely used algorithms in soil studies have been discussed in a nutshell without clinging much to their technical specifications. Simplest among them is Multi Linear Regression (MLR) which is used to find interrelationships between a target variable and a number of different covariates (James et al. 2013). The output of the target variable can be predicted with MLR, using two or more covariates by fitting a linear equation with the model parameters. Next is Cubist (Cu), which is a rule-based model which creates a tree structure initially from a dataset containing multiple covariates and the terminal trees create a linear regression model. Later using boosting training, where the weak learners are converted to strong learners by assigning more weightage to the stronger learners, it creates rules by collapsing paths through the tree. The final tree contains MLR models as the terminal nodes for calculating predicted values (Quinlan 1992). Support Vector Machine (SVM) which works on the principle of minimization of structural risk is yet another commonly used algorithm. SVM defines a maximal margin to all the maintained covariates using the support vector observations and fit the data linearly or separate it. Due to its improved generalization ability, SVM can overcome the over-fitting problems and make the results more applicable. Neural Network (NN) is yet another advanced algorithm and is inspired from the functioning of the human brain through a network of neurons. NN can be trained to perform specific functions and obtain certain outputs by adjusting the connection weights assigned to the components. A training dataset containing a large amount of information helps NN to learn abundantly. There are several types of NN like Multilayered Feedforward network, Multi Layered perceptrons (MLP), and Convolutional Neural Network (CNN). The most common algorithm in contemporary soil studies is Random Forest (RF). It consists of a group of tree predictors generated with the help of random vectors which are sampled independently from the fed input vectors. Tree predictor takes on numerical values in regression, whereas in classification it assumes class labels (Breiman 1999). In the process, multiple decision trees are constructed. On the basis average of results obtained from the total number of decision trees used, the predictions are made. The final number of trees to be grown, the random covariates selection at every node, and the minimum number of terminal sample 'leaves' for noise reduction in the training data are some of the important determinants of the model performance of RF.

2 Era of Soil Science Before Machine Learning—Issues and Constraints

Soil studies encompassing the collection of spatial soil information, fertility assessments and mapping, and pedological developments were carried out even before the integration of ML with them. Undoubtedly, the conventional methodologies have played a significant role in helping us to attain the knowledge about soils and are inseparable even today. However, using conventional methods alone, several hurdles are hindered in the progress of comprehensive study. Some of the major constraints faced with the traditional soil studies are as follows:

- ***Time and resource limitations***: Assessment of the vast areas of soil demands a huge number of human resources along with other inputs. In addition, the conventional methods are time consuming and destructive in nature
- ***Lack of continuity in the study***: Due to discrete studies without any holistic database, the traditional studies often lacked continuity. There is very little scope to build on and improvise the previous assessments.
- ***High scope of human error***: With great dependence on manual assessments and predictions, there always exists a window for manual errors. As a result, the focus

of the surveyors revolves around reducing those errors and thereby confining the scope of new explorations in the pedological domain.

- ***Lack of multidimensional approach***: Traditional studies are mainly concentrated on soil aspects with less importance being given to water, socio-economic, and environmental attributes.
- ***Limited scope of application***: Due to the missing linkage between soil and other aspects, knowledge of soil functions as an ecosystem service provider is least understood to achieve the goal of ecosystem sustainability.

Therefore, in the quest for a holistic understanding with improved efficacy, the research community started experimenting and implementing technological advancements in the arena of soil resource assessment.

3 Machine Learning in Soil Studies

Among the various technological options available to the soil scientists, ML provides the immense scope for diversity of application in soil resource studies. Padarian et al. (2020a) have categorized twelve different ML assisted soil related studies, namely, remote sensing based soil assessment (Grunwald et al. 2015; Zhang et al. 2019), soil organic carbon and carbon stocks mapping (Rial et al. 2017; Safari et al. 2017; Liu et al. 2018a, b; Keskin et al. 2019; Heuvelink et al. 2020; Dharumarajan et al. 2021a), soil water (Ahmad et al. 2010; Coopersmith et al. 2014; Attarzadeh et al. 2018; Greifeneder et al. 2021), soil pollution and contamination (Jiaet al. 2019; Hu et al. 2020), erosion and parent material (Du Plessis et al. 2020), methodologies deploying different algorithms like NN and RF (Farfani et al. 2015); spectroscopy (Morellos et al. 2016; Butler et al. 2018), modelling soil classes (Brungard et al. 2015; Dharumarajan and Hegde 2020; Camera et al. 2017), soil-crops interactions (Ivushkin et al. 2018; Khanal et al. 2018), soil physical properties (Bondi et al. 2018), and soil mapping (Pouladi et al. 2019). Digital soil mapping and NIR (Near- Infrared)/MIR (Mid-Infrared) spectroscopic soil analysis are two of the novel interventions, which have been mingling well with ML. It has been discussed further in the following section.

3.1 Digital Soil Mapping (DSM)—A ML Assisted Solution

Mapping of the soil properties has a long history. Soil maps have been used by various stakeholders for agriculture planning, framing policies, taking decisions on infrastructure projects, and landslides and other hazard assessments. However, most of the soil maps produced by conventional techniques are not scale uniform across spatial boundaries (McBratney et al. 2003). As soil properties are dynamic in nature, such maps cannot represent the present soil status and are incompatible with geospatial data in various soil surface process models. Soil resource monitoring in particular intervals through conventional soil survey is very laborious, expensive, and time consuming (Grunwald 2010). Moreover, the conventional mapping utilizes cartographic models, which degrade the quality of soil information represented by the maps. Also, the variations occurring at the boundaries of the delineated soil polygons are often missed out.

The advancements in computational power and technologies like geo-information systems have created immense potential for improvement and improvisation in techniques of soil mapping. The initiatives in the form of the SOTER (Soil and Terrain) program (1986–2016) by FAO, Italy and SoLIM (Soil Land Interface Model) (Zhu et al. 1997) assisted in changing the trajectory of soil mapping technologies in a completely new direction. SoLIM was a model which was directed to acquire and represent spatial soil information using a fuzzy logic interface. The SOTER program on the other hand was aimed at developing a global dataset at 1:1,000,000 scale. Even though these initiatives did not completely accomplish the desired goals, they paved the path of better technologies which possess the potential to meet the set targets in the true spirit. Digital soil mapping is one such advanced technology which deals with the generation of geographically referenced soil databases on the basis of relationships between environmental covariates and analyzed data from field and laboratory (McBratney et al. 2003). It can be defined as the creation of spatial soil information systems using numerical models which infer the soil spatial and temporal variations from soil observations and knowledge from derived environmental variables (Lagacherie et al. 2006).

SCORPAN (Soil, Climate, Organism, Relief, Parent material, Age, and Spatial location) is the principal model on which working of DSM is based (McBratney et al. 2003). It is a derivative of fundamental CLORPT (Climate, Organism, Relief, Parent material, and Topography) model (Jenny 1941). The inter-relationship between soil properties or soil classes and the environmental covariates is fitted using a suitable model to predict the soil properties and of unknown locations. Figure 1 shows the major components of DSM. Furthermore, DSM does not only focus on mapping soil properties, but also deals with issues such as land degradation and climate change

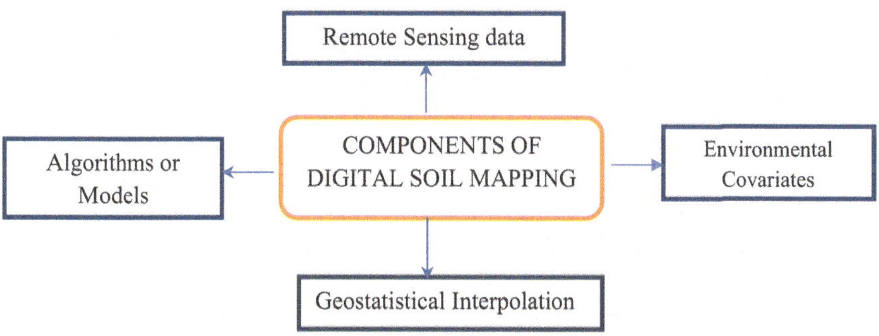

Fig. 1 Components of digital soil mapping

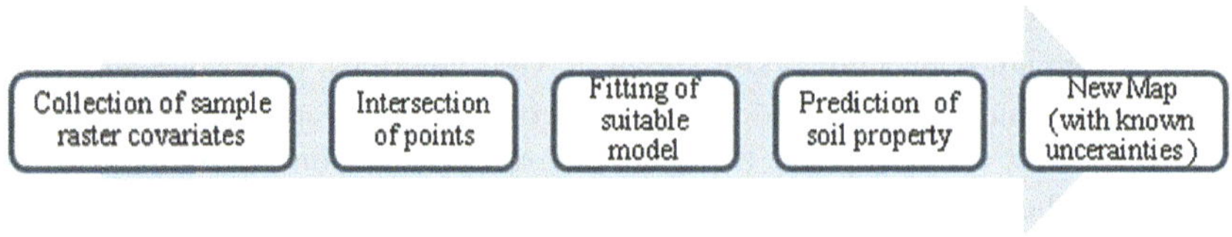

Fig. 2 Flowchart depicting the digital soil mapping methodology

which requires soil assessments in the context of environmental stressors (Grunwald 2009).

Arrouays et al. (2020) crisply pointed out the advantages of digital soil maps over the conventional soil maps, which includes quantitatively prediction of soil attributes, give uncertainty estimations of the predictions, easily incorporate and reproduce the updates, enable cost-effective and quicker map production and allow ranking and quantification of the factors influencing soil attribute distribution. Realizing its advantages, the need for the adoption of DSM at a larger scale has been emphasized time and again (Arrouays et al. 2017; Dharumarajan et al. 2019). Global Soil Map project is aimed at producing precise digital maps of different regions around the globe. Therefore, there is a growth in interest to use ML in DSM in order to accomplish the targets efficiently. The working of DSM has been depicted in a simple manner in the flow chart (Fig. 2).

Studies on the prediction of soil properties with ML algorithms with DSM can be reviewed as follows:

- **Physical properties**: ML assisted DSM has been used to predict and map soil depth (Dharumarajan et al. 2020a), soil texture (Ließ et al. 2012; Akpa et al. 2014; Vaysse and Lagacherie 2015; Dharumarajan and Hegde 2020), pH (Dharumarajan et al. 2017; Panday et al. 2018), and available water capacity (AWC) (Malone et al. 2009; Dharumarajan et al. 2021b).
- **Chemical properties**: Soil organic carbon (SOC) concentration (Kheir et al. 2010; Dai et al. 2014; Minasny et al. 2013; Mudler et al. 2016; Taghizadeh-Mehrjardi et al. 2016; Siewert 2018; Pouladi et al. 2019), and associated stocks (Grimm et al. 2008; Adhikari et al. 2014a, b; McNicol et al. 2019; Dharumarajan et al. 2021a), cation exchange capacity (Forkuor et al. 2017; Dharumarajan et al. 2020b) are being mapped with DSM. Additionally, it has been used to map soil problems including soil salinity (Taghizadeh-Mehrjardi et al. 2016).
- **Biological aspects**: Very less work has been reported in DSM with respect to the biological properties. Tajik et al. (2019) in Iran mapped the diversity of invertebrates in soil. Biological soil crust classes were digitally mapped by Brungard and Boettinger (2012). Van Den Hoogen et al. (2019) have mapped the soil nematode density at the global scale.
- **Fertility status**: Maps of soil nutrients such as nitrogen (Forkuor et al. 2017; Kalambukattu et al. 2018), phosphorus (Viscarra-Rossel et al. 2015; Hengl et al. 2017; Kalambukattu et al. 2018; Iticha and Takele, 2019; Dharumarajan et al.

2021c), potassium, calcium or magnesium (Hengl et al. 2017; Iticha and Takele, 2019), zinc (Panday et al. 2018; Dharumarajan et al. 2021c) and boron (Panday et al., 2018) among others have been prepared by DSM approach.

3.2 A Brief Look at Spectroscopic Soil Studies with ML

The spectroscopic approach for soil analysis in itself is a newly emerging field, especially in developing countries. This innovative approach is non-destructive, prevents harmful chemicals, and combines with multivariate calibrations, which can help estimate soil properties. Being sensitive to both organic and inorganic soil compositions can act as a powerful tool for assessing soil health. Moreover, ML methods and deep learning have been utilized in spectral data processing and are changing the way we model soil spectral data. This technology can prove to be an eco-friendly, economical, and reliable means for soil studies and sustainable soil management when integrated with digital soil mapping. There are several studies; however, only a few have been mentioned here to get a broad idea of diversity of applications.

CNN algorithm was trained to predict several soil properties simultaneously in one model from raw soil spectra, and it performed better compared to conventional methods such as PLS regression (Padarian et al. 2019). Ant Colony Optimization-interval Partial Least Squares (ACO-iPLS), RF, and Recursive Feature Elimination-Support Vector Machine (RF-SVM) has been used to estimate OC with VIS spectroscopy and gave promising results (Ding et al. 2018). Dangal et al. (2019) tested ML approaches in predicting soil properties, which are relevant to carbon cycle studies. Morellos et al. (2016) also investigated OC, total nitrogen, and soil moisture content based on Cubist and SVM by using VIS-NIR spectroscopy.

Additionally, Khanal et al. (2018) reported that remotely sensed data when integrated with ML algorithms offered a time and cost-effective solution for prediction of soil properties and crop yield over conventional methodology. This inter-linkage can prove to be useful in locating and demarcating the areas with potential concerns and assist in implementation of site-specific farm management practices. RF accompanied with VIS-NIR spectroscopy can be used to analyze soil salinity (Wang et al. 2018). These studies clearly prove that ML assisted spectroscopic studies can be applied diversely.

ML has also been exploited in making comparisons of the prediction accuracy based on different spectral bands. Yang et al. (2019) compared the accuracy of prediction based on bands reduced using the genetic algorithm (GA) and VIS-NIR full bands for SOM and pH prediction using Partial least squares regression (PLSR), extreme learning machines (ELM), least squares-support vector machines (LS-SVM), and the Cubist regression model (Cu) and concluded that ELM when used with GA reduced bands performed the best for SOM and pH prediction. In addition, soil sub group prediction by SVM through X-ray fluorescence and VIS spectroscopy has been experimented by Benedet et al. (2020).

3.3 A Comparative Analysis of ML Algorithms

Literature portrays that different models have performed differently under a given condition. It is not easy to designate the best suitable model for various conditions. For instance, a model performing well to predict soil texture may not give the best results for soil organic carbon prediction. The model performance is influenced by the history data fed as an input. Soil organic carbon content is a more dynamic property than soil texture, which changes slower. The covariates influencing a particular property may also have some impact on model overall performance. ML algorithms were found to outperform linear regression algorithm for most of the times in the spectroscopic soil with remote sensing data (Khanal et al. 2018). RF has gained the most popularity for DSM as they have consistently outperformed other models (Zeraatpisheh et al. 2019; Lamichhane et al. 2019; Nussbaum et al. 2018). The growing popularity of decision trees models, e.g., RF and Cubist, can be attributed to the fact that are these algorithms can deal with both linear and non-linear relationships within the data. Moreover, these approaches generally outperform classical algorithms such as linear regression, principal component analysis. Moreover, RF algorithms need minimal pre-processing and data can be used without transformation (Kuhn and Johnson 2013).

SVM has been used in soil classification (Kovačević et al. 2010), soil quality assessment (Liu et al. 2015), texture classification (Barman and Choudhury 2020), etc. ANN has been mainly used in DSM for the prediction of continuous soil variables like particle size fractions (McBratney et al. 2000; Priori et al. 2014) and soil erosion rates (Licznar and Nearing 2003). Moreover, it has also been used for prediction of discrete soil variables (Behrens et al. 2005; Silveira et al. 2013) but on a limited scale. Heung et al. (2016) evaluated ten ML algorithms, namely, ANN, MLR, SVM, CART (Classification and Regression Tree), CART with bagging, RF, *k*-NN, logistic model trees, and nearest shrunken centroid for the prediction of soil taxonomic units. Results depicted that *k*-NN and SVM showed the highest accuracy for great groups. However, due to the speed of parameterization and the interpretability of the results, algorithms like CART with bagging, RF and logistic model trees were preferred while resulting in similar accuracies. It can be inferred that choice of models and sample design has great influence on outcomes. CART, Bagged Regression Tree (BaRT), Boosted Regression Tree (BoRT), Partial Least Square Regression (PLSR), Regression Kriging (RK), Ordinary Kriging (OK) RF, and SVM were compared by Keskin et al. (2019) to predict soil organic carbon fractions and total carbon. RF was found to perform the best followed by SVM, BoRT, and BaRT. On the other hand, OK showed the poorest performance followed by CART, RK, and PLSR.

Evidently, there are several models that help in the prediction of soil properties with accuracy in a given region; however, they are inconsistent in terms of input data in form of environmental covariates, soil properties, quantitative methods, and analysis strategies (Grunwald 2009). Due to lack of consistency in the comparative studies, it is advisable to follow calibration of the algorithms using specific training datasets (Lamichhane et al. 2019). A fine balance between quantitative, geographic, and pedological expertise can ideally improve the scope of algorithm led soil understandings.

4 Digital Soil Mapping of Soil Properties Using ML Algorithms—A Case Study

A case study of the ML integrated DSM approach followed in the northern Karnataka Plateau representing Basaltic Terrain is presented in this section. Two of the key soil properties viz. pH and soil organic carbon (OC) were predicted using three ML algorithms viz. Cubist (Cu), Support Vector Machine (SVM), and Random Forest (RF) and the predicted properties were mapped.

4.1 Study Area

The study was carried out in the Bhagyanagar sub-watershed, Koppal district of Northern Karnataka Plateau in the southern part of India and located between 15° 15′ to 15°21′ N latitude and 76°5′ to 76°10′ E longitude covering with an area of 3260 ha (Fig. 3). Granite gneiss and alluvium are the major parent material and geology.

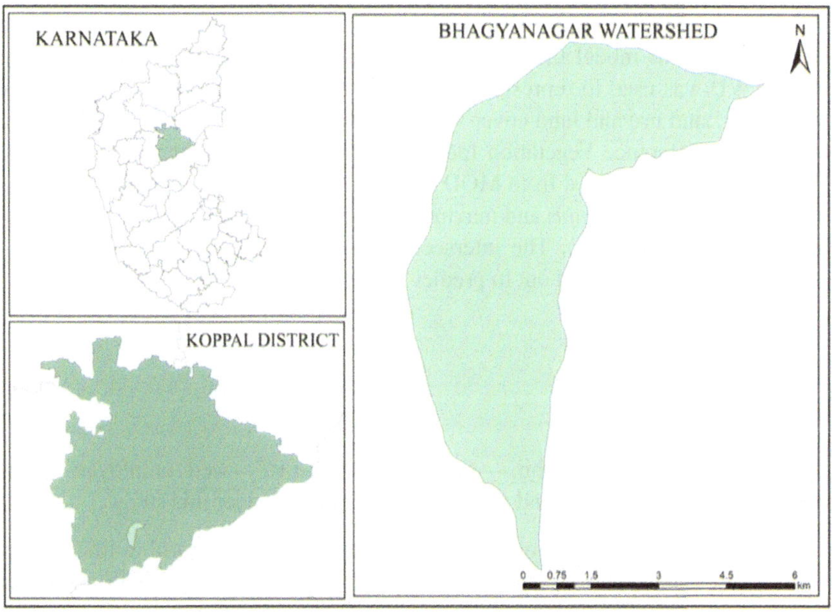

Fig. 3 Location map of the study area

The major soil orders are Vertisols and Alfisols. The sub-watershed falls under the semiarid tract of the state and is categorized as drought prone with mean annual rainfall of 662 mm. The average maximum and minimum temperature recorded for the area are 31.5 °C and 20.5 °C, respectively. The average monthly potential evapotranspiration (PET) is 145 mm. Majority of the area comes under rainfed cultivation with crops like pigeon pea, sorghum, and pearl millet.

4.2 Sampling Methodology

A total of 514 surface soil samples (0–15 cm) were collected at 250 m grid interval in the study area. Soil samples were shade dried and passed through 2 mm sieve after processing. The samples were then analyzed for soil pH (1:2.5) and soil organic carbon (SOC) (Walkley and Black, 1934).

4.3 Environmental Covariates

Digital elevation model (DEM) from Shuttle Radar Topography Mission (SRTM) data was obtained and processed in ArcGIS10 software. The terrain attributes such as aspect, slope, profile curvature, plan curvature, topographic wetness index (TWI) and topographic position index (TPI), Multi-resolution Index of Valley Bottom Flatness (MrVBF), and Multi-resolution Ridge Top Flatness (MrRTF) were used to represent relief in *SCORPAN* model and were derived using SAGA GIS. Topographic wetness index (TWI) was used to represent potential hydrological conditions on slope and catchment. Land use and land cover were represented with auxiliary variables viz., Normalized Difference Vegetation Index (NDVI), and Enhanced vegetation index (EVI) which were obtained from MOD13Q1 of NASA. Annual precipitation (*prec*) and average minimum (*tmin*) and maximum temperature (*tmax*) were used to represent the climate (Table 1). The intersection of environmental covariates for 514 sampling points was carried out to predict the soil properties.

4.4 Algorithms Used

Three supervised ML algorithms—Cubist, SVM and RF—were employed for the prediction of soil properties and uncertainties estimates under this study.

- ***Cubist model*** is an extension of the M5 model tree. This algorithm first divides the datasets into internal subsets having similar characteristics with respect to the soil property and the environmental covariates. Cubist model is an ensemble model, i.e., it allows the addition of models produced by several repetitions of

Table 1 Different covariates used for prediction

Predictor	Source	Resolution
Elevation (m)	SRTM DEM	30 m
Slope	SRTM DEM	30 m
Aspect	SRTM DEM	30 m
TPI	SRTM DEM	30 m
TWI	SRTM DEM	30 m
Plan curvature	SRTM DEM	30 m
Profile curvature	SRTM DEM	30 m
MrVBF	SRTM DEM	30 m
MrRTF	SRTM DEM	30 m
LS Factor	SRTM DEM	30 m
NDVI	MOD13Q1	250 m_16 days
EVI		
Landsat data (b1–b9)	9bands	30 m
Precipitation (mm)	Worldclim2	30 s
Average max temperature (°C)	Worldclim2	30 s
Average min temperature (°C)	Worldclim2	30 s

the particular algorithm and thereby can improve the performance of prediction using hyperparameters (Quinlan 1993). Three parameters viz. maximum number of rules in the data, number of boosted iterations of the algorithm, and extent of allowable extrapolation are to be optimized. R package 'Cubist' was used for the prediction of soil data (Kuhn et al. 2016).

- ***Support vector machine (SVM)*** was first introduced by Vapnik et al. (1995), and it is a technique that generates optimal separating hyperplane to differentiate overlapping classes which cannot be separated in a linear manner. Regression problems were solved by reduction of the structural risk with the help loss functions in SVM.
- ***Random Forest Model (RF)*** works on the principle of assembling the classification and regression trees in the forest using randomization at two levels for each tree (Breiman 1999). Like Cubist, it is also an ensemble learning algorithm that selects a set of observations from the training dataset in a random fashion by the means of bootstrapping and builds a decision tree related with those selected observations. The decision trees produce error predictions with the help an out-of-bag (OOB) strategy which in turn is generated by bootstrapping, using part of the observation dataset for training. The remaining observations are used for testing model error. This algorithm lacks sensitivity to missing data and has the ability to handle large amount of data both categorical and quantitative (Dharumarajan et al. 2017).

Table 2 Descriptive statistics of soil properties

Properties	OC (%)	pH
Mean	0.60	8.43
Min	0.29	7.06
Max	0.91	9.45
Std	0.12	0.51
CV	20	6

4.5 Models Validation

From the total dataset, 80% of the data was used for calibration and remaining 20% for validation. The models' performance was evaluated using uncertainty indicators such as Coefficient of determination (R^2), Root Mean Square Error (RMSE), mean error (ME), and Concordance correlation coefficient (CCC) for the validation dataset.

$$\text{Coefficeint of determination}\left(R^2\right) = 1 - \frac{\sum_{i=1}^{n}(pi - oi)^2}{\sum_{i=1}^{n}\left(oi - \underline{oi}\right)^2} \quad (1)$$

$$\text{Bias(ME)} = \frac{1}{n}\sum_{i=1}^{n}(oi - pi) \quad (2)$$

$$\text{Root mean squared error(RMSE)} = \sqrt{\frac{1}{n}\sum_{i=1}^{n}(oi - pi)^2} \quad (3)$$

where *pi* and *oipi* and *pi* are predicted and observed values, $\overline{pi}$ and $\overline{oi}\,\overline{pi}$ and $\overline{oi}$ are means of predicted and observed values.

$$\text{Concordance correlation coefficient}(\rho_c) = \frac{2\rho\sigma_0\sigma_p}{\sigma_{0+}^2\sigma_p^2+\left(\mu_0 - \mu_p\right)^2} \quad (4)$$

μ_0 and $\mu_p\mu_0$ and μ_p are the means of observed and predicted values and σ_0^2 and $\sigma_p^2\sigma_0^2$ and σ_p^2 are corresponding variance. $\rho\rho$ is Pearson correlation coefficient between observed and predicted values.

The good models have R^2 and $\rho_c\rho_c$ equal or close to 1 and RMSE close to 0.

4.6 *Results and Discussion*

4.6.1 Summary Statistics of Soil Properties

Summary statistics of the soil properties are represented in Table 2. The soil pH values are between 7.06 and 9.45 with a mean and standard deviation of 8.43 and 0.51, respectively. The OC ranged from 0.29 to 0.91% with a mean of 0.60%. The variability trends in pH and OC can be related to land management practices. The correlation analysis between environmental variables and soil properties (Table 3) made it evident that elevation negatively correlated with pH and positively correlated with OC.

Table 3 Correlation between variables and soil properties

Variable	OC (%)	pH
Elevation	0.45	−0.44
EVI_MODIS	0.00	−0.26
NDVI_MODIS	−0.16	−0.09
TWI	−0.13	0.18
TPI	0.00	−0.08
Slope	0.14	−0.13
Profile_Cu	0.07	−0.03
Plan_curva	0.02	0.00
MRVBF	−0.09	0.33
MRRTF	−0.33	0.43
LS_Factor	0.12	−0.07
Aspect	−0.05	0.00
Tmin	−0.34	0.39
Tmax	−0.66	0.33
Prec	0.53	−0.38
b1_LC	0.00	−0.38
b2_LC	0.05	−0.22
b3_LC	0.11	−0.14
b4_LC	0.11	−0.36
b5_LC	0.06	−0.26
b6_LC	−0.08	−0.28
b7_LC	0.00	−0.35
b8_LC	0.04	0.45

4.6.2 Model Performance and Variable Importance

The performance of each model was investigated for both the properties using statistical parameters viz. Root Mean Square Error (RMSE), Coefficient of determination (R^2), and concordance correlation coefficient (CCC) for calculation. For the prediction of pH, the results showed (Table 4) that RF performance ($R^2 = 0.78$, RMSE = 0.23, CCC = 0.86) was on par with SVM ($R^2 = 0.78$, RMSE = 0.24, CCC = 0.87). However, the Cubist ($R^2 = 0.72$, RMSE = 0.26, CCC = 0.84 and 0.00 bias) performed comparatively low in prediction of soil pH. Figure 4 pictures the scatter plots of predicted pH versus observed pH and predicted OC versus observed OC using RF model. Dharumarajan et al. (2017) emphasized that RF can be efficiently utilized to predict the pH and an increase in sampling density can potentially increase the model performance for more accurate predictions. Hengl et al. (2015) reported that RF enhanced the pH mapping accuracy over linear regression models up to an extent of 20%. OC prediction results portrayed that RF ($R^2 = 0.72$, RMSE = 0.06, CCC = 0.82) outperformed SVM ($R^2 = 0.66$, RMSE = 0.07, CCC = 0.78) and Cubist ($R^2 = 0.64$, RMSE = 0.07, CCC = 0.79) for the prediction of soil OC, with higher R^2 and CCC and relatively low RMSE value. The prediction of OC by RF was relatively comparable to results of most of the past studies (Wiesmeier et al. 2019, $R^2 = 76\%$; Sreenivas et al. 2016, $R^2 = 82\%$).

Table 4 Performance of models for prediction of pH and OC

Model	pH				OC (%)			
	R^2	CCC	RMSE	Bias	R^2	CCC	RMSE	Bias
RF	0.78	0.86	0.23	0.00	0.72	0.82	0.06	0.00
SVM	0.78	0.87	0.24	0.02	0.66	0.78	0.07	0.00
Cubist	0.72	0.84	0.26	0.00	0.64	0.79	0.07	0.00

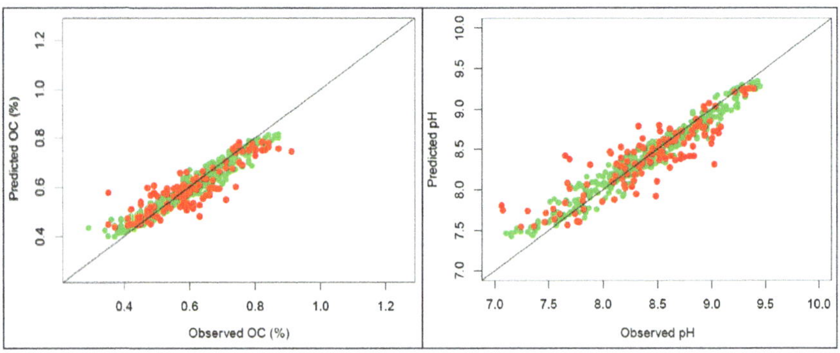

Fig. 4 Scatter plots of predicted pH and observed pH and predicted OC and observed OC using random forest model (calibration data-green points; validation data-red points)

Figure 5 depicts the variable importance rankings of RF model for OC and pH. Elevation came out to be the most significant covariate for prediction of OC followed by maximum temperature and precipitation. Dharumarajan et al. (2021a) found elevation as the top predictor in prediction of SOC since the areas on high elevations are generally under vegetation cover. The influence of topography on distribution of SOC has also been highlighted in several studies (Malone et al. 2009; Adhikari et al. 2014a, b). For pH prediction, band 8 of satellite imagery followed by band 5, elevation, MrVBF, and MrRTF emerged as important predictors.

RF model was used to map the pH and OC along with uncertainty at 30 m resolution (Figs. 6 and 7). The mean predicted pH is 8.47, and SOC is 0.60%. The uncertainty in prediction of pH and OC was more in the northern portion of the watershed compared to the southern counterpart. These detailed predicted maps could

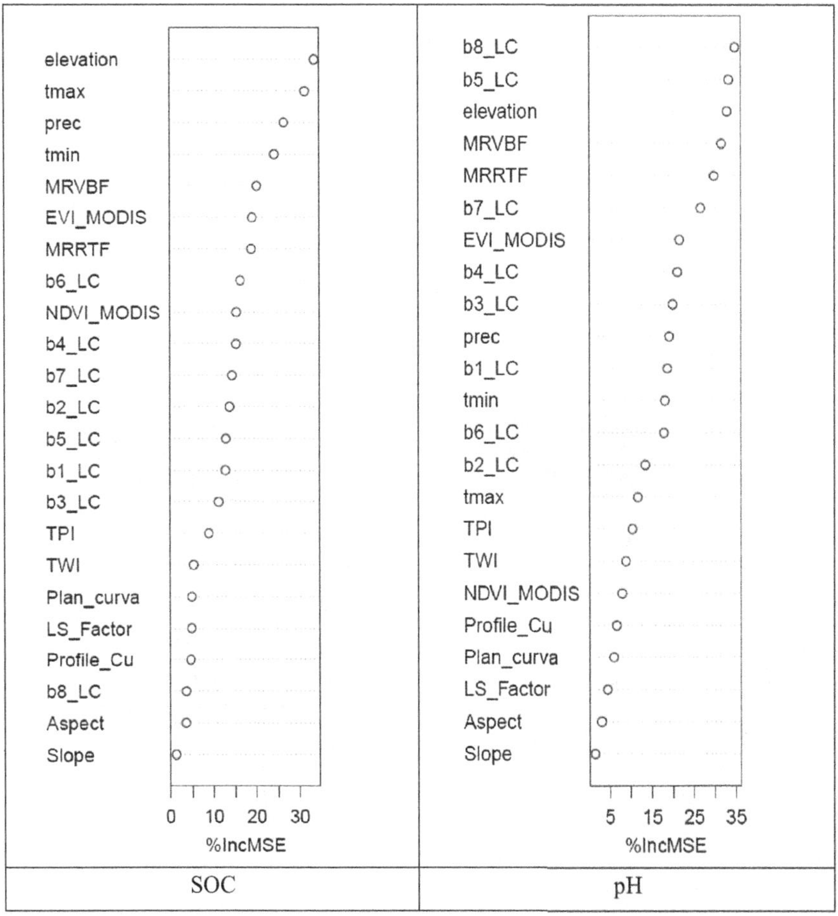

Fig. 5 Variable importance ranking of RF model for prediction of OC and pH

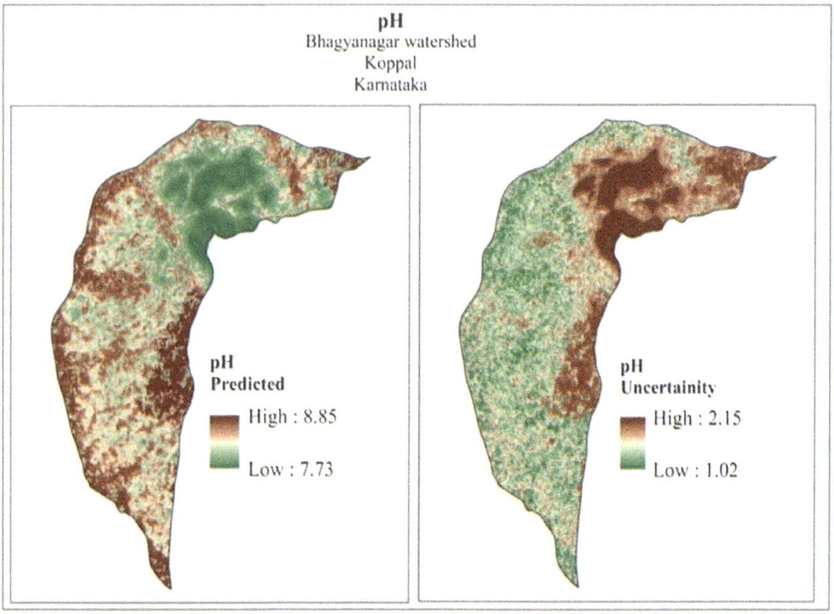

Fig. 6 Predicted pH map

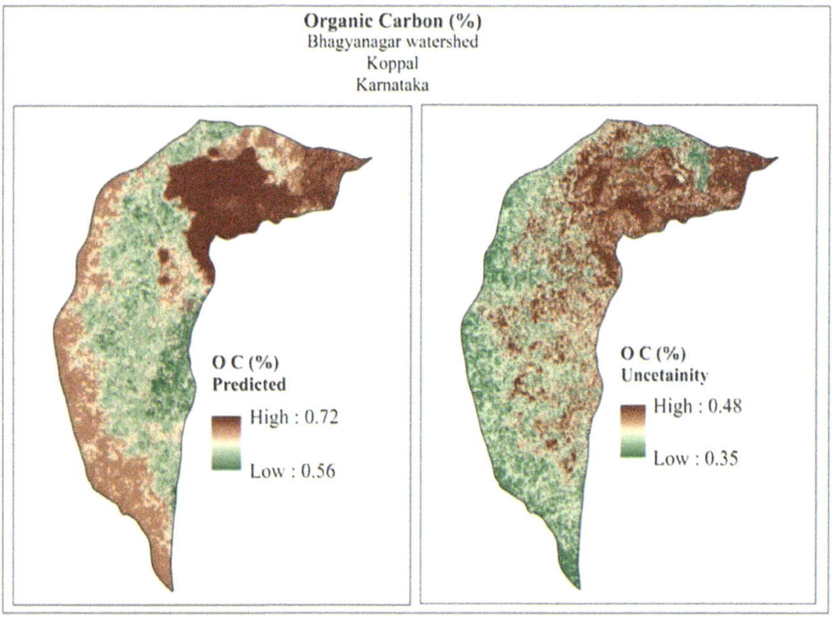

Fig. 7 Predicted OC map

help in identifying nutrients status spatially for making agronomic interventions for sustainable agricultural production and variable-rate fertilizer application.

5 Prevailing Challenges and Research Gaps

ML appears to be a promising technology which has the potential to completely reshape our knowledge and understanding about natural resources and soil as an ecosystem provider. However, currently there are certain challenges which are limiting us to realize its full potential. Some of them have been discussed here.

- ***Geographical confinements***: Though the ML methodologies are gaining popularity across the globe, they are concentrated majorly in developed countries in the present scenario (Paradian et al. 2020). This gap can be attributed to the missing link between science, development, and economic status for technology adoption in other regions. Moreover, under DSM, the models are usually being developed for a specific region. Further research is needed to apply it on a larger scale and not just confined to a small area in order to check the map accuracy.
- ***Interpretability issues***: Based on current findings, more advanced algorithms such as neural networks (NN) are generally considered as "black box" models by the researchers due to their low interpretability. RF also mostly falls in the same category (Brungard et al. 2015; Taghizadeh-Mehrjardi et al. 2016). Since, measurement of the interpretability is generally not defined appropriately and there exist numerous contradictions in views as well. However, the use of complex algorithms is justified by the use of important variables and covariates (Padarian et al. 2020b).
- ***Selection of suitable algorithm***: It is yet another challenge for the scientific community. Due to different pros and cons in the design of different ML algorithms, selection of the best among available options depends on the research objectives, data, and resource availability (Khaledian and Miller 2020). Selecting improper ones can produce poor or less reliable results. Some algorithms need specific assumptions, and the data needs to be transformed to meet those assumptions or the data normalization to the same attribute scale should be done for proper functioning of the algorithm. However, algorithms like decision tree lack such specific assumptions and do not need data transformations.
- ***Dynamic nature of soil***: Soil is a dynamic system owing huge diversity over time and space. Therefore, the search for the *perfect model* is a tough nut to crack. The studies related to development of a spatio-temporal algorithm for mapping of soils are very limited (Wadoux et al. 2020).
- ***Missing integration***: The improving mapping accuracy seems to be the most important standard to assess quality of the maps in most of the contemporary studies. It appears at times that any prediction can be turned into a soil map, neglecting the fact whether it possesses or lacks soil knowledge. There

exists a missing link which needs to be established between ML approaches and pedological knowledge.

- ***Others***: Skilled human resources to draw valuable inferences from the models and maps are not sufficiently available. Moreover, the research linking soil with ecological management via ML is still in its natal stage. Much work needs to be done in the field of updating the legacy data by taking into consideration important variables.

6 Potential Opportunities for Bridging the Gap

Amid several challenges, there are numerous potent opportunities and less explored dimensions of soil that have to be explored through the lens of ML. Opportunities encompassing the inclusion of pedological knowledge at various stages of modeling, innovative model designing improvisation of the present dataset, and output analysis of the predicted soil map using advanced parameters can aid the soil scientists to bridge the existing gap. The more a technology penetrates a particular field, the better understanding it gains. With this simple logic, the impregnation of ML in soil studies at the academic level in universities can enhance its operational interpretability. Increasing interpretability, in turn, can serve the purpose of obtaining a better understanding of soil processes (Wadoux et al. 2020). The new generation can develop innovative solutions to eliminate the problem of 'black-box' with a focus on understanding the soil processes. The well-established inter-institutional collaboration plans may accelerate this process further.

A potential opportunity also exists to explore different algorithms—both supervised and unsupervised—to further assess the sampling designs. It could be accompanied by investigating model consistency and prediction accuracy to evaluate the performance of the mapping approach as a whole. Consequently, this would enable us to generate accurate maps and link them to our existing scientific knowledge in a harmonious fashion.

7 Conclusions

Technology is an inseparable unit of smart and precision agriculture. Innovative integration of novel technologies such as ML with mature pedological understandings can help us gain crucial information about the soils. That information, in turn, can be applied in multiple dimensions to address the challenges of the twenty-first century effectively and sustainably. The case study presented above highlights that different models can help in assessing the key soil properties. Models developed may be potentially deployed in the similar regions and aid the stakeholders in better soil management practices. However, soil database must be built upon and regularly updated to enhance the model performance. Additionally, more complex and advanced models like ANN should be included with improved interpretability. Therefore, the future studies should focus on gaining deeper scientific insights from the obtained soil data by using better explainable and more interpretable ML models. Further, they should aim at meeting the unachieved targets of the SOTER program, i.e., create a global soil database. The ultimate of the research communities should be to come up with innovative solutions for sustainable management of our natural resources.

References

Adhikari K, Hartemink AE, Minasny B, Kheir RB, Greve MB, Greve MH (2014) Digital mapping of soil organic carbon contents and stocks in Denmark. PloS One 9(8):e105519

Adhikari K, Kheir RB, Greve M, Greve MH, Malone MB, Minasny B, McBratney A (2014) Mapping soil pH and bulk density at multiple soil depths in Denmark. In: GlobalSoilMap: basis of the global spatial soil information system. Taylor & Francis, London, pp.155–160.

Ahmad S, Kalra A, Stephen H (2010) Estimating soil moisture using remote sensing data: a machine learning approach. Adv Water Resou 33(1):69–80

Akpa SI, Odeh IO, Bishop TF, Hartemink AE (2014) Digital mapping of soil particle-size fractions for Nigeria. Soil Sci Soc Am J 78(6):1953–1966

Arrouays D, Lagacherie P, Hartemink AE (2017) Digital soil mapping across the globe. Geoderma Reg 9:1–4

Arrouays D, McBratney A, Bouma J, Libohova Z, Richer-de-Forges AC, Morgan CL, Roudier P, Poggio L, Mulder VL (2020) Impressions of digital soil maps: the good, the not so good, and making them ever better. Geoderma Reg 20:e00255

Attarzadeh R, Amini J, Notarnicola C, Greifeneder F (2018) Synergetic use of Sentinel-1 and Sentinel-2 data for soil moisture mapping at plot scale. Remote Sens 10(8):1285

Barman U, Choudhury RD (2020) Soil texture classification using multi class support vector machine. Inf Process Agric 7(2):318–332

Behrens T, Förster H, Scholten T, Steinrücken U, Spies ED, Goldschmitt M (2005) Digital soil mapping using artificial neural networks. J Plant Nutr Soil Sci 168(1):21–33

Benedet L, Faria WM, Silva SHG, Mancini M, Guilherme LRG, Demattê JAM, Curi N (2020) Soil subgroup prediction via portable X-ray fluorescence and visible near-infrared spectroscopy. Geoderma 365:114212

Bondi G, Creamer R, Ferrari A, Fenton O, Wall D (2018) Using machine learning to predict soil bulk density on the basis of visual parameters: tools for in-field and post-field evaluation. Geoderma 318:137–147

Breiman L (1999) Random forests. UC Berkeley TR567

Brungard CB, Boettinger JL (2012) July. Spatial prediction of biological soil crust classes; value added DSM from soil survey. In Digital soil assessments and beyond: proceedings of the 5th global workshop on digital soil mapping. CRC Press, Sydney, pp 57–60

Brungard CW, Boettinger JL, Duniway MC, Wills SA, Edwards TC Jr (2015) Machine learning for predicting soil classes in three semi-arid landscapes. Geoderma 239:68–83

Butler BM, O'Rourke SM, Hillier S (2018) Using rule-based regression models to predict and interpret soil properties from X-ray powder diffraction data. Geoderma 329:43–53

Camera C, Zomeni Z, Noller JS, Zissimos AM, Christoforou IC, Bruggeman A (2017) A high resolution map of soil types and physical properties for Cyprus: a digital soil mapping optimization. Geoderma 285:35–49

Coopersmith EJ, Minsker BS, Sivapalan M (2014) Using similarity of soil texture and hydroclimate to enhance soil moisture estimation. Hydrol Earth Syst Sci 18(8):3095–3107

Dai F, Zhou Q, Lv Z, Wang X, Liu G (2014) Spatial prediction of soil organic matter content integrating artificial neural network and ordinary kriging in Tibetan Plateau. Ecol Indic 45:184–194

Dangal SR, Sanderman J, Wills S, Ramirez-Lopez L (2019) Accurate and precise prediction of soil properties from a large mid-infrared spectral library. Soil Syst 3(1):11

Dharumarajan S, Hegde R, Singh SK (2017) Spatial prediction of major soil properties using Random Forest techniques-a case study in semi-arid tropics of South India. Geoderma Reg 10:154–162

Dharumarajan S, Hegde R, Janani N, Singh SK (2019) The need for digital soil mapping in India. Geoderma Reg 16:e00204

Dharumarajan S, Vasundhara R, Suputhra A, Lalitha M, Hegde R (2020a) Prediction of soil depth in karnataka using digital soil mapping approach. J Indian Soc Remote Sens 48:1593–1600

Dharumarajan S, Kalaiselvi B, Suputhra A, Lalitha M, Hegde R, Singh SK, Lagacherie P (2020b) Digital soil mapping of key GlobalSoilMap properties in Northern Karnataka Plateau. Geoderma Reg 20:e00250.

Dharumarajan S, Hegde R (2020) Digital mapping of soil texture classes using Random Forest classification algorithm. Soil Use Managehttps://doi.org/10.1111/sum.12668

Dharumarajan S, Kalaiselvi B, Suputhra A, Lalitha M, Vasundhara R, Kumar KA, Nair KM, Hegde R, Singh SK, Lagacherie P (2021a) Digital soil mapping of soil organic carbon stocks in Western Ghats, South India. Geoderma Reg 25:e00387

Dharumarajan S, Hegde R, Lalitha M Vasundhara R (2021b). Predicting and mapping of soil hydraulic properties in Karnataka. J Indian Soc Remote Sens 1–9

Dharumarajan S, Kalaiselvi B, Lalitha M, Vasundhara R, Rajendra Hegde (2021c) Defining fertility management units and land suitability analysis using digital soil mapping approach. Geocarto Int.https://doi.org/10.1080/10106049.2021.1926553

Ding J, Yang A, Wang J, Sagan V, Yu D (2018) Machine-learning-based quantitative estimation of soil organic carbon content by VIS/NIR spectroscopy. Peer J 6:e5714

Du Plessis C, Van Zijl G, Van Tol J, Manyevere A (2020) Machine learning digital soil mapping to inform gully erosion mitigation measures in the Eastern Cape, South Africa. Geoderma 368:114287

Farfani HA, Behnamfar F, Fathollahi A (2015) Dynamic analysis of soil-structure interaction using the neural networks and the support vector machines. Expert Syst Appl 42(22):8971–8981

Forkuor G, Hounkpatin OK, Welp G, Thiel M (2017) High resolution mapping of soil properties using remote sensing variables in south-western Burkina Faso: a comparison of machine learning and multiple linear regression models. PloS One 12(1):e0170478

Greifeneder F, Notarnicola C, Wagner W (2021) A machine learning-based approach for surface soil moisture estimations with google earth engine. Remote Sens 13(11):2099

Grimm R, Behrens T, Märker M, Elsenbeer H (2008) Soil organic carbon concentrations and stocks on Barro Colorado Island—digital soil mapping using Random Forests analysis. Geoderma 146(1–2):102–113

Grunwald S (2009) Multi-criteria characterization of recent digital soil mapping and modeling approaches. Geoderma 152(3–4):195–207

Grunwald S (2010) Current state of digital soil mapping and what is next. In: Digital soil mapping. Springer, Dordrecht, pp 3–12

Grunwald S, Vasques GM, Rivero RG (2015) Fusion of soil and remote sensing data to model soil properties. Adv Agron 131:1–109

Hengl T, Heuvelink GB, Kempen B, Leenaars JG, Walsh MG, Shepherd KD, Sila A, MacMillan RA, Mendes de Jesus J, Tamene L, Tondoh JE (2015) Mapping soil properties of Africa at 250 m resolution: random forests significantly improve current predictions. PloS One 10(6):e0125814

Hengl T, Leenaars JG, Shepherd KD, Walsh MG, Heuvelink GB, Mamo T, Tilahun H, Berkhout E, Cooper M, Fegraus E, Wheeler I (2017) Soil nutrient maps of Sub-Saharan Africa: assessment of soil nutrient content at 250 m spatial resolution using machine learning. Nutr Cycl Agroecosys 109(1):77–102

Heung B, Ho HC, Zhang J, Knudby A, Bulmer CE, Schmidt MG (2016) An overview and comparison of machine-learning techniques for classification purposes in digital soil mapping. Geoderma 265:62–77
Heuvelink GB, Angelini ME, Poggio L, Bai Z, Batjes NH, van den Bosch R, Bossio D, Estella S, Lehmann J, Olmedo GF, Sanderman J (2020) Machine learning in space and time for modelling soil organic carbon change. Eur J Soil Sci 1–17
Hu B, Xue J, Zhou Y, Shao S, Fu Z, Li Y, Chen S, Qi L, Shi Z (2020) Modelling bioaccumulation of heavy metals in soil-crop ecosystems and identifying its controlling factors using machine learning. Environ Pollut 262:114308
Iticha B, Takele C (2019) Digital soil mapping for site-specific management of soils. Geoderma 351:85–91
Ivushkin K, Bartholomeus H, Bregt AK, Pulatov A, Bui EN, Wilford J (2018) Soil salinity assessment through satellite thermography for different irrigated and rainfed crops. Int J Appl Earth Obs 68:230–237
James G, Witten D, Hastie T, Tibshirani R (2013) An introduction to statistical learning, vol 112. Springer, New York, p 18
Jenny H (1941) Factors of soil formation: a system of quantitative pedology. McGraw-Hill, New York
Jia X, Hu B, Marchant BP, Zhou L, Shi Z, Zhu Y (2019) A methodological framework for identifying potential sources of soil heavy metal pollution based on machine learning: a case study in the Yangtze Delta, China. Environ Pollut 250:601–609
Kalambukattu JG, Kumar S, Raj RA (2018) Digital soil mapping in a Himalayan watershed using remote sensing and terrain parameters employing artificial neural network model. Environ Earth Sci 77(5):1–14
Keskin H, Grunwald S, Harris WG (2019) Digital mapping of soil carbon fractions with machine learning. Geoderma 339:40–58
Khaledian Y, Miller BA (2020) Selecting appropriate machine learning methods for digital soil mapping. Appl Math Model 81:401–418
Khanal S, Fulton J, Klopfenstein A, Douridas N, Shearer S (2018) Integration of high resolution remotely sensed data and machine learning techniques for spatial prediction of soil properties and corn yield. Comput Electron Agr 153:213–225
Kheir RB, Greve MH, Bøcher PK, Greve MB, Larsen R, McCloy K (2010) Predictive mapping of soil organic carbon in wet cultivated lands using classification-tree based models: the case study of Denmark. J Environ Manage 91(5):1150–1160
Kovačević M, Bajat B, Gajić B (2010) Soil type classification and estimation of soil properties using support vector machines. Geoderma 154(3–4):340–347
Kuhn M, Johnson K (2013) Applied predictive modeling, vol 26. Springer, New York
Kuhn M, Weston, S, Keefer C, Coulter N, Ross Q, Cubist: rule-and instance-based regression modeling (2016) R package version 0.0: 19
Lagacherie P, McBratney A, Voltz M (2006) Digital soil mapping: an introductory perspective. Elsevier
Lamichhane S, Kumar L, Wilson B (2019) Digital soil mapping algorithms and covariates for soil organic carbon mapping and their implications: a review. Geoderma 352:395–413
Lamichhane S, Kumar L, Adhikari K (2021) Updating the national soil map of Nepal through digital soil mapping. Geoderma 394:115041
Licznar P, Nearing MA (2003) Artificial neural networks of soil erosion and runoff prediction at the plot scale. Catena 51(2):89–114
Ließ M, Glaser B, Huwe B (2012) Uncertainty in the spatial prediction of soil texture: comparison of regression tree and Random Forest models. Geoderma 170:70–79
Liu C, Li Z, Chang X, He J, Nie X, Liu L, Xiao H, Wang D, Peng H, Zeng G (2018a) Soil carbon and nitrogen sources and redistribution as affected by erosion and deposition processes: a case study in a loess hilly-gully catchment, China. Agr Ecosyst Environ 253:11–22

Liu X, Yang T, Wang Q, Huang F, Li L (2018b) Dynamics of soil carbon and nitrogen stocks after afforestation in arid and semi-arid regions: a meta-analysis. Sci Total Environ 618:1658–1664
Malone BP, McBratney AB, Minasny B, Laslett GM (2009) Mapping continuous depth functions of soil carbon storage and available water capacity. Geoderma 154(1–2):138–152
McBratney AB, Odeh IO, Bishop TF, Dunbar MS, Shatar TM (2000) An overview of pedometric techniques for use in soil survey. Geoderma 97(3–4):293–327
McBratney AB, Santos MM Minasny B (2003) On digital soil mapping. Geoderma 117(1–2):3–52
Minasny B, McBratney AB, Malone BP, Wheeler I (2013) Digital mapping of soil carbon. Adv Agron 118:1–47
Morellos A, Pantazi XE, Moshou D, Alexandridis T, Whetton R, Tziotzios G, Wiebensohn J, Bill R, Mouazen AM (2016) Machine learning based prediction of soil total nitrogen, organic carbon and moisture content by using VIS-NIR spectroscopy. Biosyst Eng 152:104–116
Nussbaum M, Spiess K, Baltensweiler A, Grob U, Keller A, Greiner L, Schaepman ME, Papritz A (2018) Evaluation of digital soil mapping approaches with large sets of environmental covariates. Soil 4:1–22
Padarian J, Minasny B, McBratney AB (2019) Using deep learning to predict soil properties from regional spectral data. Geoderma Reg 16:e00198
Padarian J, Minasny B, McBratney AB (2020) Machine learning and soil sciences: a review aided by machine learning tools. Soil 6(1):35–52
Padarian J, McBratney AB, Minasny B (2020) Game theory interpretation of digital soil mapping convolutional neural networks. Soil 6(2):389–397
Panday D, Maharjan B, Chalise D, Shrestha RK, Twanabasu B (2018) Digital soil mapping in the Bara district of Nepal using kriging tool in ArcGIS. PloS One 13(10):e0206350
Pouladi N, Møller AB, Tabatabai S, Greve MH (2019) Mapping soil organic matter contents at field level with Cubist, Random Forest and kriging. Geoderma 342:85–92
Priori S, Bianconi N, Costantini EA (2014) Can γ-radiometrics predict soil textural data and stoniness in different parent materials? A comparison of two machine-learning methods. Geoderma 226:354–364
Quinlan JR (1992 November) Learning with continuous classes. In: 5th Australian joint conference on artificial intelligence, vol 92, pp 343–348
Quinlan JR (1993 June) instance-based and model-based learning. In: Proceedings of the tenth international conference on machine learning, pp 236–243
Rial M, Cortizas AM Rodríguez-Lado L (2017) Understanding the spatial distribution of factors controlling topsoil organic carbon content in European soils. Sci Total Environ 609:1411–1422
Safari A, Sohrabi H, Powell S, Shataee S (2017) A comparative assessment of multi-temporal Landsat 8 and machine learning algorithms for estimating aboveground carbon stock in coppice oak forests. Int J Remote Sens 38(22):6407–6432
Silveira CT, Oka-Fiori C, Santos LJC, Sirtoli AE, Silva CR, Botelho MF (2013) Soil prediction using artificial neural networks and topographic attributes. Geoderma 195:165–172
Sreenivas K, Dadhwal VK, Kumar S, Harsha GS, Mitran T, Sujatha G, Suresh GJR, Fyzee MA, Ravisankar T (2016) Digital mapping of soil organic and inorganic carbon status in India. Geoderma 269:160–173
Taghizadeh-Mehrjardi R, Nabiollahi K, Kerry R (2016) Digital mapping of soil organic carbon at multiple depths using different data mining techniques in Baneh region. Iran. Geoderma 266:98–110
Tajik S, Ayoubi S, Shirani H, Zeraatpisheh M (2019) Digital mapping of soil invertebrates using environmental attributes in a deciduous forest ecosystem. Geoderma 353:252–263
Van Den Hoogen J, Geisen S, Routh D, Ferris H, Traunspurger W, Wardle DA, De Goede RG, Adams BJ, Ahmad W, Andriuzzi WS, Bardgett RD (2019) Soil nematode abundance and functional group composition at a global scale. Nature 572(7768):194–198
Vapnik V, Guyon I, Hastie T (1995) Support vector machines. Mach Learn 20(3):273–297

Vaysse K, Lagacherie P (2015) Evaluating digital soil mapping approaches for mapping GlobalSoilMap soil properties from legacy data in Languedoc-Roussillon (France). Geoderma Reg 4:20–30

Wadoux AMC, Minasny B, McBratney AB (2020) Machine learning for digital soil mapping: applications, challenges and suggested solutions. Earth Sci Rev 210:103359

Wang J, Ding J, Abulimiti A, Cai L (2018) Quantitative estimation of soil salinity by means of different modeling methods and visible-near infrared (VIS–NIR) spectroscopy, Ebinur Lake Wetland, Northwest China. PeerJ 6:e4703

Wiesmeier M, Urbanski L, Hobley E, Lang B, von Lützow M, Marin-Spiotta E, van Wesemael B, Rabot E, Ließ M, Garcia-Franco N, Wollschläger U (2019) Soil organic carbon storage as a key function of soils-A review of drivers and indicators at various scales. Geoderma 333:149–162

Yang M, Xu D, Chen S, Li H, Shi Z (2019) Evaluation of machine learning approaches to predict soil organic matter and pH using Vis-NIR spectra. Sensors 19(2):263

Zeraatpisheh M, Ayoubi S, Jafari A, Tajik S, Finke P (2019) Digital mapping of soil properties using multiple machine learning in a semi-arid region, central Iran. Geoderma 338:445–452

Zeraatpisheh M, Jafari A, Bodaghabadi MB, Ayoubi S, Taghizadeh-Mehrjardi R, Toomanian N, Kerry R, Xu M (2020) Conventional and digital soil mapping in Iran: past, present, and future. Catena 188:104424

Zhang S, Yao L, Sun A, Tay Y (2019) Deep learning-based recommender system: a survey and new perspectives. ACM Comput Surv (CSUR) 52(1):1–38

Zhu AX, Band L Vertessy R, Dutton B (1997) Derivation of soil properties using a soil land inference model (SoLIM). Soil Sci Soc Am J 61(2):523–533

Remote Sensing and Machine Learning for Identification of Salt-affected Soils

Nirmal Kumar, G. P. Obi Reddy, M. S. S. Nagaraju, and R. K. Naitam

Abstract Approximately 6.74 million hectares (Mha) are reported under salt-affected soils (SAS) in India. The accurate identification and monitoring of these soils is important for timely decision-making and assessing the success of reclamation measures. In the study, widely used three Machine Learning (ML) techniques, namely, Logistic Regression (LR), Support Vector Machine (SVM), and Random Forest (RF) were tested and compared in the identification of SAS in the Raibareli district of Uttar Pradesh, as a part of Indo-Gangetic Plains (IGP) of India by using Landsat 8 Operational Land Imager/Thermal Infrared Sensor (OLI/TIRS) data. Various salinity indices and principal components were computed through analysis of the Landsat 8 OLI/TIRS bands, and ancillary data like nearness to canals and streams were used to identify SAS. The three models were applied to object-oriented segments generated from these rasters. Total 361 segments were considered for training, and testing of RF model, of which 130 segments belonged to SAS, and remaining segments represented other features and normal soils. In the study, 70% of samples were used for training and 30% for testing. The accuracy of trained models on testing dataset was 96%, 98%, and 98% for LR, SVM, and RF, respectively. The trained model was then used for identification of SAS in the study area. The results were validated with field observations, and high-resolution Google Earth data. It was observed that the three models were able to identify 31,954, 16,679, and 14,070 ha area of SAS, respectively, in Raibareli district, IGP of India.

Keywords Indo-Gangetic plains · Random forest · Remote sensing · Salt-affected soils · Salinity indices

N. Kumar (✉) · G. P. O. Reddy · M. S. S. Nagaraju · R. K. Naitam
ICAR-National Bureau of Soil Survey and Land Use Planning, Amravati Road, Nagpur 440033, India
e-mail: Nirmal.Kumar3@icar.gov.in

G. P. O. Reddy et al. (eds.), *Data Science in Agriculture and Natural Resource Management*, Studies in Big Data 96, https://doi.org/10.1007/978-981-16-5847-1_13

1 Introduction

Soil salinity or sodicity is the major threat to agriculture, especially in arid and semi-arid regions (Oldeman et al. 1991; Maji et al. 2010; Reddy et al. 2018). Global distribution of salt-affected soils (SAS) indicates about 831 Mha is affected by salinity/sodicity in more than 100 countries (Martinez-Beltran and Manzur 2005; Szabolcs 1992). The Global Assessment of Soil Degradation (GLASOD), the first attempt to map human-induced soil degradation reported that about 76 Mha area was affected by salinization alone (Oldeman et al. 1991). Of the total 76 Mha, most areas under salinization were reported in Asian (53 Mha) and African (13 Mha) continents. Out of the 76 Mha, about 45 Mha comes under irrigated area, which is about 20% of the world's total irrigated area. In India, first nationwide estimate of SAS was reported by Raychaudhuri (1966). A conventional approach was applied, and the area under SAS was reported to be 6.0 Mha. Based on the GLASOD methodology by using soil map of 1:250 K, ICAR-National Bureau of Soil Survey and Land Use Planning (ICAR-NBSS&LUP) reported an area of 10.1 Mha as SAS in India (Sehgal and Abrol1994), which was revised to 5.89 Mha in 2004 (NBSS&LUP 2004). Mandal et al. (2009) prepared a Geographic Information System (GIS)-based spatial database on SAS of India and estimated about 6.73 Mha of area under SAS in the country. The spatial distribution showed a wide spread of SAS in Indo-Gangetic plains, especially in the state of Uttar Pradesh, followed by the arid and semi-arid regions of western India, especially in the states of Gujarat and Maharashtra. The coastal regions of the country were also shown to have a significant area under SAS. The SAS in the IGP and peninsular plains are primarily sodic, whereas, these are mainly saline in deltaic, coastal, and lowlands of mud flats/mangrove swamps. Various other estimates have shown SAS in India varying from approximately 1 Mha to 26 Mha (Kumar et al. 2021). These variations reflect the differences in adopted methodologies and data used.

The reliable estimates of SAS with their extent and spatial distribution assume a greater importance in achieving the land degradation neutrality goals. The conventional methods of soil survey and experts' opinion are resource consuming, subjective, and hard to reproduce. On the other hand, remote sensing has been proven successful in mapping, and monitoring of SAS. Wide range of advancements in space-borne satellite sensors in terms of spatial, temporal, and spectral resolutions have been made in recent times (Sahu et al. 2016; Kumar et al. 2018). These advancements in the sensors and the advanced digital image analysis techniques have made the assessment, mapping, and monitoring of SAS more accurate, rapid, and quantitative.

All forms of remote sensing data, including aerial photographs, thermal infrared images, PAN (panchromatic), multispectral, hyperspectral, and microwave images have been utilized in mapping of SAS (Kumar and Singh 2018; Kumar et al. 2021). However, multispectral data have been used the most. A false color composite (FCC) is preferred over a single band. The standard FCC has been used in visual interpretation (Sharma and Bhargava 1988; Sujatha et al. 2000; Sethi et al. 2006) and digital image analyses to predict SAS (Dwivedi 2001; Abbas et al. 2013). Various combinations of Landsat data have been found effective in identifying SAS. Dwivedi and Rao (1992) found the standard FCC of Landsat TM as well as band combination of 1, 3, and 5 are effective in mapping SAS in the IGP. Saha et al. (1990) found TM bands 2, 4, 5, and 7 suitable for identification of SAS and waterlogged areas. MSS bands 3, 4, and 5 were recommended for salt detection in addition to TM bands 3, 4, 5, and 7 (Naseri 1998). Metternicht and Zinck (2003) found that a combination of six Landsat TM bands (1, 2, 4, 5, 6, and 7) provides the highest separability among salt- and sodium-affected soil classes. The incorporation of the Landsat TM thermal band improves the separability of sodic soils from saline–sodic, saline, and neutral soils. Dogan and Kilic (2013) found all bands of Landsat ETM$^+$, except bands 4 and 6, suitable for salinity assessment based on correlation of digital number (DN) values with EC and pH.

Of late, image transformation techniques like Principal Components (PC) and Tasseled Cap Transformation (TCT) proved effective in assessment of SAS. PC3 developed from the bands of Landsat ETM$^+$ (Ding et al. 2011) and TM (Dwivedi and Sreenivas 1998) were found most effective to identify SAS. Peng (1998) suggested that the brightness band from the TCT can be used to identify saline soils owing to higher reflectance from them. It enhances the detection of saline features (Masoud and Koike 2006; Mitchell 2014). TCT has been found to perform better than PCA in classification and assessment for degraded lands with higher overall accuracies (Patterson and Stephen 1998; Phua and Saito 2003; Toomey and Vierling 2005). Based on transformed divergence values, Gutierrez (2002) found better separability among SAS categories using TC transformation composite than using different band combinations. Apart from these transformations, various salinity indices (SIs) have been identified to map SAS (Kumar et al. 2021), which enhances the saline patches while suppressing the non-SAS features.

National level maps on SAS have been mostly developed by visual interpretations of multispectral satellite data such as AWiFS (SAC 2007, 2016). Different multispectral images have been interpreted visually for mapping SAS (Dwivedi 1994; Singh 1994; Mandal and Sharma 2011) and their categories (Rao and Venkataratnam 1991; Sujatha et al. 2000) with limited field checks. On a standard FCC, salt-affected areas are generally identified as scattered and non-contiguous patches of irregular shapes with fine to mottled texture and bright to dull white color (subjected to moisture content) (Sharma and Bhargava 1988; Narayan et al. 1989). Space Application Centre (SAC 2007) identified 5.26 Mha of SAS in India based on visual interpretation of Advanced Wide Field Sensor (AWiFS) data for the years 2003, 2004, and 2005, which was revised to 3.99 Mha based on the same data for the same period (SAC 2016). Based on the AWiFS images of the years 2011 through 2013, it was estimated to be 3.67 Mha (SAC 2016). This shows the approach to be time consuming and subjective.

A quick and objective tool for identifying and mapping SAS is application of supervised and unsupervised classification techniques on moderate resolution data such as Landsat, and Linear Imaging Self-scanning System (LISS)-III. Apart from the Maximum Likelihood Classifiers (MLC), several other classifiers like Artificial

Neural Network (ANN), LR, Decision Tree (DT), and SVM have been used for classifying SAS and non-SAS features using remote sensing data. MLC is the widely used method. However, the other methods such as DT, SVM, and LR have been found to be more accurate than the former (Rao et al. 2005; Elnaggar and Noller 2009; Cai et al. 2010; Abbas et al. 2013; Asfaw et al. 2018; Kumar et al. 2019). The objectives of the present study are to develop and compare LR, SVM, and RF models by using various indices derived from Landsat 8 images to identify SAS in Raibareli district, IGP of India.

2 Machine Learning Techniques

2.1 Logistic Regression (LR)

Binomial logistic regression is a variation of ordinary regression, used when the dependent variable has two discrete levels. It uses the logit function which relates the independent variable to the rolling mean of the dependent variable. The LR model (Mendenhall and Sincich 1996) is expressed as

$$E(y) = \frac{\exp^{(b_0+b_1X_1+b_2X_2+\cdots+b_kX_k)}}{1+\exp^{(b_0+b_1X_1+b_2X_2+\cdots+b_kX_k)}} \tag{1}$$

where, $y = 1$, if category A occurs; 0, if category B occurs,

$E(y) = P(\text{categoryAoccurs}) = \pi$; i.e., probability that the target is SAS,

$X_1, X_2, \ldots, X_k$ are qualitative or quantitative independent variables. b_o, b_1, $b_2, \ldots,$ b_k are the estimated coefficients of the model, which are estimated by using maximum likelihood method. Kumar et al. (2019) developed LR models for identifying SAS by using Landsat 8 Operational Land Imager (OLI) in IGP and reported that bands like green, red, and SWIR 1, and salinity indices developed by using these bands, such as SI_6-SI_{10} (Table 1), and COSRI, are suitable for identifying SAS.

2.2 Support Vector Method (SVM)

The SVM is another popular ML algorithm applicable in classification. The SVM finds an optimal separating line called hyperplane between two classes, which maximizes the margins, i.e., the distance between the points closest to the hyperplane from both the classes (Vapnik 1995). It is much more effective than other conventional nonparametric classifiers (e.g., ANN, nearest neighbors, MLC methods) in terms of classification accuracy and stability to parameter setting (Feng 2006; Zhu and Blumberg 2002; Keuchei et al. 2003). Simin et al. (2010) have shown the superiority of SVM over other classification methods in identifying SAS. Muller and Niekerk

Table 1 Indices used for assessment of SAS in the study

Index	Formula	References
Normalized Difference Vegetation Index (NDVI)	(NIR-R)/(NIR+R)	Rouse et al. (1974)
Normalized Difference Salinity Index (NDSI)	(R–NIR)/(R+NIR)	Major et al. (1990)
Salinity Index (SI_T)	NIR/SWIR1	Tripathi et al. (1997)
Brightness Index	$\sqrt{}(R^2+NIR^2)$	Khan et al. (2005)
Salinity Index 1 (SI_1)	$\sqrt{}$(GxR)	Khan et al. (2005)
Salinity Index 2 (SI_2)	$\sqrt{}(G^2+R^2+NIR^2)$	Douaoui et al. (2006)
Salinity Index 3 (SI_3)	$\sqrt{}(G^2+R^2)$	Douaoui et al. (2006)
Salinity Index 4 (SI_4)	SWIR1/SWIR 2	IDNP (2002)
Salinity Index 5 (SI_5)	(NIR-SWIR1)/(NIR+SWIR1)	
Salinity Index 6 (SI_6)	(SWIR1-SWIR2)/(SWIR1+SWIR2)	
Salinity Index 7 (SI_7)	B/R	Abbas and Khan (2007) Abbas et al. (2013)
Salinity Index 8 (SI_8)	(B-R)/(B+R)	
Salinity Index 9 (SI_9)	(GxR)/B	
Salinity Index 10 (SI_{10})	$\sqrt{}$(BxR)	
Salinity Index 11 (SI_{11})	(BxR)/G	
Salinity Index 12 (SI_{12})	(RXNIR)/G	
Salinity Index 13 (SI_{13})	$\sqrt{}$(RxNIR)	Dehni and Lounis (2012)
Salinity Index 14 (SI_{14})	(R/NIR)*100	Allbed and Kumar (2013)
Combined Spectral Response Index (COSRI)	[(B+G)/(R+NIR)]*NDVI	Fernandez-Buces et al. (2006)

Note B—Blue, G—Green, R—Red, NIR—Near Infrared, SWIR1—Landsat8 OLI band 6, SWIR2—Landsat8 OLI band 7

(2016) compared six different classification algorithms (ML, NN, RF, DT, and SVM with and without textural features) at two places. RF was found to produce the best results at one place, while SVM with textural features produced the best results at second.

2.3 *Random Forest (RF)*

As an ensemble learning technique, RF constructs multiple decision trees based on training datasets and use mode of outputs of individual trees for determining the final class of an object. The RF technique was developed by Breiman (2001) and Cutler et al. (2012). This technique rectifies the error of overfitting of training data by decision tree method. Vermeulen and van Niekerk (2016) used very high-resolution WorldView-2 imagery in distinguishing salt-affected and non-salt-affected soils in

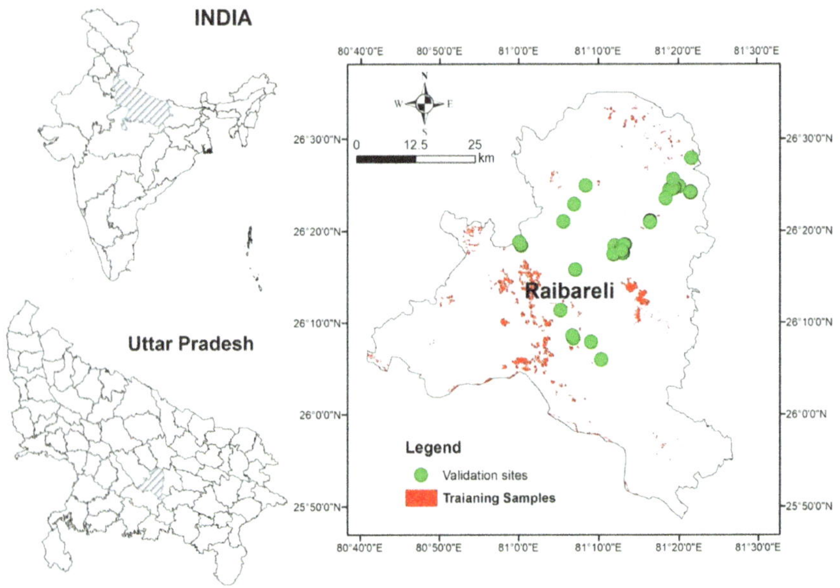

Fig. 1 The study area of Raibareli district (UP) with sampling sites for training and validation

South Africa by applying several machine-learning algorithms in which random forest method had overall accuracy of 80%.

3 Materials and Methods

3.1 Geographical Settings

In the study, the Raibareli district of state Uttar Pradesh, India was selected, which is located between 25°50′ and 26°35′ N latitude, 80°40′ and 81°27′ E longitude with total geographical area (TGA) of 307,415 Ha (Fig. 1). Raibareli district is situated in the Indo-Gangetic plains (IGP) having alluvial, calcareous, and salt-affected soils. River Ganges flows in its western part. This district experiences subtropical climate with mean annual rainfall of 840.8 mm, most of which is received during south-western monsoon period. Raibareli district has an area of 16.8% of TGA area has been reported to be under SAS. These SAS are mostly saline-sodic in nature. Most of the cultivated area in this district is irrigated through canals and tube wells. The important crops grown in the district are rice and wheat.

3.2 Data

The cloud-free Landsat 8 OLI and TIRS data were downloaded freely from https://earthexplorer.usgs.gov/ for 11 March, 2021. The pre-processing including layer stacking and sub-setting of Landsat 8 OLI and TIRS data was performed by using open-source R software by selecting 'raster' (Hijmans 2019) and 'rgdal' (Bivand et al. 2019) packages. Various salinity indices (Table 1) available in the literature were computed using Landsat 8 OLI and TIRS data in free and open-source R software. The important indices developed in the study are shown in Table 1. Further, principal components (PCs) were also developed, and first two components were used as variables.

The spatial distribution of rivers and canals was also used as they have been found as important factor in identifying SAS (Kumar 2018). Geo-referenced field photographs and high-resolution Google Earth images were used in this study for validation.

3.3 Methodology

3.3.1 Object-Based Segmentation Approach

In the study, an object-based segmentation approach was followed for the selection of training and testing sites. In the object-based segmentation approach, the pixels are grouped into polygons based on their relative homogeneity in spatial and spectral features with their neighbors. FCC of Landsat 8 OLI and TIRS data of March, 2021 were used to create image segments through SAGA (System for Automated Geo-scientific Analyses) GIS software. A total of 14,395 segments were created for the study area.

3.3.2 Training Sites

A total of 361 training segments (130 salt-affected soils and 131 normal soils or other features) were selected for the development of different models. Training segments were identified by field observations and using available legacy reports. The location of training segments selected is shown in Fig. 3. In the study, 70% data was used for training, and the remaining 30% for testing of the models.

3.3.3 Development of Models

LR, SVM, and RF models were developed with all the bands and indices as inputs in R. The models were applied to all the segments of the study area to generate the SAS maps. The RF model was developed by using 'randomForest' package (Liaw and Wiener 2002) and others by using 'caret' package (Kuhn 2014) of R software.

3.3.4 Validation

A field survey was done to collect the ground truth information. The output classified raster was validated with field verifications and with high-resolution Google Earth™ data.

4 Results and Discussion

4.1 Performance of Logistic Regression Models

The LR model was applied to each band separately. The significance and the accuracies are shown in Table 2. The results of the study indicate that the Landsat OLI and TIRS bands 5 and 10 were not able to differentiate the SAS and other features. Similarly, the various cost rasters depicting the distance to canal, natural streams, and overall drainages were also not able to differentiate the same. Among the indices, SI_T, NDVI, NDSI, SI_4, SI_5, SI_6, SI_{14}, and COSRI were not able to differentiate the two classes. The significant bands and indices were used to classify the SAS and other features by fitting a backward stepwise logistic regression. The resulting equation with the most significant variables is shown as Eq. 2. The estimates and the significance of the identified parameters in the equation are shown in Table 3. The model provided 96% accuracy on the validation set of 95 samples.

$$E(y) = \frac{\exp^{[-177-0.02\,b7+0.01\,PC1-0.35\,SI10+0.18\,SI11-0.11\,SI12+0.28\,SI13]}}{1+\exp^{[-177-0.02\,b7+0.01\,PC1-0.35\,SI10+0.18\,SI11-0.11\,SI12+0.28\,SI13]}} \quad (2)$$

where,

$$y = \{1 \text{if category is SAS } 0 \text{ if category is others}$$

$$E(y) = P(\text{category 1 occurs}) = \pi$$

Applying the equation to each segment produces a probability of being SAS, with values ranging from 0 to 1. A threshold value of probability of 0.5 was applied, so

Table 2 Model summary and accuracy of logistic regression models with different bands/indices

Variables		Coefficients	SE	Sig.	Accuracy		
					SAS	Other	Over all
DN (b1)	β0	−12.87	1.71	0.000	54.84	90.63	78.95
	β1	0.00	0.00	0.000	–	–	–
DN (b2)	β0	−11.94	1.58	0.000	61.29	92.19	82.11
	β1	0.00	0.00	0.000	–	–	–
DN (b3)	β0	−11.50	1.49	0.000	64.52	92.19	83.16
	β1	0.00	0.00	0.000	–	–	–
DN (b4)	β0	−9.06	1.19	0.000	58.06	90.63	80.00
	β1	0.00	0.00	0.000	–	–	–
DN (b5)	β0	−6.85	1.38	0.000	19.35	70.31	53.68
	β1	0.00	0.00	0.000	–	–	–
DN (b6)	β0	−11.27	1.39	0.000	67.74	93.75	85.26
	β1	0.00	0.00	0.000	–	–	–
DN (b7)	β0	−6.97	0.92	0.000	51.61	90.63	77.89
	β1	0.00	0.00	0.000	–	–	–
DN (b10)	β0	−34.78	8.13	0.000	18.92	81.08	60.36
	β1	0.00	0.00	0.000	–	–	–
PC1	β0	−8.84	1.19	0.000	54.84	90.63	78.95
	β1	0.00	0.00	0.000	–	–	–
PC2	β0	−22.54	2.95	0.000	67.74	93.75	85.26
	β1	0.00	0.00	0.000	–	–	–
Distance to canal	β0	−0.51	0.13	0.000	0.00	100.00	67.37
	β1	0.00	0.00	0.462	–	–	–
Distance to Streams	β0	−1.03	0.23	0.000	19.35	92.19	68.42
	β1	0.00	0.00	0.003	–	–	–
Distance to drainage	β0	−0.48	0.13	0.000	0.00	100.00	67.37
	β1	0.00	0.00	0.338	–	–	–
SIt	β0	4.94	1.08	0.000	29.03	85.94	67.37
	β1	−5.14	1.05	0.000	–	–	–
BI	β0	−17.55	2.52	0.000	54.84	89.06	77.89
	β1	0.00	0.00	0.000	–	–	–
NDSI	β0	0.27	0.30	0.365	0.00	92.19	62.11
	β1	4.17	1.45	0.004	–	–	–
NDVI	β0	0.27	0.30	0.366	0.00	92.19	62.11
	β1	−4.17	1.45	0.004	–	–	–
SI1	β0	−10.15	1.33	0.000	58.06	92.19	81.05

(continued)

Table 2 (continued)

Variables		Coefficients	SE	Sig.	Accuracy		
					SAS	Other	Over all
	β1	0.00	0.00	0.000	–	–	–
SI2	β0	−19.46	2.60	0.000	64.52	98.44	87.37
	β1	0.00	0.00	0.000	–	–	–
SI3	β0	−10.12	1.32	0.000	58.06	92.19	81.05
	β1	0.00	0.00	0.000	–	–	–
SI4	β0	3.31	2.34	0.158	0.00	100.00	67.37
	β1	−3.37	2.06	0.102	–	–	–
SI5	β0	−0.25	0.14	0.070	25.81	85.94	66.32
	β1	−11.33	2.22	0.000	–	–	–
SI6	β0	−0.12	0.32	0.715	0.00	100.00	67.37
	β1	−6.44	4.70	0.171	–	–	–
SI7	β0	30.44	4.42	0.000	58.06	85.94	76.84
	β1	−35.78	5.15	0.000	–	–	–
SI8	β0	−5.04	0.70	0.000	54.84	85.94	75.79
	β1	−62.16	8.90	0.000	–	–	–
SI9	β0	−8.74	1.13	0.000	61.29	92.19	82.11
	β1	0.00	0.00	0.000	–	–	–
SI10	β0	−10.32	1.36	0.000	58.06	92.19	81.05
	β1	0.00	0.00	0.000	–	–	–
SI11	β0	−9.27	1.25	0.000	58.06	90.63	80.00
	β1	0.00	0.00	0.000	–	–	–
SI12	β0	−19.85	2.80	0.000	67.74	87.50	81.05
	β1	0.00	0.00	0.000	–	–	–
SI13	β0	−19.88	2.40	0.000	70.97	98.44	89.47
	β1	0.00	0.00	0.000			
SI14	β0	−2.06	0.72	0.004	0.00	96.88	65.26
	β1	0.02	0.01	0.029			
COSRI	β0	0.22	0.34	0.531	0.00	96.88	65.26
	β1	−5.43	2.39	0.023	–	–	–

that the segments with probabilities of greater than 0.5 are classified as SAS. The identified segments are shown in (Fig. 2). A total area of 31,954 ha was found to be under SAS in the district constituting ~10% of the total geographic area of the district. SAS were reported mainly from northeastern and south-eastern part, where the canal density is higher and the natural streams are sparse.

Table 3 Logistic regression model developed with back ward stepwise method

Variables	Coefficients	SE	P value	Accuracy		
				SAS	Other	Over all
(Intercept)	−177.00	44.76	0.0001	94	100	96
B7	−0.02	0.01	0.0012	–	–	–
PC1	0.01	0.00	0.0026	–	–	–
SI10	−0.35	0.12	0.0041	–	–	–
Si11	0.18	0.06	0.0057	–	–	–
SI12	−0.11	0.04	0.0054	–	–	–
SI13	0.28	0.10	0.0037	–	–	–

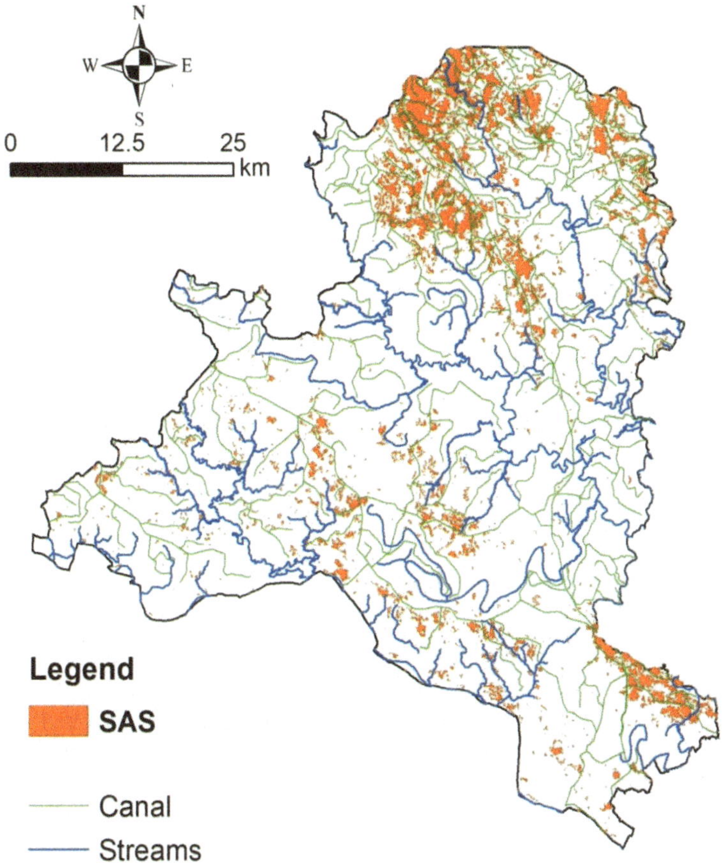

Fig. 2 SAS identified through logistic regression modeling

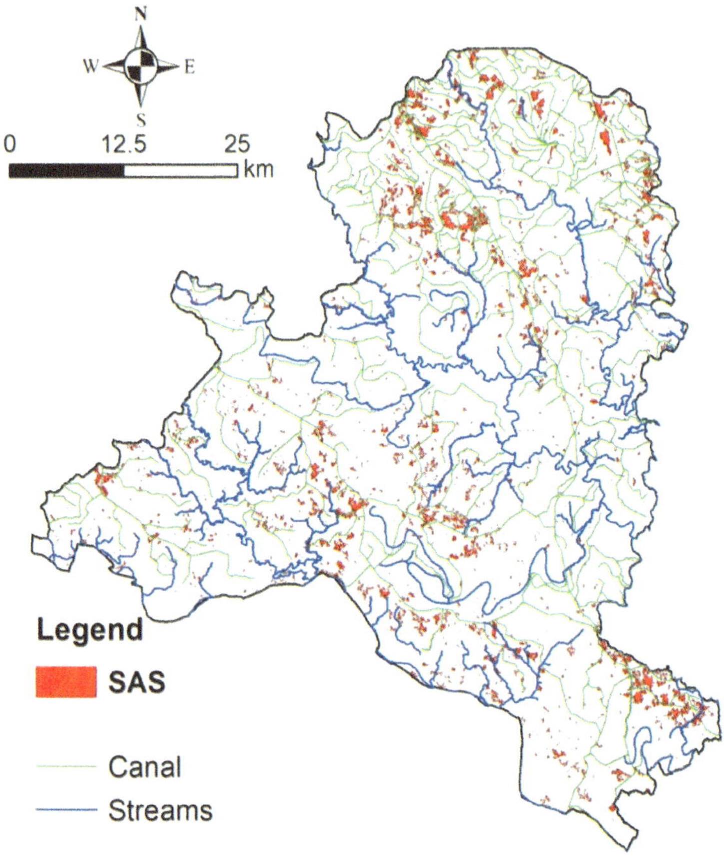

Fig. 3 SAS identified through support vector machine modeling

4.2 Performance of SVM Model

The map result from the SVM modeling is shown in Fig. 3. All the bands/PCs/indices were used for modeling. The SAS in the district were found to be 16,679 ha constituting about 5% of the TGA of the district with an accuracy of 98% and kappa of 0.96. The spatial distributions were the same as in case of LR. However, the area under SAS was found relatively less.

4.3 Performance of RF Model

4.3.1 Training of Random Forest Model

Several RF models were developed using the training dataset with different combinations of parameters N and m. Here, N represents number of decision trees and m represents number of variables, which are randomly sampled as candidates at each split in the RF model. The Out-Of-Bag (OOB) estimate of error rate was determined for each combination. The combination of N and m at which OOB error estimate was found least and stable was selected as final parameters for training random forest model. The plot of OOB estimate of error rate against the number of trees at $m = 10$ also depicts the same (Fig. 4). Therefore, the random forest model was trained with 700 number of trees ($N = 700$), and 10 variables were considered at each split ($m = 10$). The trained random forest model predicted training data with 96% accuracy, and Kappa coefficient of 0.92, whereas, it predicted testing data with 98% accuracy, and Kappa coefficient of 0.96. The confusion matrix of training and testing data is depicted in Tables 4 and 5, respectively.

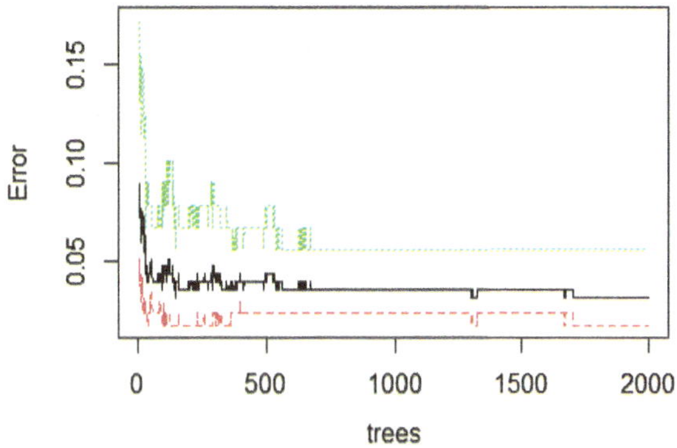

Fig. 4 Variation of Out-Of-Bag estimate of error rate with number of trees in the random forest model at $m = 10$

Table 4 Confusion matrix of training data

Prediction	Reference		
		Normal soils/other features	**Salt-affected soils**
	Normal soils/other features	154	6
	Salt-affected soils	3	87

Table 5 Confusion matrix of testing data

Prediction	Reference		
		Normal soils/other features	**Salt-affected soils**
	Normal soils/other features	74	2
	Salt-affected soils	0	35

4.3.2 Identification of Important Variables/Indices

For identifying the variables or indices which are important for the identification of salt-affected soils in the random forest model, variable importance was computed based on Mean Decrease Accuracy (MDA), and Mean Decrease Gini (MDG) measure (Fig. 5). MDA represents the amount of decrease in random forest model accuracy when a particular variable is excluded. From mean decrease accuracy measure, it can be concluded that the SI_{13} is the most important variable for identification of salt-affected soils in the random forest model followed by reflectance in Near Infrared (NIR), salinity indexes 6 and 4, and reflectance in green band. The first ten most important variables are same in case of MDG index. The MDG function is based on Gini impurity, which is the measure of probability of a particular variable being wrongly classified if it is selected randomly. Gini impurity is used in decision trees to decide which variable to be used to split at each node. MDG function is the measure of total decrease in node impurities from splitting on the variable, averaged over all

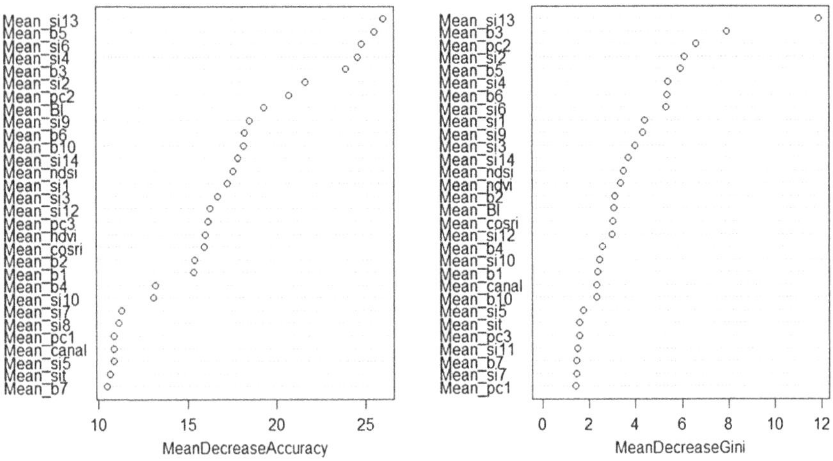

Fig. 5 Variable importance plot based on Mean Decrease Accuracy and Mean Decrease Gini in the random forest model

decision trees in the random forest model. Higher value of MDG function indicates higher importance of that variable in splitting of nodes in the random forest model. It can be concluded that the mean of salinity indices SI_{13}, band 3 (green), principal component 2, SI_2, and NIR bands had higher importance deciding the splitting of nodes in the trees of RF model.

4.3.3 Identification of Salt-affected Soils

The trained RF model with $m = 10$ and $N = 700$ was used for the prediction of salt-affected soils in the Raibareli district using the raster datasets of various computed variables mentioned before. The RF model identified 14,070 ha of land under salt-affected soils whose spatial distribution is depicted in Fig. 6.

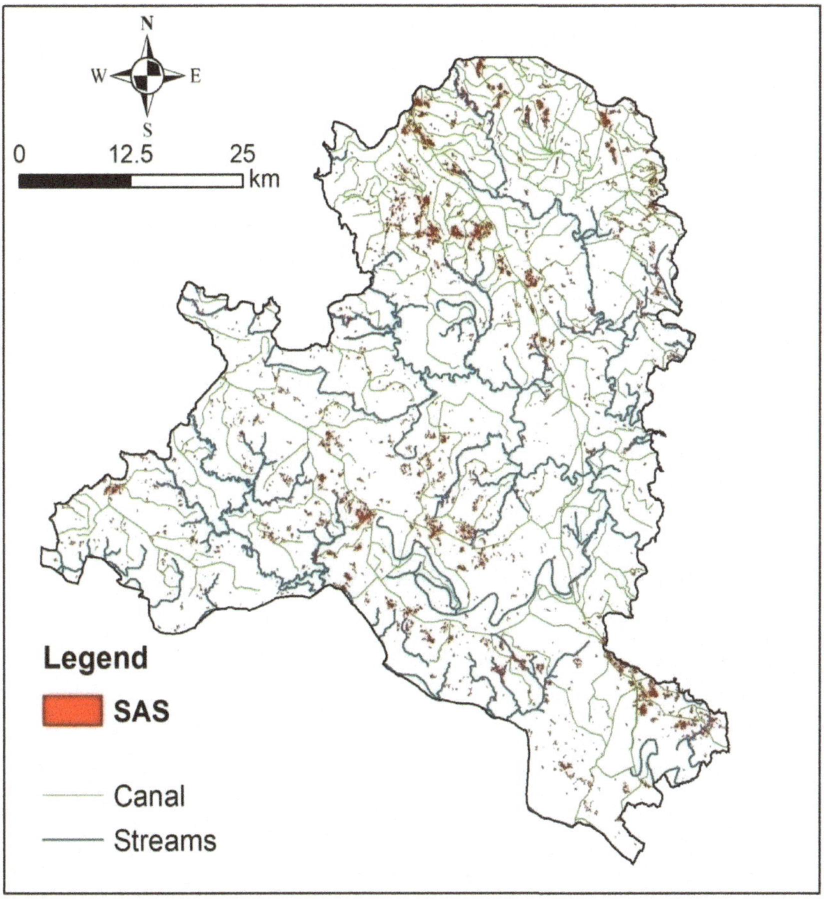

Fig. 6 Results of Random Forest modeling

4.4 Comparing the model's Output

In earlier studies, ICAR-NBSS&LUP reported 31,693 ha of SAS in the Raibareli district under slight, moderate, and severe categories with an area of 14,248, 6640, and 10,804 ha, respectively. In the study, a comparison with spatial distribution was made between the legacy map and the three new estimates reported in the study (Fig. 7). The results show the area under SAS, which is common between the legacy data and the models applied (common), and the areas under SAS, which are exclusive to the legacy data (legacy only) and the models (new). Figure 7 shows almost similar areas under common category in all the three models. However, the LR predicts larger area under SAS, which are exclusive to it. The other two models show lesser areas under new category, i.e., areas mapped by the models, which are not present in the legacy data. A closer look at the data shows that about 17,400 ha area is under severe and moderate categories in the legacy data, which matches with the area estimated with the SVM and the RF models. The maps developed were validated with Google Earth images and field verification (Fig. 8).

5 Conclusions

The present study demonstrates the potential use of LR, SVM, and RF models for identifying SAS by using Landsat 8 data. The study also determined the critical indices for the identification of salt-affected soils in the IGP. The study showed that the Landsat band 7, PC1, and salinity indices, namely, SI_{10}, SI_{11}, SI_{12}, and SI_{13}, were computed based on the blue, green, red, and NIR bands' essential parameters in the case of LR modeling. The RF model identified the salinity indices SI_{13}, SI_6 and reflectance in Near Infrared (NIR), green bands as essential parameters to identify the SAS. We understood from the study that all the models proved accurate with the training and validation samples. However, compared with the legacy data, the LR model overestimated the SAS in several new places in the district. The other two models have identified the severe and moderate types of severity well. The use of multi-season data in the assessment of SAS may provide still better results.

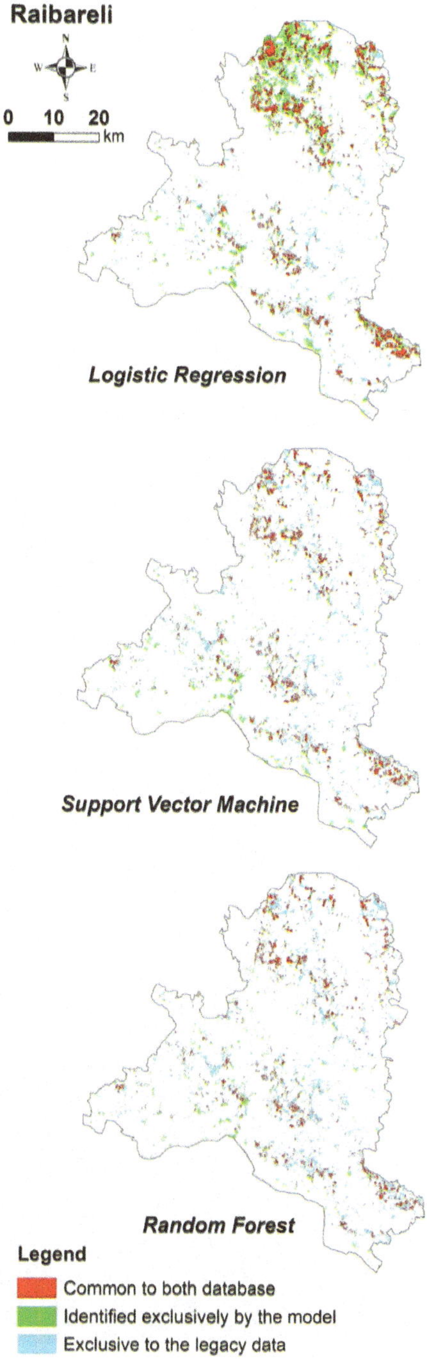

Fig. 7 Comparison of SAS estimates by the three models with legacy data

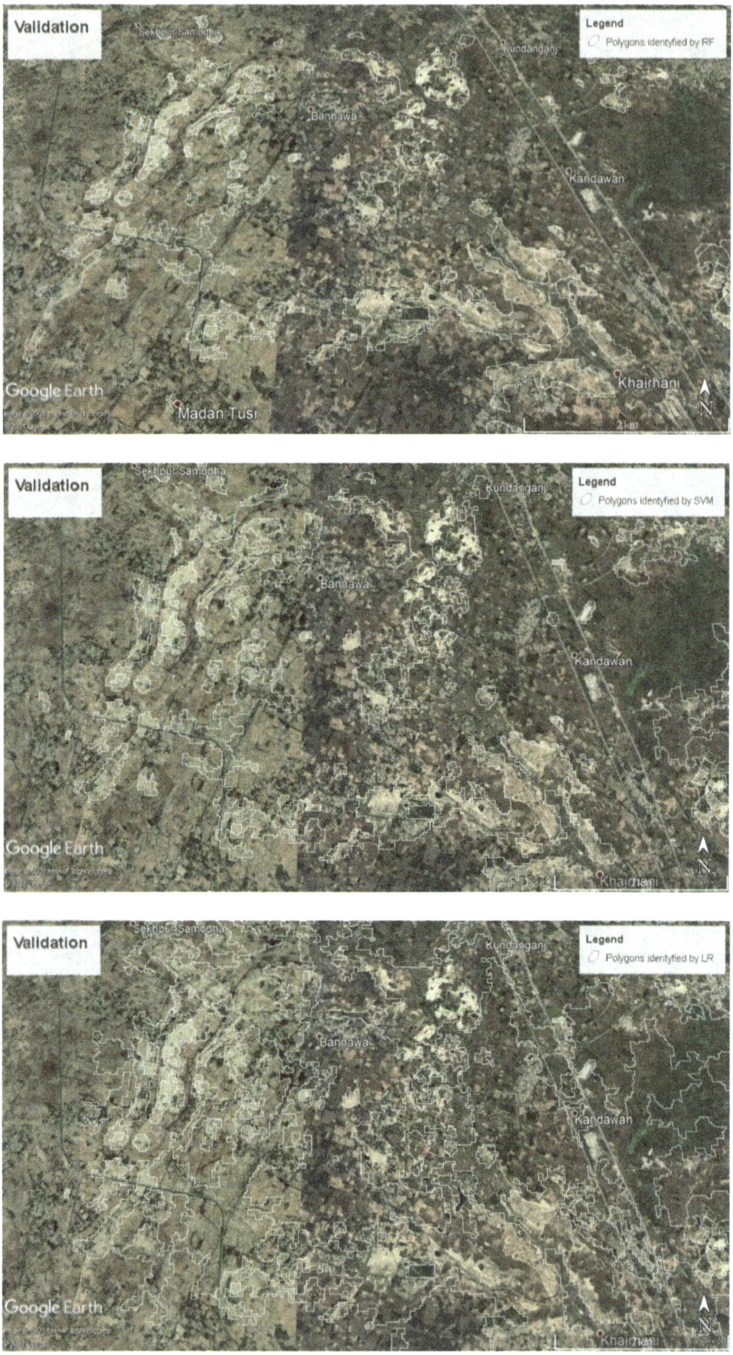

Fig. 8 Validation of SAS estimates with Google Earth images

References

Abbas A, Khan S (2007) Using remote sensing techniques for appraisal of irrigated soil salinity. In: International congress on modelling and simulation (MODSIM). Modelling and simulation society of Australia and New Zealand

Abbas A, Khan S, Hussain N, Hanjra MA, Akbar S (2013) Characterizing soil salinity in irrigated agriculture using a remote sensing approach. Phys Chem Earth, Parts A/B/C 55:43–52

Allbed A, Kumar L (2013) Soil salinity mapping and monitoring in arid and semi-arid regions using remote sensing technology: a review. Adv Remote Sens 2:373–385

Asfaw E, Suryabhagavan KV, Argaw M (2018) Soil salinity modeling and mapping using remote sensing and GIS: the case of Wonji sugar cane irrigation farm. Ethiopia J Saudi Soc Agric Sci 17:250–258

Bivand R, Keitt T, Rowlingson B (2019) rgdal: Bindings for the geospatial data abstraction library, R Package Version 1.1–10. R Foundation for Statistical Computing, Austria

Breiman L (2001) Random forest. Machine learning, vol 45. Kluwer Academic Publishers, pp 5–32

Cai S, Zhang R, Liu L, Zhou D (2010) A method of salt-affected soil information extraction based on a support vector machine with texture features. Math Comput Model 51:1319–1325

Cutler A, Cutler DR, Stevens JR (2012) Random Forest. Ensemble machine learning. Springer, Boston, MA, pp 157–175

Dehni A, Lounis M (2012) Remote sensing techniques for salt affected soil mapping: application to the Oran region of Algeria. Procedia Eng 33:188–198

Ding J, Wu M, Tiyip T (2011) Study on soil salinization information in arid region using remote sensing technique. Agric Sci China 10:404–411

Dogan HM, Kilic OM (2013) Modelling and mapping some soil surface properties of Central Kelkit Basin in Turkey by using Landsat-7 ETM+ images. Int J Remote Sens 34:5623–5640

Douaoui AEK, Nicolas H, Walter C (2006) Detecting salinity hazards within a semiarid context by means of combining soil and remote-sensing data. Geoderma 134:217–230

Dwivedi RS (2001) Soil resources mapping: a remote sensing perspective. Remote Sen Rev 20:89–122

Dwivedi RS, Rao BRM (1992) The selection of the best possible Landsat-TM band combinations for delineating salt-affected soils. Int J Remote Sens 13:2051–2058

Dwivedi RS, Sreenivas K (1998) Image transforms as a tool for the study of soil salinity and alkalinity dynamics. Int J Remote Sens 19:605–619

Dwivedi RS (1994) Study of salinity and waterlogging in Uttar Pradesh (India) using remote sensing data. Land Degrad Dev 5:191–199

Elnaggar AA, Noller JS (2009) Application of remote- sensing data and decision-tree analysis to mapping salt- affected soils over large areas. Remote Sens 2:151–165

Feng PP (2006) System reliability forecasting by support vector machines with genetic algorithms. Math Comput Model 43:262–274

Fernandez-Buces N, Siebe C, Cram S, Palacio JL (2006) Mapping soil salinity using a combined spectral response index for bare soil and vegetation: a case study in the former lake Texcoco. Mexico J Arid Environ 65:644–667

Gutierrez C (2002) A comparison of false color composites in mapping and discriminating between salt-affected soils in Kings county, California. Masters thesis. Oregon State University. p 42

Hijmans RJ (2019) raster: geographic data analysis and modelling. https://CRAN.R-project.org/package=raster, https://earthexplorer.usgs.gov/

IDNP (2002) Indo-Dutch Network Project: a methodology for identification of water-logging and soil salinity conditions using remote sensing. Central Soil Salinity Research Institute

Keuchei AJ, Naumann S, Heiler RM (2003) Automatic land cover analysis for Tenerife by supervised classification using remote sensed data. Remote Sens Environ 86:530–541

Khan NM, Rastoskuev VV, Sato Y, Shiozawa S (2005) Assessment of hydrosaline land degradation by using a simple approach of remote sensing indicators. Agric Water Manag 77:96–109

Kuhn M (2014) Contributions from Wing J, Weston S, Williams A, Keefer C, Engelhardt A, Cooper T, Mayer Z and the R Core Team. Caret: classification and regression training. R package version 6.0. 24

Kumar N, Singh SK, Mishra VN, Reddy GPO, Bajpai RK (2018) Open-source satellite data and GIS for land resource mapping. In: Reddy GPO, Singh SK (eds) Geospatial technologies in land resources mapping, monitoring and management. Geotechnologies and the environment, vol 21. Springer, Cham. pp 185–200

Kumar N, Singh SK, Reddy GPO, Mishra VN, Bajpai RK (2021) Remote sensing applications in mapping salt affected soils. Agric Rev10.18805/ag. R-2008

Kumar N (2018) Study on identification, characterization and mapping of degraded lands using time series MODIS NDVI and LANDSAT data. A dissertation for doctoral degree award in Indira Gandhi Krishi Vishvavidyala, Raipur, 256

Kumar N, Singh SK (2018) Land degradation assessment using MODIS NDVI time series data. In: Singh SK, Chattaraj S, Sharma RP (eds) ICAR-NBSS&LUP technologies, pp 33–42

Kumar N, Singh SK, Reddy GPO, Naitam RK (2019) Developing logistic regression models to identify salt-affected soils using optical remote sensing. In: Krishna AP (ed) Mukherjee AB. Interdisciplinary approaches to information systems and software engineering, IGI Global, pp 233–256

Liaw A, Wiener M (2002) Classification and regression by random forest. R News 2:18–22

Maji AK, Reddy GPO, Sarkar D (2010) Degraded and wastelands of India, status and spatial distribution. New Delhi: Indian Council of Agricultural Research and National Academy of Agricultural Science

Major DJ, Baret F, Guyot G (1990) A ratio vegetation index adjusted for soil brightness. Int J Remote Sens 11:727–740

Mandal AK, Sharma RC (2011) Delineation and characterization of waterlogged salt affected soils in IGNP using remote sensing and GIS. J Indian Soc Remote Sens 39:39–50

Mandal AK, Sharma RC, Singh G (2009) Assessment of salt affected soils in India using GIS. Geocarto Int 24:437–456

Martinez-Beltran J, Manzur CL (2005) Overview of salinity problems in the world and FAO strategies to address the problem. In: Proceedings of the international salinity forum, Riverside, California, April 2005, pp 311–313

Masoud A, Koike K (2006) Arid land salinization detected by remotely-sensed land cover changes: a case study in the Siwa Region, NW Egypt. J Arid Environ 66:151–167

Mendenhall W, Sincich T (1996) A second course in statistics: regression analysis. Prentice-Hall, Englewood Cliffs, NJ

Metternicht GI, Zinck JA (2003) Remote sensing of soil salinity: potentials and constraints. Remote Sens Environ 85:1–20

Mitchell DE (2014) Identifying salinization through multispectral band analysis, Lake Urmia, Iran. Ryerson University, Canada

Muller SJ, Niekerk AV (2016) Identification of WorldView-2 spectral and spatial factors in detecting salt accumulation in cultivated fields. Geoderma 273:1–11

Narayan LRA, Rao DP, Gautam NC (1989) Wasteland identification in India using satellite remote sensing. Int J Remote Sens 10:93–106

Naseri MY (1998) Characterization of salt-affected soils for modeling sustainable land management in semi-arid environment: a case study in the Gorgan region, Northeast Iran. PhD thesis, Ghent University, Belgium

National Bureau of Soil Survey and Land Use Planning (NBSS&LUP) (2004) Annual Report, NBSS&LUP, Nagpur, India

Oldeman LR, Hakkeling RTA, Sombroek WG (1991) World map of the status of human-induced soil degradation: an explanatory note, second (revised edition). ISRIC, Wageningen

Patterson MW, Stephen RY (1998) Mapping fire-induced vegetation mortality using Landsat Thematic Mapper data: a comparison of linear transformation techniques. Remote Sens Environ 65:132–142

Peng W (1998) Synthetic analysis for extracting information on soil salinity using remote sensing and GIS: a case study of Yanggao Basin in China. Environ Manag 22:153–159
Phua M, Saito H (2003) Estimating of biomass of a mountainous tropical forest using Landsat TM data. Canadian J Remote Sens 29:429–440
Rao BRM, Venkataratnam L (1991) Monitoring of salt-affected soils-a case study using aerial photographs, Salut-7 space photographs and Landsat-TM data. Geocarto Int 1:5–11
Rao P, Chen S, Sun K (2005) Improved classification of soil salinity by decision tree on remotely sensed images. In: Proceedings of SPIE-international society for optical engineering, 6027, 20th Congress of the international commission for optics
Raychaudhuri SP (1966) Land and soil. National Book Trust, New Delhi, India
Reddy GPO, Kumar N, Singh SK (2018) Remote sensing and GIS in mapping and monitoring of land degradation. In: Reddy GPO, Singh SK (eds) Geospatial technologies in land resources mapping, monitoring and management. Geotechnologies and the environment, vol 21. Springer, Cham, pp 401–424
Rouse JW, Haas RH, Schell JA, Deering DW (1974) Monitoring vegetation systems in the Great Plains with ERTS. In Third ERTS-1 symposium. NASA, Washington, DC
Space Application Centre (SAC) (2007) Desertification and land degradation atlas of India
Space Application Centre (SAC) (2016) Desertification and land degradation atlas of India
Saha SK, Kudrat M, Bhan SK (1990) Digital processing of Landsat TM data for wasteland mapping in parts of Aligarh District (Uttar Pradesh), India. Remote Sens 11:485–492
Sahu N, Singh SK, Reddy GPO, Kumar N, Nagaraju MSS, Srivastava R (2016) Large-scale soil resource mapping using IRS-P6 LISS-IV and Cartosat- 1 DEM in basaltic terrain of central India. J Indian Soc Remote Sens 44:811–819
Sehgal J, Abrol IP (1994) Soil degradation in India: status and impact. Oxford and IBH, Oxford
Sethi M, Dasog GS, Van Lieshout A, Salimath SB (2006) Salinity appraisal using IRS images in ShorapurTaluka, upper Krishna irrigation project, phase I, Gulbarga District, Karnataka, India. Int J Remote Sens 27:2917–2926
Sharma RC, Bhargava GP (1988) Landsat imagery for mapping saline soils and wet lands in northwest India. Int J Remote Sens 9:39–44
Simin C, Ringqun Z, Liming L, De Z (2010) A method of salt-affected soil information extraction based on a support vector machine with texture features. Math Comput Model 51:1319–1325
Singh AN (1994) Monitoring change in the extent of salt-affected soils in northern India. Int J Remote Sen 15:3173–3182
Sujatha G, Dwivedi RS, Sreenivas K Venkataratnam L (2000) Mapping and monitoring of degraded lands in part of Jaunpur district of Uttar Pradesh using temporal spaceborne multispectral data. Int J Remote Sens 21:519–531
Szabolcs I (1992) Salinization of soil and water and its relation to desertification. Desertif Control Bull 21:32–37
Toomey M, Vierling LA (2005) Multispectral remote sensing of landscape level foliar moisture: techniques and applications for forest ecosystem monitoring. Canadian J Forest Res 35:1087–1097
Tripathi NK, Rai BK, Dwivedi P (1997) Proceedings. In: 18th Asian conference in remote sensing. Kuala Lumpur, Malaysia, pp. A-8-1–A-8-6
Vapnik V (1995) The nature of statistical learning theory. Springer, New York
Vermeulen D, van Niekerk A (2016) Evaluation of a WorldView-2 image for soil salinity monitoring in a moderately affected irrigated area. J Appl Remote Sens 10:026025
Zhu GB, Blumberg DG (2002) Classification using ASTER data and SVM algorithms: the case study of BeerSheva, Israel. Remote Sens Environ 80:233–240

Geoportal Platforms for Sustainable Management of Natural Resources

G. P. Obi Reddy

Abstract Geoportals are being considered as robust and innovative platforms for spatial data assimilation, analysis, management, visualization, and dissemination of geospatial information. Web-based geoportal acts as a gateway to visualize and disseminate diversified thematic databases and information derived from various sources and field surveys along with the metadata with respect to geographic data and services. The emerging geoportal platforms have redefined the efficient planning, monitoring, and sustainable management of natural resources. Geoportals developed with diversified geospatial databases by adopting open data standards facilities to develop interoperable services. This chapter aimed to discuss the evolution of geoportal platforms with reference to Bhoomi geoportal, its services, and potential applications in natural resource management. In India, ICAR-National Bureau of Soil Survey and Land Use Planning (NBSS&LUP) has developed Bhoomi geoportal as a robust platform to provide various thematic services on land and allied resources as Web Map Service (WMS) to access, query, visualize, and disseminate the information to the users. Bhoomi geoportal offers various WMS on terrain characteristics, vegetation dynamics, agro-ecology, soils, land degradation, crop suitability, and cross-domain applications in natural resources management. The challenges associated with geoportals are development of robust functionalities, services, and applications to leverage the big data in meeting the user demands. Development of cross-domain applications, geospatial data analysis, innovative query modules, and visualization tools are some of the important future thrust areas of geoportal platforms for sustainable management of natural resources and location-specific agricultural planning.

Keywords Geoportal · Natural resources management · Service oriented architecture · Web map services

G. P. O. Reddy (✉)
ICAR-National Bureau of Soil Survey and Land Use Planning, Amravati Road, Nagpur 440033, India
e-mail: GPO.Reddy@icar.gov.in

G. P. O. Reddy et al. (eds.), *Data Science in Agriculture and Natural Resource Management*, Studies in Big Data 96, https://doi.org/10.1007/978-981-16-5847-1_14

1 Introduction

In the twenty-first century, in view of global climate change and food security, optimal utilization and sustainable management of natural resources assumes a greater importance to realize the Sustainable Development Goals (SDGs) especially no poverty, zero hunger, and life on land (FAO 2017). The accessibility of precise, consistent, and accurate information on the status of natural resources significantly helps the planners and policy makers to rationalize the decision-making processes in natural resource management (Reddy et al. 2017a). In many of the developing countries, lack of availability of spatial datasets in digital form is a primary hindrance in developing the plans, policies in optimal utilization, and management of natural resources. Abundant data on natural resources either exists in isolation or scattered at various places, making it difficult for decision makers to dovetail the decision-making processes. To overcome such challenges, there is a need for paradigm shift from the traditional data processing systems to spatial data distribution mechanisms for effective data acquisition, processing, storage, visualization, and dissemination (Guo et al. 2020). Thus, the timely availability of reliable information on natural resources plays a significant role in planning and suggestion of appropriate land uses based on their capabilities and limitations (FAO 2011; Reddy et al. 2017a). The fast-emerging geoportal platforms especially in managing of Earth Observation (EO) data (Baumann et al. 2016) getting increasing attention to efficiently store, manage, analyze, and disseminate EO data.

In India, in the field of natural resources management, prior to the concept of geoportal, many researchers developed soil information systems by using stand-alone Geographic Information System (GIS) (Maji et al. 2001, 2002a, b, c; Reddy 2012, 2016) and generated various soil based thematic maps and land use plans. The comprehensive soil information system provides the extent and spatial distribution of terrain characteristics, soil types, soil physical, and chemical properties, which immensely helps in agricultural planning (Reddy et al. 2014a, b). These GIS based soil information systems were extensively used in agricultural resources planning, especially in estimation of soil loss (Sahoo et al. 2014; Reddy et al. 2016b; Mahapatra et al. 2018), assessment of land degradation (Reddy et al. 2002; Maji et al. 2010), and crop planning (Walke et al. 2012; Kumar et al. 2017; Ramamurthy et al. 2020). Therefore, knowledge on soils with respect to their spatial extent, physical, and chemical characteristics, potential, and limitations play an important role in sustained agricultural development (Yagi et al. 2015). The geoportal capabilities were further augmented by the advanced capabilities of GIS in effectively collect, store, analyze, retrieve, transform, and visualize the spatial data (Reddy 2018a).

In recent years, the emerging capabilities of geospatial technologies have greatly enhanced the capabilities in developing robust geoportals and manage voluminous geospatial data on natural resources for their effective integration, analysis, and visualization (Reddy et al. 2017a, b; Reddy and Singh 2018a). Geoportal acts as an information aggregator and provides a platform to discover, visualize, retrieve, and disseminate diversified information and services. Globally, the growing movement for development of geoportals was observed in spatial data assimilation from different sources, its integration, analysis, and visualization through data analytics tools (Yang et al. 2007; Reddy and Singh 2018b). Geoportals become reliable platforms to access and utilize spatial information by the researchers, planners, and policy makers in decision-making processes (Reddy and Singh 2018b). Open Geospatial Consortium (OGC) (2004) describes the geoportal as 'a human interface to a collection of online geospatial information resources, and it includes spatial and non-spatial datasets and services'.

The recent advancements in geospatial technology augmented the efforts of many institutes and organizations in establishment of spatial database systems and geoportals (Reddy and Singh 2018b) at national, regional, and local levels. Geoportals have the capabilities to offer numerous functions to the users to explore and analyze the spatial data, thematic mapping, visualization, and reporting. The advanced capabilities of web-based geoportal platforms are widening the scope for geospatial data assimilation, integration, analysis, and dissemination. The initiatives in development of web-based geoportals provide a robust digital platform to discover, collect, integrate, analyze, and disseminate geospatial information and services with reference to natural resource management and agriculture (Arozarena et al. 2016). Many geoportals have designed the architecture of their platforms, and data integration, management, and dissemination of information as per the systems performance, and spatial data policies in place (Sinnott et al. 2011). Based on functionalities, geoportals can be grouped into catalog geoportals, application geoportals, and enterprise geoportals (Tang and Selwood 2005). The core aim of catalog geoportals is to provide metadata records of data; by using it users can search, access, and visualize the data and services provided by the approved data providers. The application geoportals platforms are designed primarily to meet the needs of a specific group of users. This category of geoportals can combine the necessary tools to offer effective data access mechanisms and various functional services needed by the stakeholders (Maguire and Longley 2005). Whereas, the enterprise portals are designed with a comprehensive infrastructure that allows speedy access and perform recurring transactions in large industries and service sectors. The core aim of this chapter is to discuss the evolution of various geoportal platforms, their core functionalities with special emphasis on design and development of Bhoomi geoportal, its architecture, services, and potential applications in sustainable management of natural resources.

2 Geoportal Platforms—Functionalities and Services

The standard principles of findable, accessible, interoperable, and reusable (FAIR) guide the data producers and publishers for sustainable data management and geoportal services (Wilkinson et al. 2016). The core functionalities of geoportals can be grouped into four categories namely search, mapping, publishing, and administration. The search functions can be used to execute sequential steps in searching for geographic as well as attribute data. These functionalities can be used to obtain the spatial and non-spatial content directly or content metadata. The mapping functionality can be deployed to enrich the value of the search process. This functionality helps the user's community to fully access and visualize the published content by using the functions like pan and zoom. The feature detect capabilities of geoportal can be used to evaluate the published content and visualize several map services either in single or combination of maps. The publishing function performs various functions to add, delete, and modify the metadata content. Depending on the architecture of the geoportal, the process of publishing metadata can be either manual or automated through the customized web service interface. Administration functions enable the administrator of geoportal to review and approve the submitted data or metadata content by the data providers to publish on the geoportal. The database server stores the metadata files, which can be retrieved by the users to know more about the corresponding spatial data or data of geoportal. Metadata services provide the descriptive information about the web services of geoportal to increase their usability in terms of interoperability. Administrative privileges at different levels help to edit and validate the published content.

Services and standards developed by using OGC and International Standardization Organization (ISO) open standards form as critical components of geoportals to allow the devices to communicate with services in interoperable mode. The common protocols set by the standards facilities to enable the important functions of the interoperability across the geoportal platforms. The important services provided by the geoportal platforms are summarized in Table 1.

3 Geoportal Architecture

3.1 Spatial Data Infrastructure (SDI)

With the rapid advancements in spatial data acquisition systems as well as information and communication technologies, the development of SDI assumes a greater importance for spatial data assimilation, analysis, and effective use at various levels (Williamson et al. 2006). The core objectives of SDIs are to access, distribute, and share geographic information for the users' community (Tang and Selwood 2005). SDIs are basically internet-based information systems designed and developed in a collaborative mode to provide geospatial data and customized services with harmonized data specifications for widest possible accessibility at a minimum cost (Masser 2005). SDI must have an interoperability component for spatial data management and dissemination of information at various scales to meet the aspirations and demands of users (Cömert 2004). The geoprocessing capabilities of SDI in distributed applications offer various geoprocessing services. SDI architecture helps to reduce the duplication of efforts among the agencies/institutes, minimize the costs while making geographic data accessible, enhance the benefits in using of offered spatial data, and establish key partnerships.

Table 1 Important services of geoportals and their functions

Geoportal services	Functions
Web Catalog Services for the Web (CSW)	CSW offers the abilities and functionalities to publish, discover, and query the metadata content of data, services, and associated information. Metadata provided through CSW explains the characteristics of service that can be queried, visualized, and processed either by humans or software. Maguire and Longley (2005) described that the CSW as a core feature of SDI geoportal provides the ability to publish and access the metadata about geographic data and services.
Web Map Service (WMS)	WMS is one of the standard protocols of geoportals primarily followed in publishing georeferenced maps and data in the form of image format like JPG, PNG, TIF, GIF, SVG, Geotiff, and XML. WMS offers the functionalities to view and query the attributes depending upon the geoportal architecture and data access policies. This interface is mainly used to query and visualize the registered WMS either from standalone or network of distributed geoportal platforms.
Web Map Tile Service (WMTS)	WMTS offers the functionality of pre-cached images, which are basically georeferenced. Service providers can adopt WMTS to keep the geoportal platform stable and efficient in handling the large volume of datasets. OGC describes the WMTS as the advanced version of WMS that enables it to improve the service delivery performance. Here, instead of generating a new image for every request of the user, WMTS serves small pre-generated images or reuses similar previous requests, which follows a discrete set of tile systems. The cache mechanism of WMTS supports the actions like scanning, and access of numerous raster/image data, and visualizing them in a shortest possible time by the users.
Web Coverage Service (WCS)	WCS is elucidated as retrieval of geospatial data in coverage format. WCS is an appropriate form of service to respond to diverse requirements of users from a single data source. WCS facilitates the extraction of data, and its properties by the users either fully or partially depend on the available coverage properties adopted by the geoportal platform. WCS provides the facility to the users to select part or full of server's information depending upon the spatial constraints and query criteria adopted.

(continued)

Table 1 (continued)

Geoportal services	Functions
Web Feature Service (WFS)	WFS offers a robust interface to allow the requests from the users to access and use geographical features in analysis across the web by using platform-independent requests. WFS describes the interface to perform operations like analysis and manipulation on geographic features data. WFS also allows the users to query, generate, update, and delete feature instances depending on the spatial and non-spatial constraints adopted.
Web Processing Service (WPS)	WPS interface provides geoprocessing functionalities for spatial analysis of features, wherein users can perform spatial analysis and functions on web. WPS interfaces define the standards and rules to standardize the inputs and outputs of service to perform geospatial processing operations. WPS provides various tools for the users to serve the data required by the WPS across distributed geoportal platforms. WPS is an effective web-based service to perform various geoprocessing operations on provided data by the users to obtain the desired results instead of reaching the whole data provided by the data provider.

3.2 *Service Oriented Architecture (SOA)*

SOA approaches are usually adopted to build distributed systems to deliver various application functionalities as services to meet the requirements of end-user applications (Colan 2004). Web service can be developed as an application and deployed across the web to accomplish a specific task. Web services usually build by adopting open internet standards and protocols, like HTTP for connection and communication, Uniform Resource Identifier (URI) for identification, eXtensible Markup Language (XML) for contents specification, Web Services Definition Language (WSDL) for expression of service descriptions, and Universal Description, Discovery, and Integration (UDDI) protocol for directory services. Web service ecosystem can be designed and developed, where service providers can present their services through a catalogue service, and deploy them on geoportal to access the services by clients in catalogue. Web services of SOA architecture are able to perform the functions like publish, find, and dynamic binding between service providers and users.

3.3 Geoportal Interoperability

Interoperability refers to the ability by which different applications of geoportal can communicate with each other. The distributed computing systems provide the foundational standards, protocols, and technology in design and development of web-based distributed geoportals. The main aim of the distributed geoportal platform and interoperability infrastructure is to provide services by using web-based technology. Geoportal platforms aimed to provide interoperability, must adopt standards, and definitions in geoportal design and development. Web service interoperability can be achieved through adopting OGC open data specifications, which includes CSW, WMS, WMTS, WCS, WFS, and WPS. Many researchers followed ontology-based architecture to enhance service discovery and interoperability in the network of distributed geoportals (Lacasta et al. 2007). However, the challenges like semantic-based extraction, enrichment, search, analysis, and semantic interoperability of geospatial services still remain (Lutz 2007; Kokla and Guilbert 2020).

3.4 Cloud Computing Architecture

It refers to the utilization of networked infrastructure, databases, software capabilities, and capacity to offer resources to users' on-demand. Cloud computing facilitates to catch information temporarily on users' devices and stored in central servers (Gurjar and Rathore 2013). Cloud platforms are efficient to store, access, and analyze large datasets by deploying high-end servers. These platforms provide resources like infrastructure, platform, storage of services, and software packages in various ways for the users' community (Ma et al. 2015). Cloud computing architecture has five characteristics. Firstly, providing cloud computing as on-demand service, where accessing various computing capabilities as and when needed by the users in automatic mode; secondly, the availability of wide network access where the required services are accessible over the network; thirdly, the capability of resource pooling where the service providers pooled resources to serve multiple clients; fourthly, offering the rapid elasticity, it gives the capability to quickly scale-in and scale-out of services; and the fifth characteristics is the ability to measure services, where services are controlled and optimized based on metering (Chappell 2010). The front-end architecture of cloud computing platforms provides a robust ecosystem to develop algorithms and interactively access and visualize data. It allows users to add and arrange their own data and collections as well as mine the massive warehouse of data for various applications. Google Engine Earth (GEE) is the robust cloud computing platform, and it facilitates process of big geodata of over large geographical areas of different time scales. Cloud computing models offer suitable on-demand network access with a wide range of shared computing resources like network and servers' infrastructure, storage space, applications as well as services, which can be quickly accessed and used with minimum interaction of service providers (Mell and Grance 2011). Cloud computing deployment models can be public, private, community, and hybrid.

3.5 Cloud Computing Services

Services provided by the cloud computing models can be grouped into three categories namely software as a service (SaaS), platform as a service (Paas), and infrastructure as a service (IaaS). On SaaS cloud, applications run on the cloud, and it offers an architecture that supports running of multiple instances irrespective of location. SaaS is also known as Web-based software where users can access software as service through the web instead of installing and maintaining software on their machines. SaaS cloud services offer high scalability, which gives customers the different options to access more and more services or features on demand and provide easy access irrespective of the geographic location. Major advantages of this model include flexible transactions, scalable usage, automatic updates, accessibility, and persistence. The Paas cloud model offers the capabilities to the developers to write and run the applications on the cloud. Several application services are available for the developers to use for quick deployment. IaaS shares the computing infrastructure, and it can be accessed on the web. The SaaS cloud model consists of servers, databases, storage, security, and other peripherals. Web-based software provides remote access to the system, and using this kind of software service user can access the system and perform the computing tasks remotely.

4 Evolution of Geoportal Platforms

In recent years, many organizations/institutions across the globe are adopting geoportal technology to facilitate the geospatial data storage, discovery, and dissemination of thematic services and applications (Maguire and Longley 2005; Fritz et al. 2012). Web-based geoportals provide services to find and access the geospatial data and various web services to display, edit, analyze, and publish geographic information. Geoportal-based integration of WPS, spatial analysis, query, and visualization has potential to support various applications in natural resources management and agriculture. Many international organizations/institutions developed geoportals to share various EO datasets and services. The scripting languages such as R and Python greatly augmented the developments of data analysis architecture in geoportals. The open-source spatial packages (Shao et al. 2012) also provide a very useful workspace in development of a wide variety of applications and user interfaces. Globally, Infrastructure for Spatial Information in Europe (INSPIRE) (http://inspire-geoportal.ec.europa.eu/), Global Earth Observations System of Systems (GEOSS) (https://www.geoportal.org/) (Anderson et al. 2017), and Google Earth Engine

(GEE) (https://earthengine.google.com) are the important geoportal platforms in the field of EO. However, in India, Bhuvan geoportal (https://bhuvan.nrsc.gov.in/nhp/) of Indian Space Research Organization, India Geoportal (https://www.nsdiclearinghouse.gov.in/GeoportalNSDI/Home.aspx) of DST-National Spatial Data Infrastructure (NSDI), and Bhoomi geoportal (http://www.bhoomigeoportal-nbsslup.in/) of ICAR-National Bureau of Soil Survey and Land Use Planning (NBSS & LUP) are some of the important geoportal platforms, which provides various thematic services in the field of natural resource management.

4.1 INSPIRE Geoportal

INSPIRE geoportal was developed by the European Union with an aim to provide various geospatial datasets and services to the member states. It provides harmonised sources of spatial information to formulate, implement, and evaluate various community policies (Annoni 2006). INSPIRE collect and collate the base information generated by the member states of EU with an aim to respond and provide inputs in a wide range of policy initiatives at various levels. INSPIRE geoportal provides metadata to describe the existing information resources and allows to discover, view, and download information resources. The coordination and monitoring mechanisms are core of INSPIRE geoportal. Periodically, INSPIRE assesses the current status and needs of spatial information in Europe to fill the gaps and eliminate the duplication of information collection. INSPIRE geoportal allows to monitor the availability of INSPIRE datasets, discovery of suitable datasets based on their metadata, and access the selected datasets or download services by the users.

4.2 GEOSS Geoportal

The Group on Earth Observations (GEO) was developed by the GEOSS Portal to address various areas of critical importance to the people and society, called as Societal Benefit Areas (SBAs). GEOSS addresses thematic areas like climate, water, agriculture, disasters, ecosystems, energy, health, and biodiversity (Nativi et al. 2015). GEOSS offers the capabilities for the user of EO to search, access, and consume the data, information, tools, and services. GEOSS infrastructure includes web portal, clearinghouse, components, services registry, standards, and interoperability registry. The GEO agriculture SBA supports the applications in agriculture and allied sectors like forestry and fishery for sustainable agriculture. The flagship program of GEO GLobal Agricultural Monitoring (GEOGLAM) under GEOSS provides a platform to bring the key players together in the field of global agricultural monitoring and information sharing. It addresses critical areas like mapping and monitoring of global croplands, monitoring of global agricultural crop production, and food supply and provides effective early warning of famines in ensuring global food security.

4.3 GEE Geoportal

GEE as a cloud computing platform provides robust functionalities to effectively store, manage, and process enormous geospatial datasets to perform various data analysis activities to derive information for policy inputs and decision making (Mutanga and Kumar 2019). It provides time-series satellite imagery of global coverage, cloud computing infrastructure, and access to software, and algorithms to process large datasets (Gorelick et al. 2017). GEE platform provides enormous computational capabilities for data preparation, data creation, training of numerous classifiers, and algorithms (Gorelick 2013). By using ML algorithms, GEE allows for efficient testing of various base classifiers, as well as training of the ensemble with a wide range of spatial, spectral, and temporal data derived from satellite sensors (Aguilar et al. 2018). GEE provides facilities in sharing of computer codes and obtains active online support to the users. GEE made a paradigm shift in big data storage, process, and analysis of various remotely sensed data products, which is not feasible by using simple desktop systems. By virtue of its speed, flexibility, and accessibility, GEE provides ample opportunities to make use of this platform for various EO applications in natural resources management and agriculture.

4.4 India Geoportal

In India, to accomplish its national mandate, National Spatial Data Infrastructure-India (NSDI India) has designed, developed, and being implemented India Geoportalas a network of spatial data nodes established with various data providing nodal agencies in central, and state governments to improve the access of metadata of geospatial data on various resources provided by data providers. The nodal agencies under the NSDI framework adopted OGC and ISO standard specifications in building the geospatial databases in their respective domain to facilitate an interoperable mechanism for dissemination of information among the participating organizations' data nodes and end-users. Standards-based WMS are accessible from data providing agencies for use in development of various GIS applications. Efforts are on to reengineer and maintain available datasets for provision of WFS and improve the access to processable feature datasets with stakeholders. Several state-level SDIs were also in place to provide the geospatial information required to India geoportal under governance framework. Data cleaning, metadata update, data registry, and catalogue services are continuous efforts of India Geoportal to support search and discovery of services. Developed National Data Registry (NDR) framework supports registration, search, and discovery of data services and application development on a cloud-based India geoportal platform.

4.5 Bhuvan Geo-Portal

ISRO's Bhuvan Geoportal provides a variety of EO data and visualization services to the stakeholders. Many government agencies use Bhuvan platform to use its services, share, and host their data as service to meet the specific domain applications. Bhuvan geoportal provides Indian satellite data of various spatio-temporal resolutions and derived products. These services are being used in various applications and programs conducted by the researchers, organisations, students, etc. Bhuvan provides crowd sourcing services for field data collection and geo-tagging under various government programs. Many sectors are consuming the services of Bhuvan in developing domain specific applications, which includes natural resources management and agriculture. 'Bhuvan Panchayats' a web portal module facilitates decentralized planning through sharing high resolution spatial and non-spatial information to augment developmental activities at grassroot level. Bhuvan geoportal allows the users to query and visualize various services by using the customized tools like pan, zoom, search, overlays, online editing, etc.

4.6 Bhoomi Geo-Portal

To fill the gap in availability of dedicated geoportal on soil and allied resources, ICAR-NBSS & LUP has developed Bhoomi Geoportal as web-based platform to deploy various thematic services of land and allied resources as WMS and facilitates to visualize, access, query, and disseminate the land and allied resource information. The WMS of Bhoomi Geoportal encompasses socio-economic profiles at district and block level, SRTM DEM (30 and 90 m) and ALOES DEM (12.5 m) of India, mean MODIS NDVI (250 m) for *kharif* season (June–September) of 2000–2018, physiography, sub-physiography and landform units, agro-ecological regions (1992 and 2015), agro-ecological sub-regions (1999) and aridity index (1992 and 2018) of India, benchmark soils, soil series data at states level, soil based services at various scales, soil organic carbon, soil loss, land degradation status of India, crop suitability for various crops at national, and state level. Keeping its importance and its services on various soil and allied resources in natural resource management, the details on design and development of Bhoomi Geoportal platform and its services are summarized in the following sections.

5 Design and Development of Bhoomi Geoportal

ICAR-NBSS & LUP, Nagpur India has designed and developed Bhoomi geoportal by using open source Geoserver for efficiently organizing, cataloging, storing, managing, searching, and visualizing various thematic services on land and allied resources (Reddy and Singh 2018b). GeoServer was used as a browser-based web administration interface application to configure the needful function of Bhoomi geoportal to add and publish various geospatial services. The individual spatial layers were processed in GIS, developed legends, and published as WMS on Bhoomi geoportal. Geoserver, and PostgreSQL were used in design and development of user-friendly interface of Bhoomi geoportal to visualize various thematic layers, services, and applications. HTML and JavaScript were used to develop a user-friendly interface of Bhoomi geoportal. The style and legend properties of thematic services of geoportal were developed by using Style Layer Descriptor (SLD) by using open-source QGIS software.

The services of Bhoomi geoportal platform were grouped depending on the data content and nature of applications while publishing the services. Application Programming Interface (API) was used to build geoportal applications to support various capabilities like access, query, and visualize the services. On Bhoomi geoportal, various WMS on land and allied resources were cataloged and deployed (Reddy and Singh 2018b). Various utility tools for data access, query, extract, visualize, and legend development were developed on geoportal. The postgre Structured Query Language (SQL) data management system was used to store, manage, and query geospatial data for Bhoomi geoportal. OGC metadata standards were adopted in development of metadata for various services and provided service metadata for the majority of the services on Bhoomi geoportal, which facilities to query and abstract information on various WMS. Bhoomi geoportal platform supports the development of various web applications by using libraries, packages, and templates. Bhoomi geoportal provides access to WMS, data discovery, and dynamic visualization tools for querying the geospatial data. Developed data query modules and data visualization tools as dashboards enable to query and visualize spatial and non-spatial services. The integration of cloud computing and process models with internal architecture of Bhoomi geoportal services provides impetus to the interoperability of services. Integration of Machine Learning (ML) algorithms with geospatial data on a geoportal platform offers the ability to build predictive models to perform various classification and clustering tasks. Three-tier architecture adopted in design and development of Bhoomi geoportal and its front-end are shown in Figs. 1 and 2.

5.1 Spatial Data Visualization on Bhoomi Geoportal

Visualization interfaces and dashboards of any geoportals are powerful tools to access and visualize various geospatial data, services, and associated information. Bhoomi geoportal offers the functionalities to the user to access, explore, visualize, and examine the published various thematic services and their metadata content. The data visualization functions like pan, zoom-in, and zoom-out were provided to identify and visualize the thematic services effectively by the users and evaluate the published content. Data query modules and dashboards were developed on Bhoomi

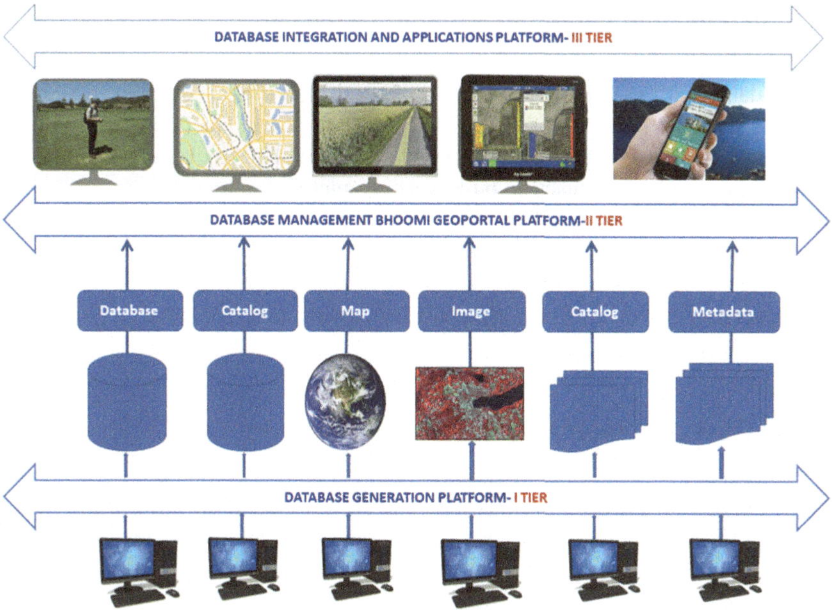

Fig. 1 Three-tier architecture adopted in design and development of Bhoomi geoportal

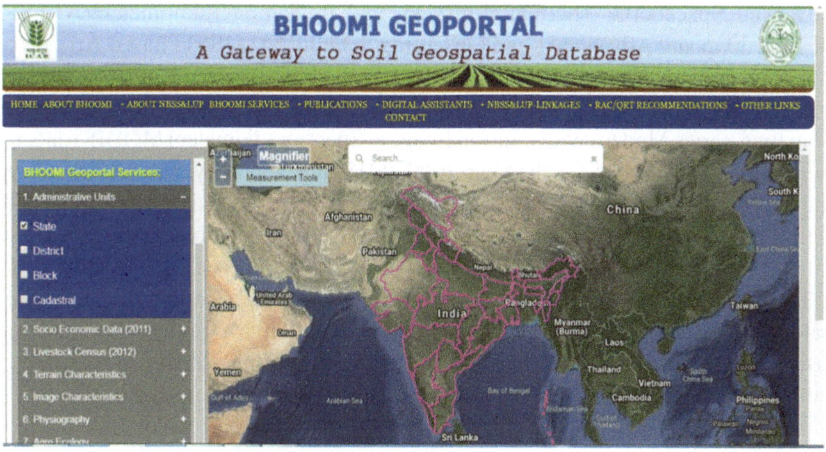

Fig. 2 The front-end interface of Bhoomi geoportal

geoportal to query spatial and non-spatial databases and various services. These functions support queries and visualize various WMS either in single or combination and derive useful information for its use in decision-making processes in sustainable management of natural resources and agriculture.

5.2 Spatial Data Query on Bhoomi Geoportal

Spatial data queries of geoportals can be spatial, non-spatial, or even mixed. On Bhoomi geoportal platform, various WMS services can be queried and visualized based on administrative units like national, state, district, and block level. The advance capabilities of SQL were used to build various query modules to query and access the information and services. Data query functions developed as an aggregations of building block tools on Bhoomi geoportal, which can be executed through selection of thematic service or selection of place name or selection of location name from the list to query the selected service, its attributes, and content metadata.

6 Services of Bhoomi Geoportal Platform

Bhoomi geoportal provides various thematic services as WMS on land and allied resources to visualize, access, query, and disseminate the information. The WMS of Bhoomi geoportal platform covers terrain characteristics, vegetation dynamics, agro-ecology, soils, land degradation, crop suitability, and cross-domain applications in natural resources management (Reddy and Singh 2018b). Bhoomi geoportal offers WMS on socio-economic profiles at district level, processed Digital Elevation Models (DEMs) of Shuttle Radar Topography Mission (SRTM) at 90 and 30 m, Terra Advanced Spaceborne Thermal Emission, and Reflection Radiometer (ASTER) DEM at 30 m and Advanced Land Observing Satellite (ALOES) Phased Array type L-band Synthetic Aperture Radar (PALSAR) DEM at 12.5 m for India, processed mean Normalized Difference Vegetation Index (NDVI) at 250 m for India derived from time-series Moderate Resolution Imaging Spectroradiometer (MODIS) data for *kharif* and *rabi* seasons for the year 2000–2018, physiography, sub-physiography, and landform units, agro-ecological regions (1992 and 2015), agro-ecological sub-regions (1999), aridity index (1992 and 2018) of India, benchmark soils, soil series data for selected states. Bhoomi geoportal also provides status on soil organic carbon, soil erosion, and land degradation, crop suitability for various crops at national and state level. The services provided by Bhoomi geoportal have immense potential in sustainable management of natural resources and agricultural land use planning in the country. The important services of Bhoomi geoportal are depicted in Fig. 3.

6.1 Digital Terrain Analysis

The detailed analysis of terrain and landforms is an important aspect in natural resources management, agriculture, and environmental planning (Reddy et al. 1999, 2001; Hengl and Reuter 2008). In recent years, DEMs are being widely used in the field of digital terrain analysis (Reddy 2018b), geomorphological, and pedological

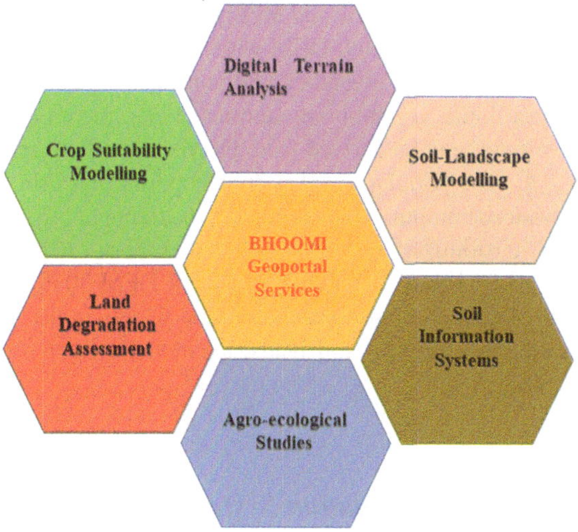

Fig. 3 Important web map services of Bhoomi geoportal

studies (Zhang et al. 2020). Terrain parameters are extensively used in mapping of primary and secondary variables in spatial prediction of various soil properties (McKenzie and Ryan 1999; Li et al. 2020). The advanced geoprocessing capabilities of GIS and availability of open-source global coverage DEMs at various spatial resolutions like at 90, 30, and 12.5 m revolutionized the field of digital terrain analysis (Bogaart and Troch 2006). However, the selection of suitable resolution and quality DEMs are important in digital terrain analysis and modeling. ICAR-NBSS&LUP processed and developed mosaic of SRTM DEMs at 90 and 30 m for India and are being widely used in analysis and characterization of various geomorphometric parameters in varied terrain conditions (Reddy et al. 2012) especially in land resource inventories and soil-landscape modeling. The processed and deployed SRTM DEM data at 30 m, ASTER DEM at 30 m, and ALOES-PALSAR DEM at 12.5 m on Bhoomi geoportal have immense potential to develop applications in conjunction with high resolution satellite data especially in the field of large-scale digital terrain analysis, landform mapping, land resources inventory, and soil-landscape modeling.

6.2 Soil-Landscape Modeling

Soil-landscape modeling based on geomorphometric parameters derived from DEMs and legacy soils data are important inputs to predict soil characteristics (Reddy and Sarkar 2012a; Reddy et al. 2013a). The recent developments in digital terrain analysis and application of machine learning (ML) algorithms in digital soil mapping broaden the scope of DEMs in the field of soil-landscape modeling and applications (Wadoux et al. 2021). DEMs in conjunction with soil information systems and satellite data significantly enhance the capabilities in regional to local scale soil mapping (Dobos et al. 2000; Lagacherie and McBratney 2006). Reddy et al. (2013a) demonstrated the potential of integrated DEM with satellite data in delineation of distinct landforms for soil resource inventory and soil-landscape modeling on basaltic landscape of central India. By applying automated classification algorithms various topographic covariates can be derived from DEMs to predict various soil parameters (Guo et al. 2019). The popular cloud-based geoportal platform like GEE supports performing advanced computational tasks for processing of big geospatial data in terrain analysis with customized algorithms (Gorelick et al. 2017). The processed and deployed SRTM DEMs at 90 and 30 m, ASTER DEM 30 m, and ALOES DEM at 12.5 m for India on Bhoomi geoportal are being widely in the fields of digital terrain analysis, landforms mapping, soil-landscape modeling, soil resources inventory, and land use planning.

6.3 Soil Information Systems

Soil information system is an integrated, centralized georeferenced information system that combines information on all soil properties, and these systems assumed a greater importance in natural resource management and agricultural land use planning (Maji et al. 2002b, 2012; Reddy 2012). Globally, some of the available soil information systems are Australian Soil Resource Information System (ASRIS), Canadian Soil Information Service (CanSIS), State Soil Geographic (STATSGO), Soil Survey Geographic (SSURGO), Soil Terrain Databases (SOTER), Global Soil Map.net, and World Inventory of Soil Emission Potential (WISE). The recent advances in the field of geospatial technology have greatly enhanced the capabilities in storing, cataloging, manipulating, analyzing, and visualizing the soil data in the form of soil information system (Reddy et al. 2016a). Availability of georeferenced soil resources databases on geoportal platforms play a key role in sustainable management of natural resources and agriculture planning (Reddy et al. 2017a). Geoportal platforms provide a host of powerful analytical tools to analyze and visualize soil information and integrate with advance remote sensing datasets and DEMs found to be robust approach in developing various applications in natural resources management and agriculture (Reddy et al. 2017b; Balla et al. 2020). Bhoomi geoportal offers WMS on soil information systems at 1:1 m and 1:250 k for India and 1:50 k for selected districts of India to query, find, and visualize geographic information on various soil parameters to derive information for use in variety of applications in natural resource management and agriculture planning. The available legacy spatial database on benchmark soils of India, soil series, and soil fertility for selected states on Bhoomi geoportal is of immense help to integrate with advanced remote sensing, DEMs, and ML algorithms in developing applications in natural resource management and agricultural planning. The WMS of soils of India (1:1 m) on Bhoomi geoportal is shown in Fig. 4.

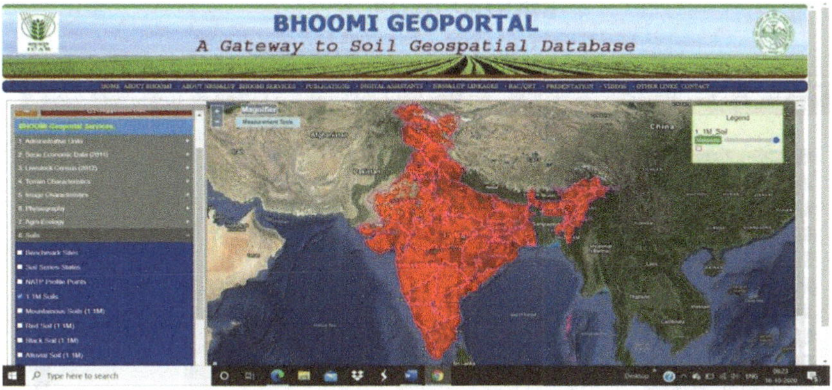

Fig. 4 WMS of soil information of India (1:1 m) on Bhoomi geoportal

6.4 Agro-Ecological Studies

Agro-ecological approach (FAO 1996), it includes identification of homogeneous units by integrating the relevant climatic parameters like length of growing period (LGP), bio-climate conditions, terrain characteristics like physiography/landforms and selected soil parameters form a strong foundation for agricultural land use planning and agro-technology transfer. Besides the information on bio-climatic resources, the available DEMs, physiographic units, and information on soil and allied resources on Bhoomi geoportal immensely helps to derive the necessary inputs in characterization of agro-ecological settings. Earlier, the spatial information on physiography, bioclimate, LGP, soil moisture, and temperature regimes was integrated in GIS environment and delineated 20 Agro-Ecological Regions (AER's) (Sehgal et al. 1992) and 60 Agro-Ecological Sub Regions (AESR's) (Velayutham et al. 1999) of India. These thematic layers and allied agro-ecological parameters of India were processed, catalogued, and deployed on Bhoomi geoportal platform. The information on AERs and AESRs can be effectively used in sustainable agricultural resources planning and improve the existing land use systems by maintaining soil health with no or minimum land degradation. The information on agroecology in conjunction with the soil information system on Bhoomi geoportal immensely helps in delineation of efficient crop productivity zones, agro-technology transfer, and resources-based agricultural land use planning. In the context of climate change, the readily available soil information systems, temporal remote sensing data, and advance capabilities of web-based Bhoomi geoportal provide a robust mechanism to revisit the boundaries of AER's and AESR's for their refinement and use in adoption of suitable land use systems to achieve higher productivity, especially in rainfed ecosystem of the country. Further, vegetative indices like NDVI derived from time-series remote sensing data can be effectively used in assessment, mapping, and monitoring of agricultural droughts (Reddy et al. 2020; Sandeep et al. 2021a; Kumar et al. 2021). Bhoomi geoportal

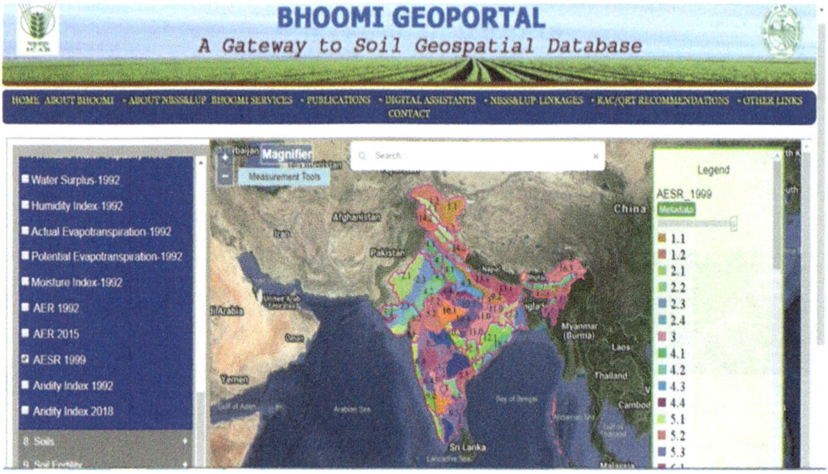

Fig. 5 WMS of AESRs of India (1999) on Bhoomi geoportal

provides WMS on AERs and AESRs of India, mean annual rainfall, water storage, water deficit, potential, and actual evapo-transpiration for the specific periods. These available thematic services on agro-ecology immensely help in developing agricultural plans and agro-technology transfer at various levels in India. The WMS of AESRs of India (1999) on Bhoomi geoportal is shown in Fig. 5.

6.5 *Land Degradation Assessment*

Land degradation assessment is a comprehensive approach that includes the study of soil erosion, soil salinity, acidification, soil nutrients, and soil structure (Maji et al. 2010; Minhas and Reddy 2017). In many regions of the world, water erosion is the prominent degradation process in the forms of sheet, rill, and gully erosion. In India, many authors applied Universal Soil Loss Equation (USLE) and GIS capabilities in assessment of soil erosion on different terrain conditions (Sahoo et al. 2005; Reddy and Sarkar 2012b; Reddy et al. 2013c; Mahapatra et al. 2018; Reddy et al. 2018) and subsequently used it in assessment of land degradation. Maji et al. (2010) harmonized degraded and wastelands datasets of India in GIS environment and reported that about 120.72 Mha area is under different categories of degraded and wastelands in India. Assessment of land degradation involves analysis of geospatial datasets like terrain parameters obtained from DEM, historical climatic data, soil characteristics, and time-series vegetation indices derived from various satellite-based products (Sandeep et al. 2021b). Bhoomi geoportal platform offers various geospatial datasets and WMS on DEMs at various resolutions, soil information, soil erosion, and time-series vegetation indices could be effectively used in assessment of land degradation at finer scales in India to address various emerging issues in natural resources management and agricultural planning in India. Precise site-specific databases on land degradation immensely help the planners and policy makers to develop a comprehensive strategy to realize land degradation neutrality (LDN) by 2030 as mooted by the United Nations.

6.6 Land Suitability Modeling

Land suitability assessment offers the information on the degree of usefulness of land resources through analysis of its potentials and limitations for potential land use (FAO 1976). Under the changing climate and availability of limited land resources, assessment of crop suitability assumed a greater importance in sustainable agricultural development in countries like India to feed the growing population and ensure food security (Reddy and Singh 2014). In land suitability modeling, the spatial information on various terrain characteristics, climate, soil physical, and chemical properties of an area play a significant role in defining its potentials for growing specific crops (Vishakha et al. 2016; Ramamurthy et al. 2020). GIS-based soil information system in conjunction with climatic variables and satellite datasets allows us to build robust models to assess land suitability (Harasheh 1994). The advanced tools and techniques of GIS offer powerful tools in integrating the diversified datasets in crop suitability analysis and modeling (Kumar et al. 2017). The advance tools like multi-criteria decision analysis (MCDA) and GIS modeling become a powerful approach in land suitability assessment for various crops (Malczewski 1996; Joerin et al. 2001; Maji et al. 2005; Walke et al. 2012; Reddy et al. 2013b). The geospatial databases coupled with fuzzy logic, Artificial Neural Network (ANN), and Analytic Hierarchy Process (AHP) models were also used under diversified climatic and soil conditions in crop suitability modeling (Ramamurthy et al. 2020; Pilevara et al. 2020). The available WMS on DEMs, agro-ecological data, and soil information system at various scales on Bhoomi geoportal in conjunction with suitable ML algorithms could be effectively used in building the models for crop suitability mapping at national, state, and district level. The suitable areas for maize cultivation in India as deployed on Bhoomi geoportal are shown in Fig. 6.

7 Geoportal Platforms and Challenges

There are many challenges associated with design and development of geoportal platforms in providing effective, sustained geoportal services, and applications. The development and offering of standardized web services depends on the flexible data integration and reuse of services and computational power of geoportal platforms. The geo-big data has immense potential to grow at a much faster rate in future, especially for in-depth scientific inventories and discoveries in natural

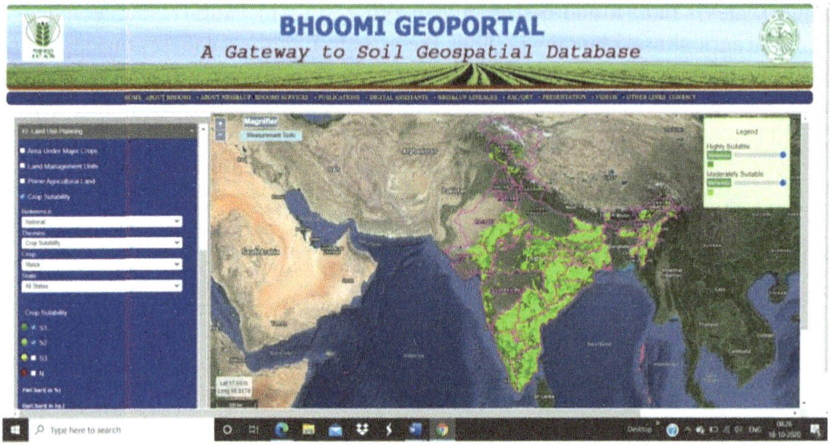

Fig. 6 Suitable areas for maize cultivation in India on Bhoomi geoportal

resources management and environment planning (Guo et al. 2016). At the same, geo-big data poses many challenges like its volume, velocity, variety, veracity, and value (Nativi et al. 2015). The availability of updated metadata of services is a yet another challenge often faced by many geoportal platforms. The adoption of a distributed geospatial information processing mechanism is a challenge to provide geo-processing capabilities in a distributed geoportal ecosystem. The recent developments in geoportal architecture provide solutions to develop distributed geospatial information processing mechanisms. However, the challenges like data quality, uncertainty, reliability, scalability, standards, interoperability, and data security of geoportals remain pertinent.

8 Way Forward

- Adoption of advanced geoportal technologies is necessary to provide innovative geospatial applications to meet the emerging needs and requirements of users in natural resources management and agriculture planning.
- Implementing global standards as per the open data standards in design and development of geoportals enhance the value of services offered by geoportals in consuming the available digital resources and sustained geoportal interoperable ecosystems.
- Cloud's computing capabilities and potential benefits must be leveraged for sustained geoportals services and applications in distributed architecture.
- Development of user-friendly Graphical User Interface (GUI) will enhance the visualization capabilities and obtain the user feedback to improve the services.

- Development of ontologies to ensure semantic interoperability and deployment of advanced cybersecurity tools enhance the geoportal capabilities in the distributed geoportal architecture.
- Decentralization and development of cross-domain hybrid geoportal architectures under a loosely coupled environment to be encouraged through adoption of SOA to enhance the capabilities of geoportal infrastructure.
- The emerging cloud platforms and trending technologies like machine learning/deep learning and data analytics techniques need to be integrated in development of interoperability-based services and applications.
- Detection of cyber security risks and development of necessary contingency plans along with mitigation strategies should be an integral part of geoportal development to provide uninterrupted geoportal services and applications.
- Collaborations among the geoportal platforms are required for integration and interoperability of geoportals to provide cross-domain services and applications to address the emerging issues and meet the stakeholders' aspirations and demands.

9 Conclusions

To summarise the study on evolution of geoportals, their core functionalities, and services reveals that globally, INSPIRE, GEOSS, and GEE are the important geoportal platforms in the field of EO. Whereas, in India, NSDI India geoportal, ISRO Bhuvan, and ICAR-NBSS&LUP Bhoomi are the important geoportal platforms in natural resources management. Bhoomi geoportal offers the functionalities to effectively catalogue, store, manage, analyze, and visualize large volumes of databases on land and allied resources. Bhoomi geoportal provides various thematic services along with attribute information to facilitate the decision-making process at various levels in natural resource management and agricultural planning. This platform also provides capabilities to leverage geospatial technology, avoid the duplication of efforts in providing cost-effective, and time efficient web-based services and applications to the users. The implemented SOA of Bhoomi geoportal platform helps to offer various WMS to the planners and policy makers in land resource management and land use planning. The user-friendly Bhoomi geoportal platform facilitates the users to query and visualize various cross-domain applications in natural resource management with reference to India. The important advantages of geoportal platforms include improvement of efficiency and effectiveness in offering services and applications across the organizations, develop collaborations, leverage geospatial resources to maintain quality datasets and improve the efficiency in query, and visualization of relevant geospatial resources in natural resource management and agriculture planning. The major challenges like data quality and uncertainty, reliability, scalability, standards and interoperability, and data securityof geoportals remain pertinent in providing robust functionalities, services, and applications. The emerging cloud platforms and machine learning/deep learning algorithms and data analytics tools need to be integrated in development of interoperability-based services and applications.

Development of cross-domain applications, online geospatial data processing capabilities, and advanced dashboards will enhance the capabilities of geoportal platforms to integrate, analyse, and visualize various services and applications in sustainable management of natural resources.

Acknowledgements The author would like to thank the scientific staff, technical officers, and project staff of ICAR-NBSS&LUP for their contributions in Bhoomi geoportal. The author is grateful to the Director, ICAR-NBSS&LUP, Nagpur, for his encouragement, and extending the facilities to carry out part of the presented work.

References

Aguilar R, Zurita-Milla R, Izquierdo-Verdiguier E, de By RA (2018) A cloud-based multi-temporal ensemble classifier to map smallholder farming systems. Remote Sens 10:729
Anderson K, Ryan B, Sonntag W, Kavvada A, Friedl L (2017) Earth observation in service of the 2030 agenda for sustainable development. Geo-Spat Infor Sci 20:77–96
Annoni A (2006) JRC and INSPIRE interoperability. GIM Int 20:3
Arozarena A, Villa G, Valcárcel N, Pérez B (2016) Integration of remotely sensed data into geospatial reference information databases. UN-GGIM national approach. Int Arch Photogramm Remote Sen Spatial Infor Sci 41:721–725
Balla D, Novák TJ, Kiss E, Mester T, Karancsi G, Zichar M (2020) Classification and geovisualization process of soil data using a web-based spatial information system. Open Agri 5:638–655
Baumann P, Mazzetti P, Ungar J, Barbera R, Barboni D, Beccati A, Wagner S (2016) Big data analytics for earth sciences: the earthserver approach. Int J Digit Earth 9(1):3–29
Bogaart PW, Troch PA (2006) Curvature distribution within hillslopes and catchments and its effect on the hydrological response. Hydrol Earth Syst Sci 10:925–936
Chappell D (2010) GIS in the cloud. The ESRI Example, pp 1–12
Colan M (2004) Service-oriented architecture expands the vision of web services, part 2. IBM Developer Works
Cömert Ç (2004) Web services and national spatial data infrastructures. In: Proceedings of geo-imagery bridging continents, XXth ISPRS congress 2004, vol 4. Spatial Data Infrastructure, Istanbul, Turkey
Dobos E, Micheli E, Baumgardner MF, Biehl L, Helt T (2000) Use of combined digital elevation model and satellite radiometric data for regional soil mapping. Geoderma 97(3–4):367–391
FAO (1976) A framework for land evaluation. Soil bulletin, vol 32. Food and agriculture organization of the United Nations, Rome
FAO (1996) Agro-ecological zones, guidelines. FAO soils bulletin, vol 73. Rome, Italy
FAO (2011) The state of the world's land and water resources for food and agriculture (SOLAW)—managing systems at risk. Food and Agriculture Organization of the United Nations, Rome and Earthscan, London
FAO (2017) The future of food and agriculture—trends and challenges. Rome
Fritz S, McCallum I, Schill C, Perger C, See L, Schepaschenko D, Van Der Velde M, Kraxner F, Obersteiner M (2012) Geo-Wiki: an online platform for improving global land cover. Environ Model Softw 31:110–123
Gorelick N (2013) Google earth engine. In: EGU general assembly conference abstracts; American geophysical union. Vienna, Austria, p 11997
Gorelick N, Hancher M, Dixon M, Ilyushchenko S, Thau D, Moore R (2017) Google earth engine: planetary-scale geospatial analysis for everyone. Remote Sens Environ 202:18–27

Guo H, Goodchild MF, Annoni A (2020) Manual of digital earth. Springer, Singapore, pp 593–622
Guo H, Wang L, Liang D (2016) Big earth data from space: a new engine for earth science. Sci Bull 61(7):505–513
Guo Z, Adhikari K, Chellasamy M, Grevec MB, Owensd PR, Greve MH (2019) Selection of terrain attributes and its scale dependency on soil organic carbon prediction. Geoderma 340:303–312
Gurjar YS, Rathore VS (2013) Cloud business intelligence—is what business need today. Int J Recent Technol Eng 1(6):81–86
Harasheh EH (1994) Agricultural applications of remote sensing and geographic information system in land use and land suitability mapping: AARS, agriculture/soil. GIS development.net, pp 1–4
Hengl T, Reuter HI (2008) Geomorphometry: concepts, software, applications, 33rd edn. Elsevier, Amsterdam, p 772
Joerin F, Theriault M, Musy A (2001) Using GIS and outranking multi-criteria analysis for land-use suitability assessment. Int J Geogr Inf Sci 15:153–174
Kokla M, Guilbert E (2020) A review of geospatial semantic information modeling and elicitation approaches. Int J Geoinf 9:146
Kumar KCA, Reddy GPO, Masilamani P, Sandeep P, Turkar SY (2021) Integrated drought monitoring index: a tool to monitor agricultural drought by using time series space-based earth observation satellite datasets. Adv Space Res 67(1):298–315
Kumar N, Reddy GPO, Chatterji S, Srivastava R, Singh SK (2017) Land suitability evaluation for soybean using temporal satellite data and GIS a case study from central India. In: Reddy GPO, Patil NG, Chaturvedi A (eds) Sustainable management of land resources: an Indian perspective. Apple Acade Press, USA, pp 387–410
Lacasta J, Nogueras-Iso J, Béjar R, Muro-Medrano PR, Zarazaga-Soria FJ (2007) A web ontology service to facilitate interoperability within a spatial data infrastructure: applicability to discovery. Data and Knowl Eng 63(3):947–971
Lagacherie P, McBratney A (2006) Spatial soil information systems and spatial soil inference systems: perspectives for digital soil mapping. Dev Soil Sci 31:3–22
Li X, McCarty GW, Du L, Lee S (2020) Use of topographic models for mapping soil properties and processes. Soil Syst 4:32
Lutz M (2007) Ontology-based descriptions for semantic discovery and composition of geoprocessing services. Geoinformatica 11(1):1–36
Ma Y, Wu H, Wang L, Huang B, Ranjan R, Zomaya A, Jie W (2015) Remote sensing big data computing: challenges and opportunities. Future Gener Comput Syst 51:47–60
Maguire DJ, Longley PA (2005) The emergence of geoportals and their role in spatial data infrastructures. Comput Environ Urban Syst 29:3–14
Mahapatra SK, Reddy GPO, Nagdev R, Yadav RP, Singh SK, Sharda VN (2018) Assessment of soil erosion in fragile Himalayan ecosystem of Uttarakhand using USLE and GIS for sustainable productivity. Current Sci 115(1):108–121
Maji AK, Nayak DC, Krishna NDR, Srinivas CV, Kamble K, Reddy GPO, Velayutham M (2001) Soil information system of Arunachal Pradesh in a GIS environment for land use planning. Int J Appl Earth Obser Geoinfor 3:69–77
Maji AK, Reddy GPO, Sarkar D (2010) Degraded and wastelands of India. Status and spatial distribution, directorate of information and publications of agriculture, Indian Council of Agricultural Research, New Delhi, p 158
Maji AK, Reddy GPO, Sarkar D (eds) (2012) Acid soils of India—their extent and spatial variability, vol 145. NBSS Publication NBSS & LUP, Nagpur, p 147
Maji AK, Reddy GPO, Tamgadge DB, Gajbhiye KS (2005) Spatial modeling for crop suitability analysis using AGROMA GIS software. Asian J Geoinfor 5(3):47–56
Maji AK, Srinivas CV, Dubey PN, Reddy GPO, Kamble KH, Velayutham M (2002) Soil resources information of Nagaland in GIS for land use planning in mountainous region. GIS India 11(4):13–16

Maji AK, Srinivas CV, Thamgadge DB, Reddy GPO, Velayutham M (2002b) Soil information of baster district in GIS for land use planning. In: Patil VC, et al (eds) Proceedings of the first national conference on agro-informatics. NCAI, Dharwad, pp 30–33

Maji AK, Tamgadge DB, Reddy GPO, Gajbhiye KS (2002c) GIS based soil information system of Bilaspur district for land use planning, In: Navalgund RR, et al (eds) Proceedings of the ISPRS commission VII symposium on resources and environmental monitoring, December 3–6. NRSA, Hyderabad, pp 746–749

Malczewski J (1996) A GIS-based approach to multiple-criteria group decision making. Int J Geogr Inf Syst 10:955–971

Masser I (2005) The future of spatial data infrastructures, ISPRS workshop on service and application of spatial data infrastructure, XXXVI (4/W6). Hangchou, China, pp 14–16

McKenzie NJ, Ryan PJ (1999) Spatial prediction of soil properties using environmental correlation. Geoderma 89:67–94

Mell P, Grance T (2011) The NIST definition of cloud computing: recommendations of the national institute of standards and technology. National Institute of Standards and Technology U.S Department of Commerce, Gaithesburg, pp 800–145

Minhas PS, Reddy GPO (2017) Edaphic stresses and agricultural sustainability: an Indian perspective. Agri Res 6(1):8–21

Mutanga O, Kumar L (2019) Google earth engine applications. Remote Sens 11:591

Nativi S, Mazzetti P, Santoro M, Papeschi F, Craglia M, Ochiai O (2015) Big data challenges in building the global earth observation system of systems. Environ Model Softw 68:1–26

OGC (2004) Geospatial portal reference architecture, a community guide to implementing standards-based geospatial portals, version:0.2. Discussion paper, OGC, pp 04–039

Pilevara AR, Matinfara HR, Sohrabia A, Sarmadian F (2020) Integrated fuzzy, AHP and GIS techniques for land suitability assessment in semi-arid regions for wheat and maize farming. EcolInd 110:105887

Ramamurthy V, Reddy GPO, Kumar N (2020) Assessment of land suitability for maize (Zea mays L) in semi-arid ecosystem of Southern India using integrated AHP and GIS approach. Comput Electron Agri 179:105806

Reddy GPO (2012a) Conceptual design of soil information system—a geoportal. In: Reddy GPO, Sarkar D (eds) Remote sensing and GIS in digital terrain analysis and soil-landscape modeling, vol 152. NBSS & LUP Publication, pp 217–226

Reddy GPO (2016) Soil geoportal developmental and its applications. In: Singh SK, Bhatta RK, Chatterji S (eds) Land use planning for arresting land degradation, combating climate change and ensuring food security—a training manual, vol 171. NBSS & LUP Publication, pp 50–56

Reddy GPO (2018a) Spatial data management, analysis, and modelling in GIS: principles and applications. In: Reddy GPO, Singh SK (eds) Geospatial technologies in land resources mapping, monitoring and management. Geotechnologies and the environment, vol. 21. Springer, Cham, pp 127–142

Reddy GPO (2018b) Remote sensing and GIS in digital terrain modeling, In: Reddy GPO, Singh SK (eds) Geospatial technologies in land resources mapping, monitoring and management. Geotechnologies and the environment, vol 21. Springer, Cham, pp 201–222

Reddy GPO, Kothare RS, Sena DR, Harindranath CS, Naidu LGK Sarkar D, Sharda VN (2013c) Soil erosion of Goa, vol 155. NBSS Publication, NBSS & LUP (ICAR), Nagpur, p 54

Reddy GPO, Kumar N, Sahu N, Srivastava R, Singh SK, Naidu LGK, Chary GR, Biradar CM, Gumma MK, Reddy BS, Kumar JN (2020) Assessment of spatio-temporal vegetation dynamics in tropical arid ecosystem of India using MODIS time-series vegetation indices. Arab J Geosci 13:704

Reddy GPO, Kumar N, Singh SK (2018) Remote sensing and GIS in mapping and monitoring of land degradation. In: Reddy GPO, Singh SK (eds) Geospatial technologies in land resources mapping, monitoring and management. Geotechnologies and the environment, vol 21. Springer, Cham, pp 401–424

Reddy GPO, Kurothe RS, Sena DR, Harindranath CS, Niranjana KV, Naidu LGK, Singh SK, Sarkar D, Mishra PK, Sharda VN (2016) Assessment of soil erosion in tropical ecosystem of Goa, India using universal soil loss equation, geostatistics and GIS. Ind J Soil Conser 44(1):1–7

Reddy GPO, Maji AK, Das SN, Srivastava R (2012) Development of GIS based seamless mosaic of SRTM elevation data of India to analyze and characterize the selected geomorphic parameters, project report. NBSS & LUP, Nagpur, p 54

Reddy GPO, Maji AK, Srinivas CV, Thayalan S, Velayutham M (2001) Landscape ecological planning in a basaltic terrain, central India, Using remote sensing and GIS techniques. J Ind Sco Remote Sens 29(1 & 2):3–16

Reddy GPO, Maji AK, Srinivas CV, Velayutham M (2002) Geomorphological analysis for inventory of degraded lands in a river basin of basaltic terrain, using remote sensing data and geographical information gystems. J Ind Soc Remote Sen 30(1&2):15–31

Reddy GPO, Mondal C, Srivastava R, Bhattacharyya T, Naidu LGK, Sidhu GS, Baruah U, Singh RS, Kumar N, Singh SK, Sarkar D (2014a) Development of Indian soil information system–a geoportal, project report. NBSS & LUP, Nagpur, p 34

Reddy GPO, Nagaraju MSS, Ramteke IK, Sarkar D (2013) Terrain characterization for soil resource mapping using IRS-P6 data and GIS - a case study from basaltic terrain of Central India. J Ind Soc Remote Sen 41:331–343

Reddy GPO, Patil NG, Chaturvedi A (eds) (2017) Sustainable management of land resources—an Indian perspective. Apple Academic Press Inc., Canada, p 796

Reddy GPO, Sarkar D (2012b) Assessment of soil loss for prioritization of sub-watersheds - A Remote sensing and GIS approach, vol 137. NBSS Publication, NBSS & LUP, Nagpur, pp 55

Reddy GPO, Sarkar D (eds) (2012a) Remote sensing and GIS in digital terrain analysis and soil-landscape modeling, vol 152. NBSS & LUP Publication, pp 300

Reddy GPO, Sarkar D, Mandal C, Srivastava R, Bhattacharyya T, Naidu LGK, Sidhu GS, Baruah U, Singh SK, Singh RS, Nair KM, Sen TK, Chandran P, Sahoo AK, Srinivas S, Kumar N, Sapana C (2016) Digital soil resource database and information system. In: Dwivedi RS, Roy PS (eds) Geospatial technology for integrated natural resources management. Yes Dee Publishing Pvt Ltd., Chennai, pp 321–351

Reddy GPO, Sarkar D, Prasad J, Ramamurthy V (2013) Geospatial modeling in assessment of biophysical resources for sustainable land resource management. Trop Ecol 54(2):227–238

Reddy GPO, Shekinah DE, Maurya UK, Thayalan S, Prasad J, Ray SK, Bhasker BP (1999) Landscape-soil relationship in part of Bazargaon plateau Maharashtra. Geog Rev India 61(3):280–291

Reddy GPO, Singh SK (2014) Achieving food security through land management. Geospatial Today, pp 14–20

Reddy GPO,Singh SK (2018a). Geospatial technologies in land resources mapping, monitoring, and management. Geotechnologies and the environment, vol 21. Springer, p 638

Reddy GPO, Singh SK (2018b) Land resources information system and NBSS BHOOMI geoportal for land resource monitoring and management. In: Reddy GPO, Singh SK (eds) Geospatial technologies in land resources mapping, monitoring and management. Geotechnologies and the environment, vol 21. Springer, Cham, pp 451–468

Reddy GPO, Singh SK, Kumar N, Mondal C, Srivastava R, Maji AK, Sarkar D (2017) Geospatial technologies in development of soil information system and prototype geoportal. In: Reddy GPO, Patil NG, Chaturvedi A (eds) Sustainable management of land resources-an Indian perspective. Apple Academic Press, Waretown, pp 411–442

Reddy GPO, Singh SK, Mondal C, Srivastava R, Bhattacharyya T, Naidu LGK, Sidhu GS, Baruah U, Singh RS, Kumar N, Sarkar D (2014b) Development of district soil information system (DSIS) on 1:50,000 scale (50 districts), project report. NBSS & LUP, Nagpur, p 160

Sahoo AK, Sarkar D, Sah KD, Maji AK, Dhyani BL (2005) Soil erosion of Orissa, vol 126. NBSS Publication. NBSS & LUP, Nagpur, p 48

Sahoo AK, Sarkar D, Singh SK, Reddy GPO, Dhyani BL, Mishra PK, Sharda VN (2014) Soil erosion in Jharkhand, vol 159. NBSS Publication, NBSS & LUP (ICAR), Nagpur, p 39

Sandeep P, Reddy GPO, Jegankumar R, Kumar KCA (2021a) Monitoring of agricultural drought in semi-arid ecosystem of Peninsular India through indices derived from time series CHIRPS and MODIS datasets. Ecol Indic 121:107033

Sandeep P, Reddy GPO, Jegankumar R, Kumar KCA (2021) Modeling and assessment of land degradation vulnerability in semi-arid ecosystem of southern India using temporal satellite data. AHP and GIS Environ Model Assess 26(2):143–154

Sehgal J, Mandal DK, Mandal C, Vadivelu S (1992) Agro-ecological regions of India, 2nd edn, vol 24. NBSS & LUP, Publcation, Nagpur, India, pp 130

Shao J, Kuk G, Anand S, Morley JG, Jackson MJ, Mitchell T (2012) Mapping collaboration in open-source geospatial ecosystem. Trans GIS 16:581–597

Sinnott RO, Galang G, Tomko M, Stimson R (2011) Towards an e-infrastructure for urban research across Australia. In: 2011 IEEE seventh international conference on e-science. Stockholm, Sweden, pp 295–302

Tang W, Selwood J (2005) Spatial portals: adding value to spatial data infrastructures. In: ISPRS workshop on service and application of spatial data infrastructure. Hangzhou, China, pp XXXVI(4/W6)

Velayutham M, Mandal DK, Mandal C, Sehgal JL (1999) Agro-ecological sub regions of India for planning and development, vol 35. NBSS & LUP Publication, Nagpur, India, pp 372

Vishakha D, Maji AK, Reddy GPO, Ramteke IK (2016) Land suitability evaluation for rice (*Oryza sativa* L.) in Tirora tehsil of Gondia district, Maharashtra—A GIS approach. Agropedology 26(01):69–78

Wadoux AMJ-C, Minasny B, McBratney AB (2021) Machine learning for digital soil mapping: applications, challenges and suggested solutions. Earth-Sci Rev 210:103359

Walke N, Reddy GPO, Maji AK, Thayalan S (2012) GIS based multi-criteria overlay analysis in soil- suitability valuation for cotton (Gossypium spp.): a case study in black soil region of Central India. Comp Geosci 41:108–118

Wilkinson MD, Dumontier M, Aalbersberg IJ, Appleton G, Axton M, Baak A, Blomberg N, Boiten J-W, da Silva Santos LB, Bourne PE et al (2016) The FAIR guiding principles for scientific data management and stewardship. Sci Data 3:160018

Williamson I, Rajabifard A, Binns A (2006) Challenges and Issues for SDI development. Int J Spat Data Infrastruct Res 1:24–35

Yagi K, Agus F, Arao T, Aulakh MS, Bai Z, Carating R, Jung K, Kadono A, Kawahigashi M, Lee SH, Ma L, Reddy GPO, Sidhu GS, Takata Y, Tien TM, Xu R, Yan X, Yokoyama K, Zhang F, Zhou D (2015) Regional assessment of soil changes in Asia. In: Status of the world's soil resources (SWSR)–main report. FAO and ITPS, Rome, Italy, pp 287–329

Yang P, Evans J, Cole M, Marley S, Alameh N, Bambacus M (2007) The emerging concepts and applications of the spatial web portal. Photogramm Eng Remote Sen 73(6):691–698

Zhang B, Fan Z, Du Z, Zehng J, Luo J, Wang N, Wang Q (2020) A geomorphological regionalization using the upscaled DEM: the Beijing-Tianjin-Hebei area. China Case Study Sci Rep 10:10532

Author Index

G. P. O. Reddy et al. (eds.), *Data Science in Agriculture and Natural Resource Management*, Studies in Big Data 96, https://doi.org/10.1007/978-981-16-5847-1

The manufacturer's authorised representative in the EU is Springer Nature Customer Service Centre GmbH, Europaplatz 3, 69115 Heidelberg, Germany. If you have any concerns regarding our products, please contact ProductSafety@springernature.com

Printed and bound by CPI Group (UK) Ltd, Croydon, CR0 4YY
30/09/2026
02225294-0014